Frontiers of Fundamental Physics 4

Frontiers of Fundamental Physics 4

Edited by

B. G. Sidharth

B. M. Birla Science Centre
Adarsh Nagar, Hyderabad, India

and

M. V. Altaisky

Laboratory of Information Technologies
Dubna, Russia

Springer Science+Business Media, LLC

Library of Congress Cataloging-in-Publication Data

Frontiers of fundamental physics 4/edited by B.G. Sidharth and M.V. Altaisky.
p. cm.
Proceedings of the Fourth International Symposium on Frontiers of Fundamental Physics, held December 9-13, 2000, in Hyderabad, India—T.p. verso.
Includes bibliographical references and index.
ISBN 978-1-4613-5505-2 ISBN 978-1-4615-1339-1 (eBook)
DOI 10.1007/978-1-4615-1339-1
1. Physics—Congresses. 2. Astrophysics—Congresses. 3. Geophysics—Congresses. I. Sidharth, B. G. (Burra Gautam), 1948- II. Altaisky, M. V. III. International Symposium on Frontiers of Fundamental Physics (4th: 2000: Hyderabad, India).

QC1 .F826 2001
530—dc21

2001038292

Proceedings of the Fourth International Symposium on Frontiers of Fundamental Physics, held December 9-13, 2000, in Hyderabad, India

ISBN 978-1-4613-5505-2

Originally published by Kluwer Academic / Plenum Publishers, New York in 2001
Softcover reprint of the hardcover 1st edition 2001

http://www.wkap.nl

10 9 8 7 6 5 4 3 2 1

A C.I.P. record for this book is available from the Library of Congress

Preface

The B.M. Birla Science Centre, Hyderabad, India has been organizing the series of International Symposia on "Frontiers of Fundamental Physics". The object has been to provide a platform for frontier physicists from slightly different areas in Theoretical Physics, Astrophysics, Cosmology and so on to meet and present the latest developments in these fields, and interact amongst themselves. Indeed this has proved to be a very fruitful exercise because in the last few years dramatic new developments have been surfacing, calling for revisions and reviews. These symposia have been attended by physicists from all over the world. Young physicists have also been encouraged to participate, to benefit from the exercise. The topics covered in this series have ranged from Quantum Superstrings and Particle Physics to Cosmology, presented by leading world authorities.

The Fourth International Symposium of the series was held from December 9-13, 2000. Once again it attracted physicists from USA, Japan, South Africa, different European countries, Russia and the CIS states, India and elsewhere. A feature of the Symposium were the talks by Nobel Laureates Prof. G't Hooft of the University of Utrecht and Prof. Steven Chu of Stanford University. These apart there have been invited talks by a number of other eminent physicists, touching upon apart from conventional theories different interesting perspectives which could come to fruition in the new century.

The present volume is a collection of many of these articles. The compilation had to be made under stringent deadlines, and any lapses may be forgiven.

Thanks are due to Ms. Susan Safren and Kluwer Academic/Plenum Publishers, but for whose persistence and interest, this volume would not have become a reality.

B.G. Sidharth
Hyderabad.

CONTENTS

A Confrontation With Infinity

Gerard 't Hooft
Institute for Theoretical Physics, University of Utrecht, 3584 CC Utrecht, The Netherlands

1 Appetizer

Early attempts at constructing realistic models for the weak interaction between elementary particles were off-set by the emergence of infinite, hence meaningless, expressions when one tried to derive the radiative corrections. When models based on gauge theories with Higgs mechanism were discovered to be renormalizable, the bothersome infinities disappeared-they cancelled out. If this success seemed to be due to mathematical sorcery, it may be of interest to explain the physical insights on which it is actually based.

2 Introduction

In this lecture I intend to reflect on the efforts that were needed to tame the gauge theories, the reasons for our successes at this polint, and the lessons to be learned. I realize the dangers of that. Often in the past, progress was made precisely because lessons from the past were being ignored. Be that as it may, I nevertheless think these lessons are of great importance, and if researchers in the future should choose to ignore them, they must know what they are doing.

When I entered the field of elementary particle physics, no precise theory for the weak interactions existed[1]). It was said that any theory one attempted to write down was nonrenormalizable. What was meant by that? In practice, what it meant was that when one tried to compute corrections to scattering amplitudes, physically impossible expressions were encountered. The result of the computations appeared to imply that these amplitudes should be infinite. Typically, integrals of the following form were found:

$$\int d^4k \frac{Pol(k_\mu)}{(k^2+m^2)[(k+q)^2+m^2]} = \infty, \tag{1}$$

where $Pol(k_\mu)$ stands for some polynomial in the integration variables k_μ. Physically, this must be nonsense. If, in whatever model caculation, the effects due to some obscure secondary phenomenon appear to be infinitely strong, one knows what this means: the so-called secondary effect is not as innocent as it might have appeared-it must have been represented incorrectly in the model; one has to improve the model by paying special attention to the features that were at first thought to be negligible. The infinities in the weak-interaction theories were due to interactions from virtual particles at extremely high energies. High energy also means high momentum, and

Frontiers of Fundamental Physics 4, Edited by Sidharth and Altaisky,
Kluwer Academic/Plenum Publishers, New York, 2001

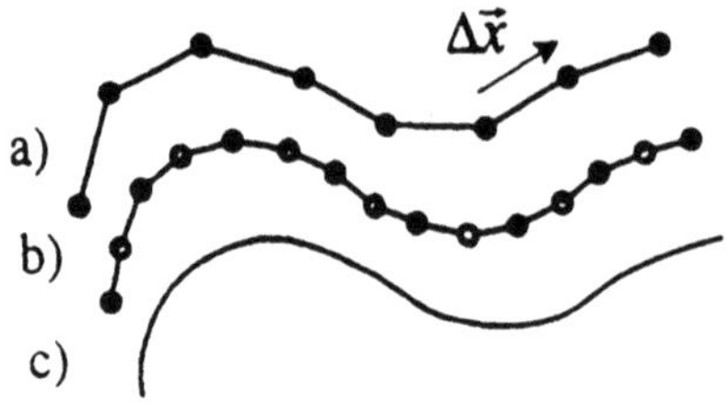

Figure 1: Differentiation.

in quantum mechanics this means that the waves associated with these particles have very short wavelengths. One had to conclude that the short distance structure of the existing theories was too poorly understood.

Short distance scales and short time intervals entered into theories of physics first when Newton and Leibniz introduced the notion of differentiation. In describing the motion of planets and moons, one had to consider some small time interval Δt and the displacement $\Delta\vec{x}$ of the object during this time interval [see Fig.1(a)]. The crucial observation was that, in the limit $\Delta t \to 0$, the ratio

$$\frac{\Delta\vec{x}}{\Delta t} = \vec{v} \tag{2}$$

makes sense, and we call it "velocity." In fact, one may again take the ratio of the velocity change $\Delta\vec{v}$ during such a small time interval Δt, and again the ratio

$$\frac{\Delta\vec{v}}{\Delta t} = \vec{a} \tag{3}$$

exists in the limit $\Delta t \to 0$; we call it "acceleration." Their big discovery was that it makes sense to write equations relating accelerations, velocities, and positions, and that in the limit where Δt goes to zero, you get good models describing the motion of celestial bodies[Fig. 1(c)]. The mathematics of differential equations grew out of this, and nowadays it is such a central element in theoretical physics that we often do not realize how important and how nontrivial these observations actually were. In modern theories of physics we send distances and time intervals to zero all the time, also in multidimensional field theories, assuming that the philosophy of differential equations applies. But occasionally it may happen that everything goes wrong. The limits that we thought to be familiar with do not appear to exist. The behavior of our model at the very tiniest time and distance scales then has to be reexamined.

Infinite integrals in particle theory were not new. They had been encountered many times before, and in some theories it was understood how to deal with them[2]. What had to be done was called "renormalization." Imagine a particle such as an electron to be something like a little sphere, of radius R and mass m_{bare}. Now attach an electric charge to this particle, of an amount Q. The electric-field energy would be

$$U = \frac{Q^2}{8\pi R}, \tag{4}$$

and, according to Einstein's special theory of relativity, this would represent an extra amount of mass, U/c^2, where c is the speed of light. Particle plus field would carry a mass equal to

$$m_{\text{phys}} = m_{\text{bare}} + \frac{Q^2}{8\pi c^2 R}. \tag{5}$$

It is this mass, called "physical mass," that an experimentor would measure if the particle were subject to Newton's law, $\vec{F} = m_{\text{phys}}\vec{a}$. What is alarming about this effect is that the mass correction diverges to infinity when the radius R of our particle is sent to zero. But we want R to be zero, because if R were finite it would be difficult to take into account that forces acting on the particle must be transmitted by a speed less than that of light, as is demanded by Einstein's theory of special relativity. If the particle were deformable, it would not be truly elementary. Therefore, finite-size particles cannot serve as a good basis for a theory of elementary objects.

In addition, there is an effect that alters the electric charge of the particle. This effect is called "vacuum polarization." During extremely short time intervals, quantum fluctuations cause the creation and subsequent annihilation of particle-antiparticle pairs. If these particles carry electric charges, the charges whose signs are opposite to our particle in question tend to move towards it, and this way they tend to neutralize it. Although this effect is usually quite small, there is a tendency of the vacuum to "screen" the charge of our particle. This screening effect implies that a particle whose charge is Q_{bare} looks like a particle with a smaller charge Q_{phys} when viewed at some distance. The relation between Q_{bare} and Q_{phys} again depends on R, and, as was the case for the mass of the particle, the charge renormalization also tends to infinity as the radius R is sent to zero (even though the effect is usually rather small at finite R).

It was already in the first half of the 20th century that physicists realized the following. The only properties of a particle such as an electron that we ever measure in an experiment are the physical mass m_{phys} and the physical charge Q_{phys}. So, the procedure we have to apply is that we should take the limit where R is sent to zero while m_{phys} and Q_{phys} are kept fixed. Whatever happens to the bare mass m_{bare} and the bare charge Q_{bare} in that limit is irrelevant, since these quantities can never be measured directly.

Of course, there is a danger in this argument. If, in Eq.(5), we send R to zero while keeping m_{phys} fixed, we notice that m_{bare} tends to minus infinity. Can theories in which particles have negative mass be nevertheless stable? The answer is no, but fortunately Eq.(??) is replaced by a different equation in a quantized theory. m_{bare} tends to zero, not minus infinity.

3 The Renormalization Group

The modern way to discuss the relevance of the small distance structure is by performing scale transformations, using the renormalization group[3, 4], and we can illustrate this again by considering the equation of motion of the planets. Assume that we took definite time intervals Δt, finding equations for the displacements Δx. Imagine that we wish to take the limit $\Delta t \to 0$ very carefully. We may decide first to divide all Δt's and all Δx's by 2 [see Fig.1(b)]. We observe that, if the original intervals are already sufficiently small, the new results of a calculation will be very nearly the same as the old ones. This is because, during small time intervals, planets and moons move along small sections of their orbits, which are very nearly straight lines, the division by 2 would have made no difference at all. planets move along straight lines if no force acts on them. The reason why differential equations were at all successful for planets is that we may ignore the effects of the forces (the "interactions") when time and space intervals are taken to be very small.

In quantized field theories for elementary particles, we have learned how to do the same thing. We reconsider the system of interacting particles at very short time and distance scales. If at sufficiently tiny scales the interactions among the particles may be ignored, then we can understand how to take the limits where these scales go all the way to zero. Since then the interactions may be ignored, all particles move undisturbedly at these scales, and so the physics is then understood. Such theories can be based on a sound mathematical footing-we understand how to do calculations by approximating space and time as being divided into finite sections and intervals and taking the limits in the end.

So, what is the situation here? Do the mutual interactions among elementary particles vanish at sufficiently tiny scales? Here is the surprise that physicists had to learn to cope with; they do not.

Many theories indeed show very bad behavior at short distances. A simple prototype of these is the so-called chiral model[5].

In such a model, a multicomponent scalar field is introduced which obeys a constraint: its total length is assumed to be fixed,

$$\sum_i |\phi_i|^2 = R^2 = \text{fixed} \tag{6}$$

At large distance scales, the effects of this constraint are mild, as the quantum fluctuations are small compared to R. At small distance scales, however, the quantum fluctuations are large compared to R, and hence the nonlinear effects of the constraint are felt much more strongly there. As a consequence, such a theory has large interactions at small distance scales and vice versa. Therefore, at infinitesimally small distance scales, such a theory is ill-defined, and the model is unsuitable for an accurate description of elementary particles. Other examples of models with bad small-distance behavior are the old four-fermion interaction model for the weak interactions and most attempts at making a quantum version of Einstein's gravity theory.

But some specially designed models are not so bad. Examples are: a model with spinless particles whose fields ϕ interact only through a term of the form $\lambda\phi^4$ in the Lagrangian, and a model in which charged particles interact through Maxwell's equations (quantum electrodynamics, QED). In general, we choose the distance scale to be a parameter called $1/\mu$. A scale transformation by a factor of 2 amounts to adding In 2 to In μ, and if the distance scale is Δx, then

$$\frac{\mu d}{d\mu}\Delta x = -\Delta x. \tag{7}$$

During the 1960's, it was found that in all theories existing at the time, the interaction parameters, being either the coefficient λ for $\lambda\phi^4$ theory, or the coefficient e^2 in quantum electrodynamics for electrons with charge e, the variation with μ is a positive function[6], called the β function:

$$\frac{\mu d}{d\mu}\lambda = \beta(\lambda) > 0, \tag{8}$$

so, comparing this with Eq.(7), λ is seen to increase if Δx decreases.
In the very special models that we just mentioned, the function $\beta(\lambda)$ behaves as λ^2 when λ is small, which is so small that the coupling only varies very slightly as we go from one scale to the next. This implies that, although there are still interactions, no matter how small the scales at which we look, these interactions are not very harmful, and a consequence of this is that these theories are "renormalizable." If we apply the perturbation expansion for small λ then, term by term, the expansion coefficients are uniquely defined, and we might be seduced into believing that there are no real problems with these theories.
However, many experts in these matters were worried indeed, and for good reason: If β is positive, then there will be a scale where the coupling strength among particles diverges. The solution to Eq.(8) is [see Fig.2(a)]

$$\lambda(\mu) = 1/(C - \beta_2 ln\mu), \qquad \text{if} \quad \beta(\lambda) = \beta_2\lambda^2, \tag{9}$$

where C is an integration constant, $C = 1/\lambda(1)$if $\lambda(1)$ is λ measured at the scale $\mu = 1$. We see that at scales $\mu = 0[exp(1/\beta_2\lambda(1))]$, the coupling explodes. Since for small $\lambda(1)$ this is exponentially far away, the problem is not noticed in the perturbative formulation of the theory, but it was recognized that if, as in physically realistic theories, λ is taken to be not very small, there is real trouble at some definite scale. And so it was not crazy to conclude that these quantum field theories were sick and that other methods should be sought for describing particle theories.
I was never afflicted with such worries for a very simple reason. Back in 1971, I carried out my own calculations of the scaling properties of field theories, and the

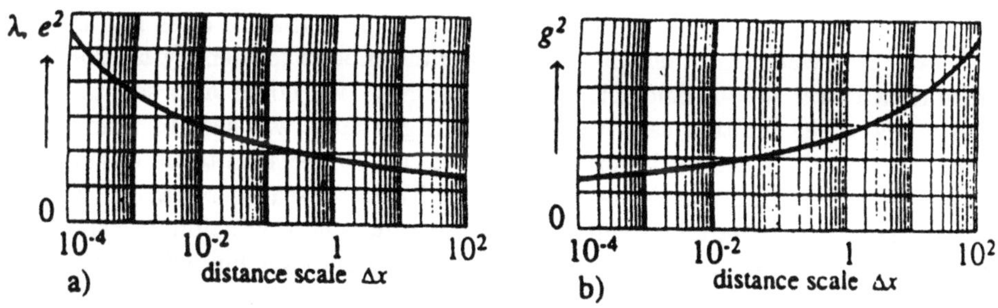

Figure 2: Scaling of the coupling strength as the distance scale varies, (a) for $\lambda\varphi^4$ theories and QED, (b) for Yang–Mills theories.

first theory I tried was Yang-Mills theory. My finding was, when phrased in modern notation, that for these theories,

$$\beta(g^2) = Cg^4 + 0(g^6) \quad \text{with} \quad C < 0 \tag{10}$$

if the number of fermion species is less than 11 [for $SU(2)$] or $16\frac{1}{2}$ [for $SU(3)$]. The calculation, which was alluded to in my first paper on the massive Yang-Mills theory [7], was technically delicate but conceptually not very difficult. I could not possibly imagine what treasure I had here or that none of the experts knew that β could be negative; they had always limited themselves to studying only scalar field theories and quantum electrodynamics, where β is positive.

4 The Standard Model

If we were to confront the infinities in our calculations for the weak-interaction processes, we had to face the challenge of identifying a model for the weak interaction that shows the correct interwining with the electromagnetic force at large distance scales but is sufficiently weakly interacting at small distances. The resolution here was to make use of spontaneous symmetry breaking. The mass generation mechanism discussed here should, strictly speaking, not be regarded as spontaneous symmetry breaking, since in these theories the vacuum does not break the guage symmetry. "Hidden symmetry" is a better phrase [8]. We simply refer to this mechanism as the "Higgs mechanism." We use a field with a quartic self-interaction but with a negative mass term, so that its energetically favored value is nonvanishing. The fact that such fields can be used to generate massive vector particles was known but not used extensively in the literature. Also the fact that one could construct reasonable models for the weak interaction along these lines was known. These models, however, were thought to be inelegant, and the fact that they were the unique solution to our problems was not realized.

Not only did the newly revived models predict hitherto unknown channels for the weak interaction, they also predicted a new scalar particle, the Higgs boson [9, 10, 11]. The new weak interaction, the so-called neutral-current interaction, could be confirmed experimentally within a few years, but as of this writing, the Higgs boson is still fugitive. Some researchers suspect that it does not exist at all. Now if this were true then this would be tantamount to identifying the Higgs field with a chiral field-a field with a fixed length. We could also say that this corresponds to the limiting case in which the Higgs mass was sent to infinity. An infinite-mass particle cannot be produced, so it can be declared to be absent. But as we explained before, chiral theories have bad small-distance behavior. We can also say that the interaction strength at small distances is proportional to the Higgs mass; if that would be taken to be infinite then we would have landed in a situation where the small-distance behavior was out of control. Such models simply do not work. Perhaps experimentalists will not succeed in producing and detecting Higgs particles, but this then would imply that entirely new theories must be found to account for the small-distance structure. Candidates for such theories have been proposed. They seem to be inelegant at present, but of course that could be due to our present limited understanding, who knows? New theories would necessarily

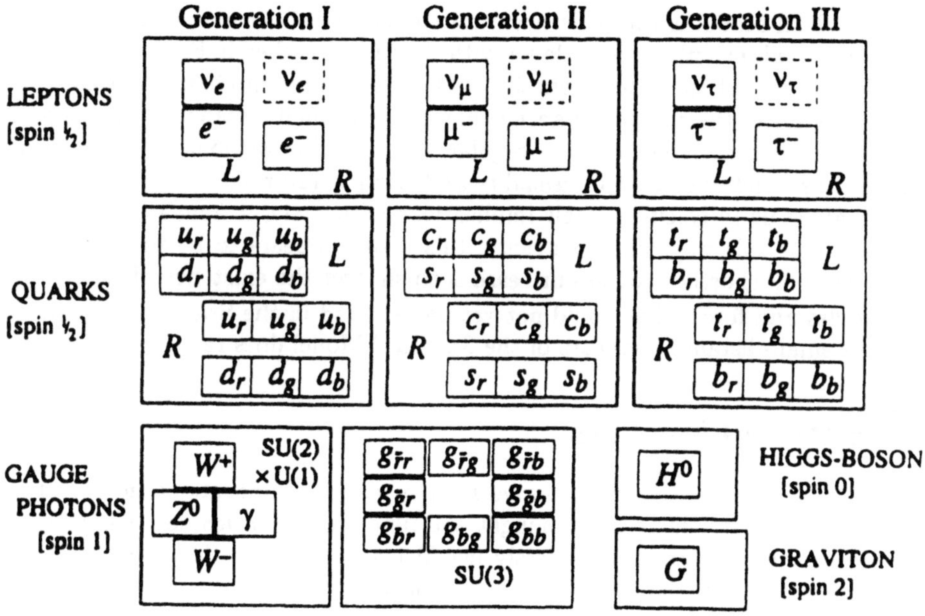

Figure 3: The standard model.

imply the existence of many presently unknown particle species, and experimenters would be delighted to detect and study such objects. We cannot lose here. Either the Higgs particle or other particles must be waiting there to be discovered, probably fairly soon[12, 13].

To the strong interactions, the same philosophy applies, but the outcome of our reasoning is very different. The good scaling behavior of pure gauge theories [see Fig.2(b)] allows us to construct a model in which the interactions at large distance scales is unboundedly strong, yet it decreases to zero (though only logarithmically) at small distances. Such a theory may describe the binding forces between quarks. It was found that these forces obtain a constant strength at arbitrarily large distances, where Coulomb forces would have decreased with an inverse square law. Quantum chromodynamics, a Yang-Mills theory with gauge group $SU(3)$, could therefore serve as a theory for the strong interactions. It is the only allowed model in which the coupling strength is large but nevertheless the small-distance structure is under control.

The weak force, in contrast, decreases exponentially as the distance between weakly interacting objects becomes large. Thus gauge theory allows us to construct models with physically acceptable behavior at short distances, while the forces at large distances may vary in any of the following three distinctive ways:

(i) The force may drop exponentially fast, as in the weak interaction;

(ii) The force may drop according to an inverse square law, as in electromagnetism, or

(iii) The force may tend towards a constant, as in the strong interactions.

The Standard Model is the most accurate model describing nature as it is known today. It is built exactly in accordance with the rules sketched in Fig. 3. Our philosophy is always that the experimentally obtained information about the elementary particles refers to their large-distance behavior. The small-distance structure of the theory is then postulated to be as regular as is possible without violating principles such as strict obedience of casuality and Lorentz invariance. Not only do such models allow us to calculate their implications accurately, it appears that Nature really is built this way. In some sense, this result appears to be too good to be true. We shall shortly explain our reasons to suspect the existence of many kinds of particles and forces that could not yet be included in the Standard Model, and that the small-distance structure of the Standard Model does require modification.

5 Future Colliders

Theoreticians are most eager to derive all they want to know about the structures at smaller distances using pure thought and fundamental principles. Unfortunately, our present insights are hopelessly insufficient, and all we have are some wild speculations. Surely, the future of this field still largely depends on the insights to be obtained from new experiments.

The present experiments at the Large Electron Positron Collider (LEP) at CERN are coming to a close. They have provided us with impressive precision measurements that not only gave a beautiful confirmation of the Standard Model, but also allowed us to extrapolate to higher energies, which means that we were allowed a glimpse of structures at the smallest distance ranges yet accessible. The most remarkable result is that the structures there appear to be smooth; new interactions could not be detected, which indicates that the mass of the Higgs particle is not so large, a welcome stimulus for further experimental efforts to detect it.

In the immediate future we may expect interesting new experimental results first from the Tevatron Collider at Fermilab, near Chicago, and then from the Large Hadron Collider (LHC) at CERN, both of which will devote much effort to finding the still elusive Higgs particle. Who will be first depends on what the Higgs mass will turn out to be, as well as other not yet precisely known properties of the Higgs. Detailed analysis of what we know at present indicates that Fermilab has a sizable chance at detecting the Higgs first, and the LHC almost certainly will not only detect these particles, but also measure many of their properties, such as their masses, with high precision. If supersymmetric particles exist, LHC will also be in a good position to be able to detect these, in measurements that are expected to begin shortly after 2005.

These machines, which will discover structures never seen before, however, also have their limits. They stop exactly at the point where our theories become highly interesting, and the need will be felt to proceed further. As before, the options are either to use hadrons such as protons colliding against antiprotons, which has the advantage that, due to their high mass, higher energies can be reached, or alternatively to use leptons, such as e^+ colliding against e^-, which has the advantage that these objects are much more pointlike, and their signals are more suitable for precision experiments[14]. Of course, one should do both. A more ambitious plan

is to collide muons, μ^+ against μ^-, since these are leptons with high masses, but this will require numerous technical hurdles to be overcome. Boosting the energies to ever-increasing values requires such machines to be very large. In particular the high-energy electrons will be hard to force into circular orbits, which is why design studies of the future accelerators tend to take the form of straight lines, not circles. These linear accelerators have the interesting feature that they could be extended to larger sizes in the more distant future.
My hope is that efforts and enthusiasm to design and construct such machines in the future will not diminish. As much international cooperation as possible is called for. A smympathetic proposal[15], is called ELOISATRON, a machine in which the highest conceivable energies should be reached in a gigantically large circular tunnel. It could lead to a hundredfold improvement of our spatial resolution. What worries me, however, is that in practice one group, one nation, takes an initiative and then asks other groups and nations to join, not so much in the planning, but rather in financing the whole thilng. It is clear to me that the best international collaborations arise when all partners are involved from the very earliest stages of the development onwards. The best successes will come from those institutions that are the closest approximations to what could be called "world machines." CERN claims to be a world machine, and indeed as such this laboratory has been, and hopefully will continue to be, extremely successful. Unfortunately, it still has an E in its name. This E should be made as meaningless as the N (after all, the physics studied at CERN has long ago ceased to be nuclear, it is subnuclear now). I would not propose to change the name, but to keep the name CERN only to commemorate its rich history.

6 Beyond the Standard Model

Other, equally interesting large scientific enterprises will be multinational by their very nature: plans are underway to construct neutrino beams that go right through the earth to be detected at the exit point, where it may be established how subtle oscillations due to their small mass values may have caused transitions from one type into another. Making world machines will not imply that competition will be eliminated; the competition, however, will not be between nations, but rather between the different collaborations who use different machines and different approaches towards physics questions.
The most interesting and important experiments are those of which we cannot guess the outcome reliably. This is exactly the case for the LHC experiments that are planned for the near future. What we do know is that the Standard Model, as it s-tands today, cannot be entirely correct, in spite of the fact that the interactions stay weak at ultrashort distance scales. Weakness of the interactions at short distances is not enough; we also insist that there be a certain amount of stability. Let us use the metaphor of the planets in their orbits once again. We insisted that, during extremely short time intervals, the effects of the forces acting on the planets have hardly any effect on their velocities, so that they move approximately in straight lines. In our present theories, it is as if at short time intervals several extremely strong forces act on the planets, but, for some reason, they all but balance out.

The net force is so weak that only after long time intervals, days, weeks, months, the velocity change of the planets become apparent. In such a situation, however, a reason must be found as to why the forces at short time scales balance out. The way things are for the elementary particles, at present, is that the forces balance out just by accident. It would be an inexplicable accident, and as no other examples of such accidents are known in Nature, at least not of this magnitude, it is reasonable to suspect that the true short distance structure is not exactly as described in the Standard Model, but that there are more particles and forces involved, whose nature is as yet unclear. These particles and forces are arranged in a new symmetry pattern, and it is this symmetry that explains why the short-distance forces balance out.

It is generally agreed that the most attractive scenario is one involving "supersymmetry," a symmetry relating fermionic particles, whose spin is an integer plus one-half, and bosonic particles, which have integral spin. (Supersymmetry has a vast literature. See, for instance, the collection of papers in [16, 17].) It is the only symmetry that can be made to do the required job in the presence of the scalar fields that provide the Higgs mechanism, in an environment where all elementary particles interact weakly. However, when the interactions do eventually become strong then there are other scenarios. In that case, the objects playing the role of Higgs particles may be not elementary objects but composites, similar to the so-called Cooper pairs of bound electronsd that perform a Higgs mechanism in ultracool solid substances, leading to superconductivity. Just because such phenomena are well known in physics, this is a scenario that cannot easily be dismissed. But, since there is no evidence at present of a new strong interaction domain at the TeV scale, the bound-state Higgs theory is not favoured by most investigators.

One of the problems with the supersymmetry scenario is the supersymmetry breaking mechanism. Since at the distance scale where experiments are done at present no supersymmetry has been detected, the symmetry is broken. It is assumed that the breaking is "soft," which means that its effects are seen only at large distances, and only at the tiniest possible distance scales is the symmetry realized. Mathematically, this is a possibility, but there is as yet no plausible physical explanation of this situation. The only situation can come from a theory at even smaller distance scales, where the gravitational force comes into play.

Until the early 1980's the most promising model for the gravitational force was a supersymmetric variety of gravity: supergravity [17]. It appeared that the infinities that were insurmountable in a plain gravity theory would be overcome in supergravity. Curiously, however, the infinities appeared to be controlled by the enhanced symmetry and not by an improved small-distance structure of the theory. Newton's constant, even if controlled by a dilaton field, still is dimensionful in such theories, with consequently uncontrolled strong interactions in the small-distance domain. As the small- distance structure of the theory was not understood, it appeared to be almost impossible to draw conclusions from the theory that could shed further light on empirical features of our world.

An era followed with even wilder speculations concerning the nature of the gravitational force. By far the most popular and potentially powerful theory is that of the superstrings[18]. The theory started out by presenting particles as made up of (either closed or open) pieces of string. Fermions living on the string provide it with

a supersymmetric pattern, which may be the origin of the approximate pattern, which may be the origin of the approximate supersymmetry that we need in our theories. It is now understood that only in a perturbative formulation do particles look like strings. In a nonperturbative formalism there seems to be a need not only of strings but also of higher dimensional substances such as membranes. But what exactly is the perturbation expansion in question? It is not the approximation that can be used at the shortest infinitesimal distances. Instead, the shortest distances seem to be linked to the largest distances by means of duality relations. Just because superstrings are also held responsible for the gravitational force, they cause curvature of space and time to such an extent that it appears to be futile to consider distances short compared to the Planck scale.
According to superstring theory, it is a natural and inevitable aspect of the theory that distance scales shorter than the Planck scale cannot be properly addressed, and we should not worry about it. When outsiders or sometimes colleagues from unrelated branches of physics attack superstring theory, I come to its defense. The ideas are very powerful and promising. But when among friends, I have this critical note. As string theory makes heavy use of differential equations it is clear taht some sort of continuity is counted on. We should attempt to find an improved short-distance formulation of theories of this sort, if only to justify the use of differential equations or even functional integrals.
Rather than regarding the above as criticism against existing theories, one should take our observations as indications of where to search for further improvements. Emphasizing the flaws of the existing constructions is the best way to find new and improved procedures. Only in this way can we hope to achieve theories that allow us to explain the observed structures of the Standard Model and to arrive at more new predictions, so that we can tell our experimental friends where to search for new particles and forces.

References

[1] R.P. Crease and C.C. Mann, "The Second Creation: Makers of the Revolution in Twentiety-century Physics", Macmillan, New York, 1986.

[2] A. Pais, "Inward Bound: Of Matter and Forces in the Physical World", Oxford University, London, 1986.

[3] K.G. Wilson and J. Kogut, Phys.Rep., Phys.Lett. 12C, 1974, 75.

[4] H.D. Politzer, Phys.Rep., Phys.Lett. 14C, 1974, 129.

[5] B.W. Lee, "Chiral Dynamics", Gordon and Breach, New York, 1972, pp.60-67.

[6] D.J. Gross, in "The Rise of the Standard Model", Cambridge University, Cambridge, 1997, p.199.

[7] G 't Hooft, Nucl.Phys.B 35, 1971, 167.

[8] S. Coleman, "Secret Symmetries", in "Laws of Hadronic Matter", Ed. A. Zichichi, Academic, New York, London, 1975.

[9] P.W. Higgs, Phys.Lett. 12, 1964a, 132.

[10] P.W. Higgs, Phys.Rev.Lett. 13, 1964b, 321.

[11] P.W. Higgs, Phys.Rev. 145, 1966, 1156.

[12] E. Accomando, et al., Phys. Rep. 299, 1998, 1.

[13] P.M. Zerwas, "Physics with an e^+e^- linear collider at high luminosity", Cargese lectures 1999, preprint DESY, 99-178.

[14] J. Ellis, "Possible Accelerators at CERN beyond the LHC", preprint CERN-TH/99-350, hep-ph/9911440, 1999.

[15] A. Zichichi, "Fifty years of subnuclear physics: From past to future and the ELN project", in "Highlights of Subnuclear Physics: 50 Years Later: Proceedings of the International School of Subnuclear Physics", Ed. A. Zichichi, World Scientific, Singapore and River Edge, London, 1999, p.161.

[16] S. Ferrara, Ed."Supersymmetry", Vol.1, North Holland, Amsterdam, 1987.

[17] S. Ferrara, Ed."Supersymmetry", Vol.2, North Holland, Amsterdam, 1987.

[18] J. Polchinski, "String Theory, Vol.1, An Introduction to the Bosonic String, Cambridge Monographs on Mathematical Physics", Ed. P.V. Landshoff et al., Cambridge University, Cambridge, 1998.

The Self-Intersecting Brane World

M. Pavšič
Department of Theoretical Physics
Jožef Stefan Institute, Jamova 39, SI-1001 Ljubljana (Slovenia)
Email:matej.pavsic@ijs.si

1 Introduction

After so many years of intensive research the quantization of gravity is still an unfinished project. Amongst many approaches followed, there is the one which seems to be especially promising. This is the so called induced gravity proposed by Sakharov [1]. His idea was to treat the metric not as a fundamental field but as one induced from more basic fields. The idea has been pursued by numerous authors [2]; especially illuminating are works by Akama, Terazawa and Naka [3]. Their basic action contains N scalar fields and it is formally just a slight generalization of the well-known Dirac–Nambu–Goto action for an n-dimensional world sheet swept by an $(n-1)$-dimensional membrane.

Here we pursue such an approach and give a concrete physical interpretation of the N scalar fields which we denote $\eta^a(x)$. We assume that the spacetime is a surface V_4, called the spacetime sheet, embedded in a higher-dimensional space V_N, and $\eta^a(x)$ are the embedding functions. An embedding model has been first proposed by Regge and Teitelboim [4] and investigated by others [5]. In that model the action contains the Ricci scalar expressed in terms of the embedding functions. In our present model [6, 7, 8], on the contrary, we start from an action which is essentially the minimal surface action[1] weighted with a function $\omega(\eta)$ in V_N. For a suitably chosen ω, such that it is singular (δ-function like) on certain surfaces $\hat{V}_m$, also embedded in V_N, we obtain on V_4 a set of world lines. It was shown that these worldlines are geodesics of V_4, provided that V_4 described by $\eta^a(x)$ is a solution to our variational procedure [6]–[8]. I will show that after performing functional integrations over $\eta^a(x)$ we obtain two contributions to the path integral. One contribution comes from all possible $\eta^a(x)$ not intersecting $\hat{V}_m$, or not self-intersecting, and the other from those $\eta^a(x)$ which do intersect the surfaces $\hat{V}_m$, or which do self-intersect. In the effective action so obtained, the first contribution gives the Einstein–Hilbert term R plus higher-order terms like R^2. The second contribution can be cast into the form of a path integral over all possible worldlines $X^\mu(\tau)$. Thus we obtain an action which contains matter sources and a kinetic term for the metric field (plus higher orders in R). So in the proposed approach both the metric field and the matter field are induced from more basic fields $\eta^a(x)$.

[1]Recently Bandos [9] has considered a string-like description of gravity by considering bosonic p-branes coupled to an antisymmetric tensor field.

Frontiers of Fundamental Physics 4, Edited by Sidharth and Altaisky,
Kluwer Academic/Plenum Publishers, New York, 2001

2 The Classical Model

2.1 The dynamics of a brane in the backround of other branes

Let us briefly summarize the results of refs. [6]–[8]. We assume that the arena where physics takes place is an N-dimensional space V_N with $N \geq 10$. Next we assume that an n-dimensional surface V_n living in V_N represents a possible spacetime. The parametric equation of such a 'spacetime sheet' V_n is given by the embedding functions $\eta^a(x^\mu)$, $a = 0, 1, 2, ..., N$, where x^μ , $\mu = 0, 1, 2, \ldots, n-1$, are coordinates (parameters) on V_n. We assume that the action is just that for a minimal surface V_n

$$I[\eta^a(x)] = \int (\det \partial_\mu \eta^a \, \partial_\nu \eta^b \gamma_{ab})^{1/2} \mathrm{d}^n x, \tag{1}$$

where γ_{ab} is the metric tensor of V_N. The dimension of the spacetime sheet V_n is taken here to be arbitrary, in order to allow for the Kaluza–Klein approach. In particular, we may take $n = 4$. We admit that the embedding space is curved in general. In particular, let us consider the case of a conformally flat V_N, such that $\gamma_{ab} = \omega^{2/n} \eta_{ab}$, where η_{ab} is the N-dimensional Minkowski tensor. Then Eq. (1) becomes

$$I[\eta^a(x)] = \int \omega(\eta) (\det \partial_\mu \eta^a \, \partial_\nu \eta^b \eta_{ab})^{1/2} \mathrm{d}^n x. \tag{2}$$

¿From now on we shall forget about the origin of $\omega(\eta)$ and consider it as a function of position in a *flat* embedding space. The indices a, b, c will be raised and lowered by η^{ab} and η_{ab}, respectively.

In principle $\omega(\eta)$ is arbitrary. But it is very instructive to choose the following function:

$$\omega(\eta) = \omega_0 + \sum_i \int \kappa_i \frac{\delta^N(\eta - \hat{\eta}_i)}{\sqrt{|\gamma|}} \, \mathrm{d}^m \hat{x} \sqrt{|\hat{f}|} \, , \tag{3}$$

where $\eta^a = \hat{\eta}_i^a(\hat{x})$ is the parametric equation of an m-dimensional surface $\widehat{V}_m^{(i)}$, called the *matter sheet*, also embedded in V_N, $\hat{f}$ is the determinant of the induced metric on $\widehat{V}_m^{(i)}$, and $\sqrt{|\gamma|}$ allows for taking curved coordinates in otherwise flat V_N. If we take $m = N - n + 1$ then the intersection of V_n and $\widehat{V}_m^{(i)}$ can be a (one-dimensional) line, i.e., a worldline C_i on V_n. In general, when $m = N - n + (p+1)$ the intersection can be a $(p+1)$-dimensional world sheet representing the motion of a p-dimensional membrane (also called p-brane). In this section we confine our consideration to the case $p = 0$, that is, to the motion of a point particle.

Inserting (3) into (2) and writing $f_{\mu\nu} \equiv \partial_\mu \eta^a \partial_\nu \eta_a$, $f \equiv \det f_{\mu\nu}$ we obtain

$$I[\eta] = \omega_0 \int \mathrm{d}^n x \sqrt{|f|} + \int \mathrm{d}^n x \sum_i \kappa_i \, \delta^n(x - X_i)(f_{\mu\nu} \dot{X}_i^\mu \dot{X}_i^\nu)^{1/2} \, \mathrm{d}\tau. \tag{4}$$

The result above was obtained by writing

$$\mathrm{d}^m \hat{x} = \mathrm{d}^{m-1} \hat{x} \, \mathrm{d}\tau \, , \qquad \hat{f} = \hat{f}^{(m-1)} (\dot{X}_i^\mu \dot{X}_{i\mu})^{1/2}$$

and taking the coordinates η^a such that $\eta^a = (x^\mu, \eta^n, ..., \eta^{N-1})$, where x^μ are (curved) coordinates on V_n. The determinant of the metric of the embedding space V_N in such curvilinear coordinates is then $\gamma = \det \partial_\mu \eta^a \, \partial_\nu \eta_a = f$.

If we vary the action (4) with respect to $\eta^a(x)$ we obtain

$$\partial_\mu \left[\sqrt{|f|} \, (\omega_0 f^{\mu\nu} + T^{\mu\nu}) \partial_\nu \eta_a \right] = 0, \tag{5}$$

where

$$T^{\mu\nu} = \frac{1}{\sqrt{|f|}} \sum_i \int \mathrm{d}^n x \, \kappa_i \, \delta^n(x - X_i) \, \frac{\dot{X}_i^\mu \dot{X}_i^\nu}{(\dot{X}_i^\alpha \dot{X}_{i\alpha})^{1/2}} \, \mathrm{d}\tau \tag{6}$$

is the stress–energy tensor of dust. Eq. (5) can be rewritten in terms of the covariant derivative D_μ on V_n:

$$D_\mu \left[(\omega_0 f^{\mu\nu} + T^{\mu\nu}) \partial_\nu \eta_a \right] = 0. \tag{7}$$

The latter equation gives

$$\partial_\nu \eta_a \, D_\mu T^{\mu\nu} + (\omega_0 f^{\mu\nu} + T^{\mu\nu}) D_\mu D_\nu \eta_a = 0, \tag{8}$$

where we have taken into account that a covariant derivative of metric is zero, i.e., $D_\alpha f_{\mu\nu} = 0$ and $D_\alpha f^{\mu\nu} = 0$, which implies also $\partial_\alpha \eta^c D_\mu D_\nu \eta_c = 0$, since $f_{\mu\nu} \equiv \partial_\mu \eta^a \partial_\nu \eta_a$. Contracting Eq.(8) by $\partial^\alpha \eta^a$ we have

$$D_\mu T^{\mu\nu} = 0. \tag{9}$$

The latter are the well known equations of motion for the sources. In the case of dust (9) implies that dust particles move along geodesics of the spacetime V_n. We have thus obtained the very interesting result that the worldlines C_i which are obtained as intersections $V_n \cap \hat{V}_m^{(i)}$ are geodesics of the spacetime sheet V_n. The same result is also obtained directly by varying the action (4) with respect to the variables $X_i^\mu(\tau)$.

A solution to the equations of motion (5) (or (7)) gives both: the spacetime sheet $\eta^a(x)$ and the worldlines $X_i^\mu(\tau)$. Once $\eta^a(x)$ is determined the induced metric $g_{\mu\nu} = \partial_\mu \eta^a \partial_\nu \eta_a$ is determined as well. But such a metric in general does not satisfy the Einstein equations. In the next section we shall see that quantum effects induce the necessary Einstein–Hilbert term $(-g)^{1/2} R$.

2.2 The dynamics of a system of many intersecting branes

Now we shall assume that V_{m_j} are dynamical too. Therefore we add a corresponding kinetic term to the action. So we obtain an action for a system of intersecting branes η_i, $i = 1, 2, \ldots$:

$$I[\eta_i] = \sum_i \int \omega_0 \sqrt{|f_i|} \, \mathrm{d}x_i + \frac{1}{2} \sum_{ij} \int \omega_{ij} \, \delta^N(\eta_i - \eta_j) \sqrt{|f_i|} \sqrt{|f_j|} \, \mathrm{d}x_i \mathrm{d}x_j \tag{10}$$

Besides the kinetic term for free branes, our action (10) contains also the interactive terms. The interactions result from the intersections of the branes.

The equations of motion for the i-th brane are

$$\partial_\mu \left[\sqrt{|f_i|} \partial^\mu \eta_i^a \left(\omega_0 + \sum_{i \neq j} \int \omega_{ij} \, \delta^N(\eta_i - \eta_j) \sqrt{|f_j|} \mathrm{d}x_j \right) \right] = 0 \tag{11}$$

The same equations (with the identification $\eta \equiv \eta_i$, $\kappa_j \equiv \omega_{ij}$) follow also from (2),(3) or (4). However, with (10) we have a self consistent system, where each brane determines the motion of all the others.

Returning now to the action (2), (3) experienced by one of the branes whose worldsheet is represented by $\eta_i^a(x_i) \equiv \eta^a(x)$, we find after integrating out x_j, $j \neq i$ that

$$I[\eta] = \omega_0 \int \mathrm{d}^n x \sqrt{|f|} + \sum_j \kappa_j \int \mathrm{d}^n x \, \mathrm{d}^{p_j+1}\xi \, (\det \partial_A X_j^\mu \partial_B X_j^\nu f_{\mu\nu})^{1/2} \, \delta^n(x - X_j(\xi)) \tag{12}$$

For various p_j, the latter expression is an action for a system of point particles ($p_j = 0$), strings ($p_j = 1$), and higher dimensional branes ($p_j = 2, 3, \ldots$), described by $X_j^\mu(\xi)$, moving in the background metric $f_{\mu\nu}$, which is the induced metric on our brane V_n (see Fig. 1)

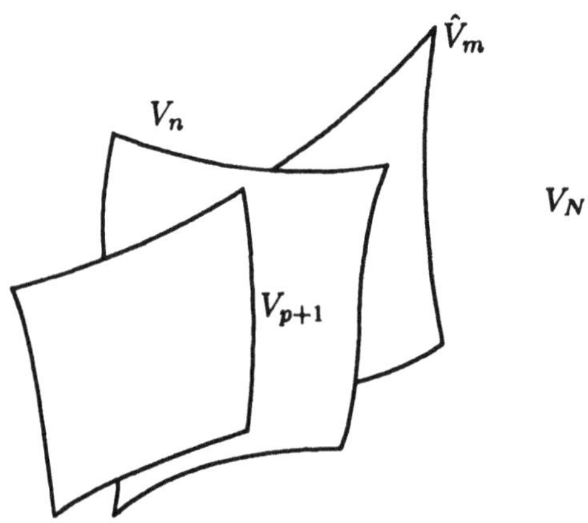

Figure 1: The intersection between two different branes V_n and $\hat{V}_m$ can be a p-brane V_{p+1}.

We see that the interactive term in (10) manifests itself in various ways, depending on how we look at it. It is a manifestation of the fact that the metric of the embedding space is curved (in particular, the metric is singular on the system of branes). From the point of view of a chosen brane V_n the interactive term becomes the action for a system of p-branes (including point particles) moving on V_n. If we now adopt the *brane world* view, where V_n is our spacetime, we see *that matter on V_n comes from other branes's worldsheets which happen to intersect our worldsheet V_n*. Those other branes, in turn, are responsible for the non trivial metric of the embedding space.

2.3 The brane interacting with itself

In (2), (3) or (10) we have a description of a brane interacting with other branes. What about self interaction? In the second term of the action (10) we have excluded the self interaction. In principle we should not exclude the self interaction, since there is no reason why a brane could not interact with itself.

Let us return to the action (2), (3) and let us calculate $\omega(\eta)$, this time assuming for simplicity that there is only one brane $V_{m_j} \equiv \hat{V}_m$ which coincides with our brane V_n. Hence the intersection is the brane V_n itself, and according to (3) we have

$$\begin{aligned}
\omega(\eta) &= \omega_0 + \kappa \int \mathrm{d}^n\hat{x}\sqrt{|\hat{f}|}\,\delta^N(\eta - \hat{\eta}(\hat{x}) \\
&= \omega_0 + \kappa \int \mathrm{d}^n\xi\,\sqrt{|\hat{f}|}\,\delta^n(x - X(\xi)) \\
&= \omega_0 + \kappa \int \mathrm{d}^n x\,\delta^n(x - X(x)) = \omega_0 + \kappa \qquad (13)
\end{aligned}$$

Here the coordinates ξ^A, $A = 0, 1, 2, ..., n-1$ cover the manifold V_n, and $\hat{f}_{AB}$ is the metric of V_n in coordinates ξ^A. The other coordinates are x^μ, $\mu = 0, 1, 2, ..., n-1$. In the last step in (13) we have used the property that the measure is invariant, $\mathrm{d}^n\xi\sqrt{|\hat{f}|} = \mathrm{d}^n x\sqrt{|f|}$.

The result (13) demonstrates that we do not need to separate a constant term ω_0 from the function $\omega(\eta)$. For a brane moving in a background of many branes we can replace (3) with

$$\omega(\eta) = \sum_j \int \kappa_j \delta^N(\eta - \eta_j)\,\sqrt{|f_j|}\mathrm{d}x_j \qquad (14)$$

where j runs over *all* the branes within the system. Any brane feels the same background, and its action for a fixed i is

$$I[\eta_i] = \int \omega(\eta_i)\sqrt{|f_i|}\mathrm{d}x_i = \sum_j \int \kappa_j \delta^N(\eta_i - \eta_j)\,\sqrt{|f_i|}\sqrt{|f_j|}\mathrm{d}x_i\,\mathrm{d}x_j \qquad (15)$$

However the background is self consistent: it is a solution to the variational principle given by the action

$$I[\eta_i] = \sum_{i \geq j} \omega_{ij}\delta^N(\eta_i - \eta_j)\sqrt{|f_i|}\sqrt{|f_j|}\mathrm{d}x_i\,\mathrm{d}x_j \qquad (16)$$

where now also i runs over *all* the branes within the system; the case $i = j$ is also allowed.

In (16) the self interaction or self coupling occurs whenever $i = j$. The self coupling term of the action is ($\kappa_i \equiv \omega_{ii}$)

$$\begin{aligned}
I_{\mathrm{self}}[\eta_i] &= \sum_i \kappa_i \int \delta^N(\eta(x_i) - \eta_i(x_i'))\sqrt{|f_i(x_i)|}\sqrt{|f_i(x_i')|}\mathrm{d}x_i\mathrm{d}x_i' \\
&= \sum_i \kappa_i \int \delta^N(\eta - \eta_i(x_i))\delta^N(\eta - \eta_i(x_i'))\sqrt{|f_i(x_i)|}\sqrt{|f_i(x_i')|}\mathrm{d}x_i\mathrm{d}x_i'\mathrm{d}^N\eta \\
&= \sum_i \kappa_i \int \delta^N(\eta - \eta_i(x_i))\delta^{n_i}(x_i - x_i')\sqrt{|f_i(x_i)|}\mathrm{d}x_i\mathrm{d}x_i'\mathrm{d}^N\eta \\
&= = \sum_i \kappa_i \sqrt{|f_i(x_i)|}\mathrm{d}^{n_i}x_i \qquad (17)
\end{aligned}$$

where we have used the same procedure which led us to Eq. (13). We see that the interactive action (16) automatically contains also the minimal surface terms, so that they do not need to be postulated separately.

2.4 A system of many branes creates bulk and its metric

We can now imagine that a system of branes (a brane configuration) can be identified with the embedding space in which a single brane moves. Here we have a concrete realization of that idea. We have a system of branes which intersect. The only interaction between the branes is due to intersection ("contact interaction"). The interaction at the intersection influences the motion of a (test) brane: it feels a potential because of the presence of other branes. If there are many branes and a test brane moves in the midst of them, then on average it feels a metric field which is approximately continuous. Our test brane moves in an effective metric of the embedding space.

A single brane or several branes give the singular conformal metric. Many branes are expected to give, on average, an arbitrary metric.

There is a close interrelationship between the presence of branes and the bulk metric. In the model we discuss here the bulk metric is singular on the branes, and zero elsewhere. Without the branes there is no metric and no bulk. Actually the bulk consists of the branes which determine its metric.

Something quite analogous occurs in string theory, more precisely, in the theory of closed superstrings. Although classically the string theory is formulated in a background spacetime with a fixed (Minkowski) metric, it turns out after quantization that the background metric in which the quantum string moves cannot be fixed. The string itself determines what are the equations of motion for the metric of the embedding space (the target space in the string theoretic jargon). This could be intuitively understood by noticing that the quantum string automatically involves many strings. A generic quantum state is a many strings state and effectively it leads to gravity in target space. What is still not quite satisfactory in string theory is its background dependent starting point, namely use of the Minkowski metric. I believe that the many intersecting brane model (which, of course, includes also strings) resolves the issue of background independence at the classical level, since the action (16), which includes also self interaction, contains no metric of a background embedding space. It is true that in (16) we have the quantity $\sqrt{|f_i|}$, $f_i \equiv \det f_{i\mu\nu}$, where $f_{i\mu\nu} \equiv \partial_\mu \eta_i^a \partial_\nu \eta_i^b \eta_{ab}$. The latter quantity is the metric on the i-th brane worldsheet, but now it can not be considered as a metric induced from an embedding space metric, since according to (14) the latter metric vanishes outside the branes. Hence in effect there is no embedding space, apart from the system of branes itself. The fixed quantity η_{ab} can even less be interpreted as a metric of an embedding space. It is the Minkowski metric of the flat space to which there corresponds a conformally flat space which, because of the singular conformal factor, is identified with the system of branes. The latter conformally flat space is our dynamical system, but the former flat space (with metric η_{ab}) is not, and therefore can hardly be considered as a background space for our dynamical system of branes.

2.5 Matter from the intersection of our brane with itself

Our model of intersecting branes allows for the possibility that a brane intersects itself, as schematically illustrated in Fig. 2.

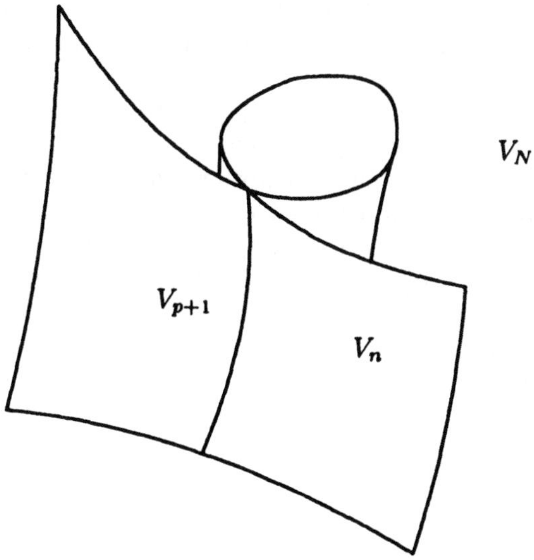

Figure 2: Illustration of a self-intersecting brane. At the intersection V_{p+1}, because of the contact interaction, the stress-energy tensor on the brane V_n is singular, and it manifests itself as matter on V_n. The manifold V_{p+1} is a worldsheet swept by a p-brane and it is a minimal surface (e.g. a geodesic, when $p = 0$) in V_n.

The analysis used so far is valid also for the situations like the one in Fig. 2, if we divide the worldsheet V_n in two pieces which are glued together at a submanifold C, situated somewhere within the "loop" region.

There is a variety of ways a worldsheet can self intersect. For instance, it may intersect itself many times to form a sort of helix or spiral. Instead of the intersection with a single loop, like in Fig. 2, the intersection may form a double or triple loop. In this respect some new possibilities occur, waiting to be explored in detail (see also [10]).

3 The Quantum Model

For the purpose of quantization we shall use the classical action [6], that is a generalization of the well known Howe–Tucker action [11], which is equivalent to (2):

$$I[\eta^a, g^{\mu\nu}] = \tfrac{1}{2} \int \mathrm{d}^n x \sqrt{|g|}\omega(\eta)(g^{\mu\nu}\partial_\mu \eta^a \partial_\nu \eta_a + 2 - n). \tag{18}$$

It is a functional of the embedding functions $\eta^a(x)$ and the Lagrange multipliers $g^{\mu\nu}$. Varying (18) with respect to $g^{\mu\nu}$ gives the constraints

$$-\frac{\omega}{4}\sqrt{|g|}\, g_{\alpha\beta}(g^{\mu\nu}\partial_\mu \eta^a \partial_\nu \eta_a + 2 - n) + \frac{\omega}{2}\sqrt{|g|}\,\partial_\alpha \eta^a \partial_\beta \eta_a = 0. \tag{19}$$

Contracting (19) with $g^{\alpha\beta}$, we find $g^{\mu\nu}\partial_\mu\eta^a\partial_\nu\eta_a = n$, and after inserting the latter relation back into (19) we find

$$g_{\alpha\beta} = \partial_\alpha\eta^a\partial_\beta\eta_a, \tag{20}$$

which is the expression for an induced metric on the surface V_n. In the following paragraphs we shall specify $n = 4$; however, whenever necessary, we shall switch to the generic case of arbitrary n.

In the classical theory we may say that a 4-dimensional spacetime sheet is swept by a 3-dimensional space-like hypersurface Σ which moves forward in time. The latter surface is specified by initial conditions, and the equations of motion then determine Σ at every value of a time-like coordinate $x^0 = t$. Knowledge of a particular hypersurface Σ implies knowledge of the corresponding intrinsic 3-geometry specified by the 3-metric $g_{ij} = \partial_i\eta^a\partial_j\eta_a$ induced on Σ $(i, j = 1, 2, 3)$. However, knowledge of the data $\eta^a(t, x^i)$ on an entire infinite Σ is just a mathematical idealization which cannot be realized in a practical situation by an observer because of the finite speed of light.

In quantum theory a state of a surface Σ is not specified by the coordinates $\eta^a(t, x^i)$, but by a wave functional $\psi[t, \eta^a(x^i)]$. The latter represents the probability amplitude that at time t an observer would obtain, as a result of measurement, a particular surface Σ.

The probability amplitude for the transition from a state with definite Σ_1 at time t_1 to a state Σ_2 at time t_2 is given by the Feynman path integral

$$K(2,1) = \langle\Sigma_2, t_2|\Sigma_1, t_1\rangle = \int \mathrm{e}^{i\,I[\eta,g]}\mathcal{D}\eta\mathcal{D}g. \tag{21}$$

Now, if in Eq.(21) we perform integration only over the embedding functions $\eta^a(x^\mu)$, then we obtain the so called ***effective action*** I_{eff}

$$\mathrm{e}^{iI_{\text{eff}}[g]} \equiv \int \mathrm{e}^{i\,I[\eta,g]}\mathcal{D}\eta, \tag{22}$$

which is a functional of solely the metric $g^{\mu\nu}$. ¿From Eq. (22) we obtain by functional differentiation

$$\frac{\delta I_{\text{eff}}[g]}{\delta g^{\mu\nu}} = \frac{\int \frac{\delta I[\eta,g]}{\delta g^{\mu\nu}}\,\mathrm{e}^{i\,I[\eta,g]}\mathcal{D}\eta}{\int \mathrm{e}^{i\,I[\eta,g]}\mathcal{D}\eta} \equiv \left\langle \frac{\delta I[\eta,g]}{\delta g^{\mu\nu}} \right\rangle = 0. \tag{23}$$

On the left–hand side of Eq. (23) we have taken into account the constraints $\delta I[\eta,g]/\delta g^{\mu\nu} = 0$ (explicitly given in Eq. (19)).

The expression $\delta I_{\text{eff}}[\eta,g]/\delta g^{\mu\nu} = 0$ gives the classical equations for the metric $g_{\mu\nu}$, derived from the effective action.

Let us now consider a specific case in which we take for $\omega(\eta)$ the expression (3). Then our action (18) splits into two terms

$$I[\eta,g] = I_0[\eta,g] + I_m[\eta,g] \tag{24}$$

with

$$I_0[\eta,g] = \frac{\omega_0}{2}\int \mathrm{d}^n x\sqrt{|g|}\,(g^{\mu\nu}\partial_\mu\eta^a\partial_\nu\eta_a + 2 - n), \tag{25}$$

$$I_m[\eta, g] = \frac{1}{2} \int \mathrm{d}^n x \sqrt{|g|} \sum_i \kappa_i \frac{\delta^N(\eta - \hat{\eta}_i)}{\sqrt{|\gamma|}} \mathrm{d}^m \hat{x}$$

$$\times \sqrt{|\hat{f}|} \, (g^{\mu\nu} \partial_\mu \eta^a \partial_\nu \eta_a + 2 - n). \tag{26}$$

The last expression can be integrated over $m-1$ coordinates $\hat{x}^\mu$, while $\hat{x}^0$ is chosen so to coincide with the parameter τ of a worldline C_i. We also split the metric as $g^{\mu\nu} = n^\mu n^\nu / n^2 + \bar{g}^{\mu\nu}$, where n^μ is a time-like vector and $\bar{g}^{\mu\nu}$ is the projection tensor, giving $\bar{g}^{\mu\nu} \partial_\mu \eta^a \partial_\nu \eta_a = n - 1$. So we obtain

$$I_m[\eta, g] = \frac{1}{2} \int \mathrm{d}^n x \sqrt{|g|} \sum_i \frac{\delta^n(x - X_i(\tau))}{\sqrt{|g|}}$$

$$\times \left(\frac{g_{\mu\nu} \dot{X}_i^\mu \dot{X}_i^\nu}{\mu_i} + \mu_i \right) \mathrm{d}\tau = I_m[X_i, g]. \tag{27}$$

Here $\mu_i \equiv 1/\sqrt{n^2}|_{C_i}$ are the Lagrange multipliers giving, after variation, the worldline constraints $\mu_i^2 = \dot{X}_i^2$. Eq. (27) is the well known Howe-Tucker action [11] for point particles.

Now let us substitute our specific action (24)–(27) into the expression (22) for the effective action. The functional integration now runs over two distinct classes of spacetime sheets V_n [represented by $\eta^a(x)$]:

(a) those V_n which either *do not intersect* the matter sheets $\hat{V}_m^{(i)}$ [represented by $\hat{\eta}_i^a(\hat{x})$] and do not self intersect, or, if they do, the intersections are just single points, and

(b) those V_n which *do intersect* $\hat{V}_m^{(i)}$ and/or self-intersect, the intersections being worldlines C_i.

The sheets V_n which correspond to the case (b) have two distinct classes of points (events):

(b1) *the points outside the intersection*, i.e., outside the worldlines C_i, and

(b2) *the points on the intersection*, i.e., the events belonging to C_i.

The measure $\mathcal{D}\eta^a(x)$ can be factorized into the contribution which corresponds to the case (a) or (b1) ($x \notin C_i$), and the contribution which corresponds to the case (b2) ($x \in C_i$):

$$\begin{aligned} \mathcal{D}\eta &= \prod_{a,x} (|g(x)|)^{1/4} \mathrm{d}\eta^a(x) \\ &= \prod_{a, x \notin C_i} (|g(x)|)^{1/4} \mathrm{d}\eta^a(x) \prod_{a, x \in C_i} (|g(x)|)^{1/4} \mathrm{d}\eta^a(x) \\ &\equiv \mathcal{D}_0 \eta \, \mathcal{D}_m. \end{aligned} \tag{28}$$

The additional factor $(|g(x)|)^{1/4}$ comes from the requirement that the measure be invariant under reparametrizations of x^μ (see Ref. [12] for details). From the very definition of $\prod_{a,x\in C_i}(|g(x)|)^{1/4}\mathrm{d}\eta^a(x)$ as the measure of the set of points on the worldlines C_i (each C_i being represented by an equation $x = X_i^\mu(\tau)$) we conclude that

$$\mathcal{D}_m\eta^a(x) = \mathcal{D}X_i^\mu(\tau). \tag{29}$$

The effective action then satisfies [owing to (24)–(29)]

$$\mathrm{e}^{iI_{\mathrm{eff}}[g]} = \int \mathrm{e}^{iI_0[\eta,g]}\,\mathcal{D}_0\eta\,\mathrm{e}^{iI_m[X_i,g]}\,\mathcal{D}X_i \equiv \mathrm{e}^{iW_0}\,\mathrm{e}^{iW_m}, \tag{30}$$

$$I_{\mathrm{eff}} = W_0 + W_m. \tag{31}$$

The measure $\mathcal{D}_0\eta$ includes all those sheets V_n that do not intersect the matter sheet and do not self-intersect [case (a)], and also all those sheets which do intersect and/or self-intersect [case (b1)], apart from the points on the intersections.

The first factor in the product (30) contains the action (25). The latter has the same form as the action for N scalar fields in a curved background spacetime with the metric $g_{\mu\nu}$. The corresponding effective action has been studied and derived in Refs. [13]. Using the same procedure and substituting our specific constants $\omega_0/2$ and $(n-1)$ occurring in Eq. (25), we find the following expression for the effective action:

$$I_{\mathrm{eff}} = \lim_{\mu^2\to 0}\int \mathrm{d}^n x\sqrt{|g|}\left(N\,\omega_0^{-1}(4\pi)^{-n/2}\sum_{j=0}^{\infty}(\mu^2)^{n/2-j}\,a_j(x)\,\Gamma(j-\frac{n}{2}) + \frac{\omega_0}{2}(2-n)\right) \tag{32}$$

with

$$a_0(x) = 1 \tag{33}$$

$$a_1(x) = R/6 \tag{34}$$

$$a_2(x) = \frac{1}{12}R^2 + \frac{1}{180}(R_{\alpha\beta\gamma\delta}R^{\alpha\beta\gamma\delta} - R_{\alpha\beta}R^{\alpha\beta}) - \frac{1}{30}D_\mu D^\mu R \tag{35}$$

where R, $R_{\alpha\beta}$ and $R_{\alpha\beta\gamma\delta}$ are the Ricci scalar, the Ricci tensor and the Riemann tensor, respectively. The function $\Gamma(y) = \int_0^\infty \mathrm{e}^{-t}\,t^{y-1}\,\mathrm{d}t$; it is divergent at negative integers y and finite at $y = \frac{3}{2}, \frac{1}{2}, -\frac{1}{2}, -\frac{3}{2}, -\frac{5}{2}, \ldots$. The effective action (32) is thus divergent in even-dimensional spaces V_n. For instance, when $n = 4$, the argument in Eq. (32) is $j-2$, which for $j = 0, 1, 2, \ldots$ is indeed a negative integer. Therefore, in order to obtain a finite effective action one needs to introduce a suitable cut-off parameter Λ and replace

$$(\mu^2)^{n/2-j}\Gamma(j-\frac{n}{2}) = \int_0^\infty t^{j-1-n/2}e^{-\mu^2 t}\mathrm{d}t$$

with

$$\int_{1/\Lambda^2}^\infty t^{j-1-n/2}e^{-\mu^2 t}\mathrm{d}t$$

Then we find

$$I_{\text{eff}} = \int \mathrm{d}^n x \sqrt{|g|} \Big(\lambda_0 + \lambda_1 R + \lambda_2 R^2 + \lambda_3 (R_{\alpha\beta\gamma\delta} R^{\alpha\beta\gamma\delta} - R_{\alpha\beta} R^{\alpha\beta})$$

$$+ \lambda_4 D_\mu D^\mu R + O(\Lambda^{n-6}) \Big) , \tag{36}$$

where[2], for $n > 4$,

$$\lambda_0 = \frac{N\omega_0}{4(4\pi)^{n/2}} \frac{2\Lambda^n}{n} + \frac{\omega_0}{2}(2-n),$$
$$\lambda_1 = \frac{N\omega_0}{4(4\pi)^{n/2}} \frac{2\Lambda^{n-2}}{6(n-2)},$$
$$\lambda_2 = \frac{N\omega_0}{4(4\pi)^{n/2}} \frac{2\Lambda^{n-4}}{12(n-4)},$$
$$\lambda_3 = \frac{N\omega_0}{4(4\pi)^{n/2}} \frac{2\Lambda^{n-4}}{180(n-4)},$$
$$\lambda_4 = -\frac{N\omega_0}{4(4\pi)^{n/2}} \frac{2\Lambda^{n-4}}{30(n-4)}. \tag{37}$$

Here λ_0 is the cosmological constant, whilst λ_1 is related to the gravitational constant G in n-dimensions according to

$$\lambda_1 \equiv (16\pi G)^{-1}. \tag{38}$$

This last relation shows how the induced gravitational constant is calculated in terms of ω_0, which is a free parameter of our embedding model, and the cutoff parameter Λ. According to Akama [3], et al., and Sugamoto [14], we consider the cutoff Λ to be a physical quantity, the inverse thickness of a membrane, because the original action (25) describes an idealized theory of extended objects with vanishing thickness, but the real extended objects have non-vanishing thickness playing a role of the ultraviolet cutoff. In the case of thin extended objects, we can ignore the $O(\Lambda^{n-4})$ terms in Eq. (36).

In the above calculation of the effective action we have treated all functions $\eta^a(x)$ entering the path integral (22) as those representing distinct spacetimes sheets V_n. However, due to the reparametrization invariance, there exist equivalence classes of functions representing the same V_n. This complication must be taken into account when calculating the entire amplitude (21). The conventional approach is to introduce ghost fields which cancel the non-physical degrees of freedom. An alternative approach, first explored in ref. [15]–[17] (see also [10]), is to assume that all possible embedding functions $\eta^a(x)$ can nevertheless be interpreted as describing physically distinct spacetime sheets $\mathcal{V}_n$. This is possible if the extra degrees of freedom in $\eta^a(x)$ describe tangent deformations of $\mathcal{V}_n$. Such a deformable surface $\mathcal{V}_n$ is then a different concept from a non-deformable surface V_n. The path integral can be performed in a straightforward way in the case of $\mathcal{V}_n$, as was done in arriving at the result (32). However, even from the standard point of view [18], gauge fixing is

[2] In the case $n = 4$ it is convenient to use Akama's regularization [3]. Similarly for the problematic coefficients in higher dimensions.

not required for the calculation of the effective action, since in the η^a integration, $g_{\mu\nu}$ is treated as a fixed background.

Let us now return to Eq. (30). In the second factor of Eq. (30) the functional integration runs over all possible worldlines $X^\mu(\tau)$. Though they are obtained as intersections of *various* V_n with $\hat{V}_m^{(i)}$, we may consider all those worldlines to be lying in *the same* effective spacetime $V_n^{(\text{eff})}$ with the intrinsic metric $g_{\mu\nu}$. In other words, in the effective theory we identify all those various V_n's, having the same induced (intrinsic) metric $g_{\mu\nu}$, as one and the same spacetime. If one considers the embedding space V_N of sufficiently high dimension N, then there is enough freedom to obtain as an intersection any possible worldline in the effective spacetime $V_n^{(\text{eff})}$.

This is even more transparent when the spacetime sheet self-intersects. Then we can have a situation in which various sheets coincide in all the points, apart from the points in the vicinity of the intersections

When the condition for the classical approximation is satisfied, i.e., when $I_m \gg \hbar = 1$, then only those trajectories X_i^μ which are close to the classically allowed ones effectively contribute:

$$e^{iW_m} = e^{iI_m[X_i,g]}\,, \qquad W_m = I_m \tag{39}$$

The effective action is then a sum of the gravitational kinetic term W_0 given in Eq. (36) and the source term I_m given in (27). Variation of I_{eff} with respect to $g^{\mu\nu}$ then gives the gravitational field equation in the presence of point particle sources with the stress–energy tensor $T^{\mu\nu}$ as given in Eq. (6):

$$R^{\mu\nu} - \frac{1}{2}g^{\mu\nu} + \lambda_0 g^{\mu\nu} + (\text{higher order terms}) = -8\pi G\, T^{\mu\nu}\,. \tag{40}$$

However, in general the classical approximation is not satisfied, and in the evaluation of the matter part W_m of the effective action one must take into account the contributions of all possible paths $X_i^\mu(\tau)$. So we have (confining ourselves to the case of only one particle, omitting the subscript i, and taking $\mu = 1$, $\kappa \equiv m$)

$$\begin{aligned} e^{iW_m} &= \int_{x_a}^{x_b} \mathcal{D}X \exp\left(\frac{i}{2}\int_{\tau_a}^{\tau_b} d\tau\, m\,(g_{\mu\nu}\dot{X}^\mu\dot{X}^\nu + 1)\right) \\ &= \mathcal{K}(x_b,\tau_b;x_a,\tau_a) \equiv \mathcal{K}(b,a) \end{aligned} \tag{41}$$

which is a propagator or a Green's function satisfying (for $\tau_b \geq \tau_a$) the equation

$$\left(i\frac{\partial}{\partial\tau_b} - H\right)\mathcal{K}(x_b,\tau_b;x_a,\tau_a) = -\frac{1}{\sqrt{|g|}}\delta^n(x_b - x_a)\,\delta(\tau_b - \tau_a)\,, \tag{42}$$

where $H = (|g|)^{-1/2}\partial_\mu((|g|)^{1/2}\partial^\mu)$. From (42) it follows that

$$\mathcal{K}(b,a) = -\left[i\frac{\partial}{\partial\tau} - H\right]^{-1}_{x_b,\tau_b;x_a,\tau_a} \tag{43}$$

where the inverse Green's function is treated as a matrix in the (x,τ) space.

Using the following relation [19] for Gaussian integration

$$\int y_m y_n \prod_{i=1}^{N} dy_i\, e^{-\sum_{ij} y_i A_{ij} y_j} \propto \frac{(A^{-1})_{mn}}{(\det|A_{ij}|)^{1/2}}\,, \tag{44}$$

we can rewrite the Green's function in terms of the second quantized field:

$$\mathcal{K}(a,b) = \int \psi^*(x_b, \tau_b)\, \psi(x_b, \tau_b)\, \mathcal{D}\psi^*\, \mathcal{D}\psi$$

$$\times \exp\left[-i\int \mathrm{d}\tau\, \mathrm{d}^n x\, \sqrt{|g|}\, \psi^*(i\partial_\tau - H)\psi\right]. \qquad (45)$$

If the conditions for the "classical" approximation are satisfied, so that the phase in (45) is much greater than $\hbar = 1$, then only those paths $\psi(\tau, x)$, $\psi^*(\tau, x)$ which are close to the extremal path, along which the phase is zero, effectively contribute to $\mathcal{K}(a, b)$. Then the propagator is simply

$$\mathcal{K}(b,a) \propto \exp\left[-i\int \mathrm{d}\tau\, \mathrm{d}^n x \sqrt{|g|}\, \psi^*(i\partial_\tau - H)\psi\right]. \qquad (46)$$

The effective one-particle "matter" action W_{m} is then

$$W_{\mathrm{m}} = -\int \mathrm{d}\tau\, \mathrm{d}^n x\, \sqrt{|g|}\, \psi^*(\tau, x)(i\partial_\tau - H)\psi(\tau, x). \qquad (47)$$

If we assume that the τ-dependence of the field $\psi(\tau, x)$ is specified[3] by $\psi(\tau, x) = \mathrm{e}^{im\tau}\, \phi(x)$, then Eq. (47) simplifies to the usual well known expression for a scalar field:

$$W_{\mathrm{m}} = \int \mathrm{d}^n x \sqrt{|g|} \phi^*(x) \left[\frac{1}{\sqrt{|g|}} \partial_\mu (\sqrt{|g|} \partial^\mu) + m^2\right) \phi(x)$$

$$= -\tfrac{1}{2} \int \mathrm{d}^n x\, \sqrt{|g|}\, (g^{\mu\nu}\, \partial_\mu \phi^*\, \partial_\nu \phi - m^2), \qquad (48)$$

where the surface term has been omitted.

Thus, starting from our basic fields $\eta^a(x)$, which are the embedding functions for a spacetime sheet V_n, we have arrived at the effective action I_{eff} which contains the kinetic term W_0 for the metric field $g^{\mu\nu}$ (see Eq. (36)) and the source term W_{m} (see Eq. (40) and (26), or (48)). Both the metric field $g_{\mu\nu}$ and the bosonic matter field ϕ are induced from the basic fields $\eta^a(x)$.

4 Conclusion

We have investigated a model which seems to be very promising in attempts to find a consistent relation between quantum theory and gravity. Our model exploits the approach of induced gravity and the concept of spacetime embedding in a higher-dimensional space and has the following property of interest: what appears as worldlines in, e.g., a 4-dimensional spacetime, are just the intersections of a spacetime sheet V_4 with 'matter' sheets $\widehat{V}_{\mathrm{m}}^{(i)}$, or self intersections of V_4. Various

[3] By doing so we in fact project out the so called physical states from the set of all possible states. Such a procedure, which employs a "fictitious" evolution parameter τ is often used [22]. When gauge fixing the action (41) one pretends that such an action actually represents an "evolution" in parameter τ. Only later, when all the calculations (e.g., path integral) are performed, one integrates over τ and thus projects out the physical quantities.

choices of spacetime sheets then give various configurations of worldlines. Instead of V_4 it is convenient to consider a spacetime sheet V_n of arbitrary dimension n. When passing to the quantized theory a spacetime sheet is no longer definite. All possible alternative spacetime sheets are taken into account in the expression for a wave functional or a Feynman path integral. The intersection points of V_n with itself or with a matter sheet $\hat{V}_m^{(i)}$ are treated specially, and it is found that their contribution to the path integral is identical to the contribution of a point particle path. We have paid special attention to the effective action which results from functionally integrating out all possible embeddings with the same induced metric tensor. We have found that the effective action, besides the Einstein–Hilbert term and the corresponding higher-order terms, also contains a source term. The expression for the latter is equal to that of a classical (when such an approximation can be used) or quantum point particle source described by a scalar (bosonic) field.

In other words, we have found that the (n-dimensional) Einstein equations (including the R^2 and higher derivative terms) with classical or quantum point-particle sources are effective equations resulting from performing a quantum average over all possible embeddings of the spacetime. Gravity —as described by Einstein's general relativity— is thus considered not as a fundamental phenomenon, but as something induced quantum-mechanically from more fundamental phenomena.

In our embedding model of gravity with bosonic sources, new and interesting possibilities are open. For instance, instead of a 4-dimensional spacetime sheet we can consider a sheet which possesses additional dimensions, parametrized either with usual or Grassmann coordinates. In such a way we expect to include, on the one hand, via the Kaluza–Klein mechanism, also other interactions besides the gravitational one, and, on the other hand, fermionic sources. The latter are expected also to result from the polivector generalization of the action, as discussed in refs. [23, 10].

It is well known that the quantum field theory based on the action (48) implies the infinite vacuum energy density and consequently the infinite (or, more precisely, the Planck scale cutoff) cosmological constant. This is only a part of the total cosmological constant predicted by the complete theory, as formulated here. The other part comes from eqs. (36), (37). The parameters N, ω_0, Λ and n could in principle be adjusted so to give a small or vanishing total cosmological constant. If carried out successfully this would be then an alternative (or perhaps a complement) to the solution of the cosmological constant problem as suggested in ref.[24].

Finally, let us observe that within the "brane world" model of Randall and Sundrum [20], in which matter fields are localized on a 3-brane, whilst gravity propagates in the bulk, it was recently proposed [21] to treat the Einstein–Hilbert term on the brane as being induced in the quantum theory of the brane. It was found that the localized matter on a brane can induce via loop correction a 4D kinetic term for gravitons. This also happens in our quantum model if we consider the effective action obtained after integrating out the second quantized field in (45). (The procedure expounded in refs. [22] is then directly applicable.) In addition, in our model we obtain, as discussed above, a kinetic term for gravity on the brane after functionally integrating out the embedding functions.

References

[1] A. D. Sakharov, Dok. Akad. Nauk. SSSR **177**, 70, 1967; Sov. Phys. JETP, **12**, 1040, 1968.

[2] S.L. Adler, Rev. Mod. Phys., **54**, 729, 1982, and references therein.

[3] K. Akama, Progr. Theor. Phys., **60**, 1900, 1978; **79**, 1299, 1988; K. Akama and H. Terazawa, Prog. Theor. Phys., **79**, 740, 1988; H. Terazawa, in Proceedings of the First International A. D. Sakharov Conference on Physics, L. V. Keldysh, et al., eds. Nova Science, New York, 1991; S. Naka and C. Itoi, Progr. Theor. Phys., **70**, 1414, 1983.

[4] T. Regge and C. Teitelboim, in "Proceedings of the Marcel Grossman Meeting, Trieste, 1975.

[5] M. Pavšič, Class. Quant. Grav., **2**, 869, 1985; Phys. Lett., **A107**, 66, 1985; V. Tapia, Class. Quant. Grav., **6**, L49, 1989; D.Maia, Class. Quant. Grav., **6**, 173, 1989.

[6] M. Pavšič, Foundations of Physics, **24**, 1495, 1994.

[7] M. Pavšič, Physics Letters A, **116**, 1, 1986.

[8] M. Pavšič, Gravitation & Cosmology, **2**, 1, 1996.

[9] I.A. Bandos, Modern Physics Letters A, **12**, 799, 1997; I. A. Bandos and W. Kummer, hep-th/9703099.

[10] M. Pavsšič, "The Landscape of Theoretical Physics: A Global View", Kluwer Academic, to appear.

[11] P.S. Howe and R.W. Tucker, J. Phys. A: Math. Gen., **10**, L155, 1977; A. Sugamoto, Nuclear Physics B, **215**, 381, 1983; E. Bergshoeff, E. Sezgin and P.K. Townsend, Physics Letter B, **189**, 75, 1987; A. Achucarro, J.M. Evans, P.K. Townsend and D.L. Wiltshire, Physics Letters B, **198**, 441, 1987; M. Pavšič, Class. Quant. Grav., **5**, 247, 1988; see also M. Pavšič, Physics Letters B, **197**, 327, 1987.

[12] K. Fujikawa, Phys. Rev. Lett., **42**, 1195, 1979; **44**, 1733, 1980; Phys. Rev. D, **21**, 2448, 1980; **23**, 2262, 1981; Nucl. Phys. B, **226**, 437, 1983; **245**, 436, 1984; for a review see M. Basler, Fortschr. Phys., **41**, 1, 1993.

[13] See e.g. B.S. De Witt, Phys. Reports, **19**, 295, 1975; S. M. Christensen, Phys. Rev., **D14**, 2490, 1976; N. D. Birrell and P.C.W. Davies, "Quantum Fields in Curved Space", Cambridge University Press, Cambridge, 1982; H. Boschi-Filho and C. P. Natividade, Phys. Rev. D, **46**, 5458, 1992.

[14] A.Sugamoto, Nuclear Physics B, **215**, 381, 1983.

[15] M. Pavšič, Found. Phys., **25**, 819, 1995;

M. Pavšič, Nuovo Cimento A, **108**, 221, 1995.

[16] M. Pavšič, Foundations of Physics, **26**, 159, 1996.

[17] M. Pavšič, Nuovo Cimento A, **110**, 369, 1997.

[18] R.Floreanini and R.Percacci, Modern Physics Letters A, **5**, 2247, 1990.

[19] See, e.g., M. Kaku, "Introduction to Superstrings" Springer–Verlag, New York, N.Y., 1988.

[20] L. Randall and R. Sundrum, Physical Review Letters, **83**, 3370, 1999; **83**, 4690, 1999.

[21] G. Dvali, G. Gabadadze and M. Porrati, Physics Letters B, **485**, 208, 2000.

[22] B.S. DeWitt, Phys. Rep., **19**, 295, 1975; S.M. Christensen, Phys. Rev. D, **14**, 2490, 1976; L.S. Brown, Phys. Rev. D **15**, 1469, 1977; T.S. Bunch and L. Parker, Phys. Rev. D **20**, 2499, 1979; H. Boschi-Filho and C.P. Natividade, Phys. Rev. D, **46**, 5458, 1992; A. Follacci, Phys. Rev. D, **46**, 2553, 1992; A. Sugamoto, Nuclear Physics B, **215**, 381, 1983.

[23] M. Pavšič, "Clifford Algebra Based Polydimensional Relativity and Relativistic Dynamics", hep-th 0011216.

[24] M. Pavšič, Phys. Lett. A, **254**, 119, 1999.

A Model of the Spacetime Foam

V. Dzhunushaliev
Phys. Dept., Kyrgyz-Russian Slavic University
Bishkek, 720000, Kyrgyz Republic
Email:dzhun@hotmail.kg

1 Introduction

A hypothesized spacetime foam is a very interesting object for investigation. The history of this question has a long time : at 50th Wheeler presupposed that a spacetime can have a very complex structure (on the topological level) on the Planck scale. By his assumption the spacetime has topological handles attached to our spacetime. These handles (we can name their as quantum wormholes) appear and disappear as quantum objects. The problem of description of such phenomenon is connected with our understanding of quantum gravity. Every physical theory has the questions answers on which has the very important meaning for understanding of this theory. Evidently such problem for quantum gravity is a hypothesized spacetime foam. Definitely a basic problem here is the variation of spacetime topology by appearing/disappearing of quantum handles. Any earlier existing physical theories had not a similar sort of problems. The problem is that by appearing/disappearing of quantum handles we have the singular points in which a time arrow is undefined. It follows from Morse theory describing the topology transition such points have name as critical points.

Hence we should try to simplify this problem : attempt to present any effective and approximate model of spacetime foam. The main goal of this paper is to submit ***an effective model of the spacetime foam.***

2 Model of a single quantum wormhole

At first we present a model of a single handle in the spacetime foam, see Fig(2). The 5D metric [3, 4, 5] for the throat is

$$
\begin{aligned}
ds^2 &= \eta_{AB}\omega^A\omega^B = \\
&\quad -\frac{r_0^2}{\Delta(r)}(d\chi - \omega(r)dt)^2 + \Delta(r)dt^2 - dr^2 - \\
&\quad a(r)\left(d\theta^2 + \sin\theta^2 d\varphi^2\right), \qquad (1)\\
a &= r_0^2 + r^2, \quad \Delta = \pm\frac{2r_0}{q}\frac{r^2 + r_0^2}{r^2 - r_0^2}, \\
&\quad \omega = \pm\frac{4r_0^2}{q}\frac{r}{r^2 - r_0^2}. \qquad (2)
\end{aligned}
$$

Frontiers of Fundamental Physics 4, Edited by Sidharth and Altaisky,
Kluwer Academic/Plenum Publishers, New York, 2001

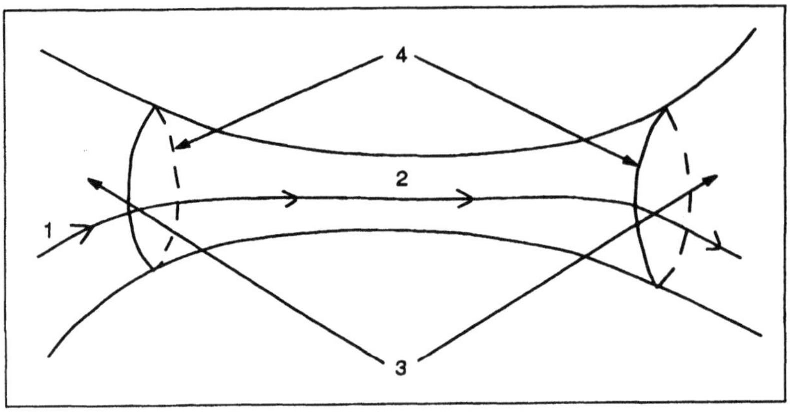

Figure 1: The model of a single quantum WH. Whole spacetime is 5 dimensional one. The external spacetime (3) has nonvariable G_{55} and the Kaluza-Klein theory is in its initial interpretation : 4D gravity + electromagnetism. In the throat (2) G_{55} component of the 5D metric is variable and Kaluza-Klein theory is equivalent to 4D gravity+electromagnetism+scalar field. Near the EH (4) the metric is the Reissner-Nordström metric and the throat is a solution of the 5D Kaluza-Klein theory. (1) is the force line of the electric field.

where χ is the 5th extra coordinate; $\eta_{AB} = (\pm, -, -, -, \mp)$, $A, B = 0, 1, 2, 3, 5$; r, θ, φ are the $3D$ polar coordinates; $r_0 > 0$ and q are some constants. We can see that there are two closed $ds^2_{(5)}(\pm r_0) = 0$ hypersurfaces at the $r = \pm r_0$. By Bronnikov designation ([6]) such hypersurfaces are named as a D-holes. On these hypersurfaces we should join [7]:

- the flux of the 4D electric field with the flux of the 5D electric field (defined by $R_{5t} = 0$ Einstein equation).
- the area of the Reissner-Nordström EH (EH) with the area of the D-hole.

The explanation of this fact that we have only these joining conditión is that in some sense on the EH holds a "holography principle" (see Ref.[8] for the more detailed explanations). This means that on the EH to we have a reduction of the amount of initial data. For example, the Einstein - Maxwell equations for the Reissner-Nordström metric

$$ds^2 = \Delta dt^2 - \frac{dr^2}{\Delta} - r^2 \left(d\theta^2 + \sin^2 d\varphi^2 \right), \tag{3}$$

$$A_\mu = (\omega, 0, 0, 0)) \tag{4}$$

(where A_μ is the electromagnetic potential, κ is the gravitational constant) is

$$-\frac{\Delta'}{r} + \frac{1-\Delta}{r^2} = \frac{\kappa}{2}\omega'^2, \tag{5}$$

$$\omega' = \frac{q}{r^2}. \tag{6}$$

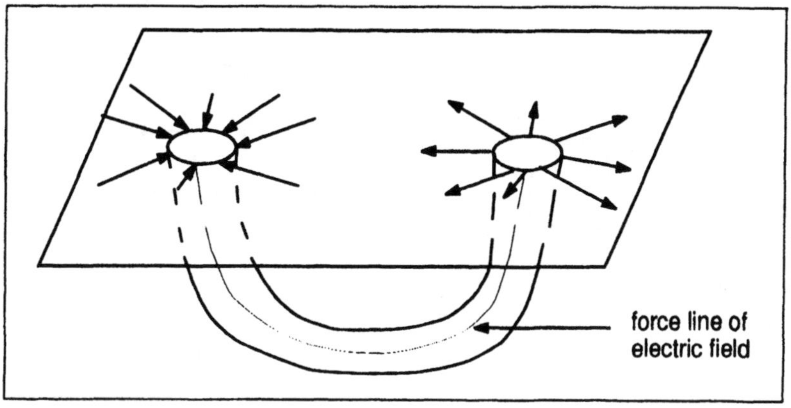

Figure 2: The left mouth of the quantum WH entraps the force lines of the electric field and looks as (-) electric charge. The force lines outcome from the right mouth of WH and looks as (+) charge.

For the Reissner - Nordström black hole the EH is defined by the condition $\Delta(r_g) = 0$, where r_g is the radius of the EH. Immediately we see that on the EH

$$\Delta'_g = \frac{1}{r_g} - \frac{\kappa}{2} r_g \omega_g'^2, \tag{7}$$

subscript (g) means that the corresponding value is taken on the EH. Thus the Einstein equation (5) is a first-order differential equations in the whole spacetime ($r \geq r_g$). The condition (7) tells us that the derivative of metric on the EH is expressed through the metric value on the EH. This is the same what we said above: the reduction of the amount of initial data takes place by such a way that we have only two integration constants (mass m and charge e for the Reissner-Nordström solution and q and r_0 for the 5D throat).

The 5D throat has an interesting property [9]. Above we have seen that the signs of the η_{55} and η_{00} can be (+) or (-). The 5D throat metric is located behind the EH therefore the 4D observer is not able to determine the signs of the η_{55} and η_{00}. Moreover it is possible that this 5D metric can fluctuate between these two possibilities. Hence the external 4D observer is forced to describe such composite WH by means of something like spinor.

Another interesting characteristic property of this model of composite WH is that we have the flux of electric field through the throat, *i.e.* each mouth can entrap the electric force lines and this leads that this mouth is like to electric charge for the external 4D observer, see Fig.2. If we reduce a cross section of the throat to zero then each mouth is point-like and we can try to describe these mouthes with help of some effective field. Taking into account the spinor-like properties of quantum handles, we assume that ***spacetime foam can be described with help of an effective spinor field.***

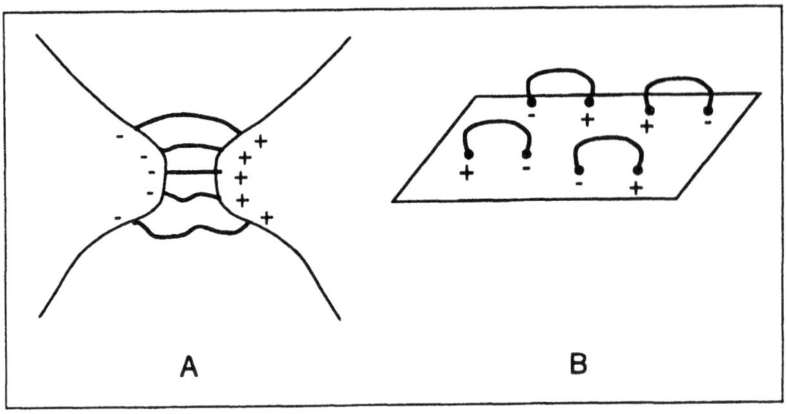

Figure 3: The figure A presents the quantum handles connecting two spaces. On the figure B the quantum handles attached to the same space are shown.

3 Approximate model of the spacetime foam

The physical meaning of the spinor field depends on the method of attaching the quantum handles to the external space, see Fig.(3).

3.1 Quantum wormholes with separated mouthes

In this case $|\psi|^2$ is a density of the mouthes in the external space and $e|\psi|^2$ is a density of the electric charge [10].

Following this way we write differential equations for the gravitational + electromagnetic fields in the presence of the spacetime foam (ψ) as follows

$$R_{\mu\nu} - \frac{1}{2}g_{\mu\nu}R = T_{\mu\nu}, \tag{8}$$

$$\left(i\gamma^\mu \partial_\mu + eA_\mu - \frac{i}{4}\omega_{\bar{a}\bar{b}\mu}\gamma^\mu \gamma^{[\bar{a}}\gamma^{\bar{b}]} - m\right)\psi = 0, \tag{9}$$

$$D_\nu F^{\mu\nu} = 4\pi e\left(\bar{\psi}\gamma^\mu\psi\right), \tag{10}$$

For our model we use the following ansatz: the spherically symmetric metric

$$ds^2 = e^{2\nu(r)}\Delta(r)dt^2 - \frac{dr^2}{\Delta(r)} - r^2\left(d\theta^2 + \sin^2 d\varphi^2\right), \tag{11}$$

the electromagnetic potential

$$A_\mu = (-\phi, 0, 0, 0), \tag{12}$$

and the spinor field

$$\tilde{\psi} = e^{-i\omega t}\frac{e^{-\nu/2}}{r\Delta^{1/4}}\left(f, 0, ig\cos\theta, ig\sin\theta e^{i\varphi}\right). \tag{13}$$

The following remark is ***very important*** : the ansatz (13) for the spinor field ψ has the $T_{t\varphi}$ component of the energy-momentum tensor and the $J^\varphi = 4\pi e(\bar{\psi}\gamma^\varphi\psi)$

component of the current. Let we remind that ψ determines the stochastical gas of the virtual WH's which can not have a preferred direction in the spacetime. This means that after substitution expression (11)-(13) into field equations they should be averaged by the spin direction of the ansatz (13). After this averaging we have $T_{t\varphi} = 0$ and $J^{\varphi} = 0$ and we have the following equations system describing our spherically symmetric spacetime

$$f'\sqrt{\Delta} = \frac{f}{r} - g\left((\omega - e\phi)\frac{e^{-\nu}}{\sqrt{\Delta}} + m\right), \tag{14}$$

$$g'\sqrt{\Delta} = f\left((\omega - e\phi)\frac{e^{-\nu}}{\sqrt{\Delta}} - m\right) - \frac{g}{r}, \tag{15}$$

$$r\Delta' = 1 - \Delta - \kappa\frac{e^{-2\nu}}{\Delta}(\omega - e\phi)\left(f^2 + g^2\right) - r^2 e^{-2\nu}\phi'^2, \tag{16}$$

$$r\Delta\nu' = \kappa\frac{e^{-2\nu}}{\Delta}(\omega - e\phi)\left(f^2 + g^2\right) - \kappa\frac{e^{-\nu}}{r\sqrt{\Delta}}fg - \frac{\kappa}{2}m\frac{e^{-\nu}}{\sqrt{\Delta}}\left(f^2 - g^2\right), \tag{17}$$

$$r^2\Delta\phi'' = -8\pi e\left(f^2 + g^2\right) - \left(2r\Delta - r^2\Delta\nu'\right)\phi', \tag{18}$$

where κ is some constant. This equations system was investigated in [11] and result is the following. A particle-like solution exists which has the following expansions at the origin $r = 0$

$$f(r) = f_1 r + \mathcal{O}(r^2), \quad g(r) = \mathcal{O}(r^2), \tag{19}$$

$$\Delta(r) = 1 + \mathcal{O}(r^2), \quad \nu(r) = \mathcal{O}(r^2), \quad \phi(r) = \mathcal{O}(r^2) \tag{20}$$

and the following asymptotical Reissner-Nordström behaviour $r \to \infty$

$$\Delta(r) \approx 1 - \frac{2m_\infty}{r} + \frac{(2e_\infty)^2}{r^2}, \quad \nu(r) \approx const, \tag{21}$$

$$\phi(r) \approx \frac{2e_\infty}{r}, \tag{22}$$

$$f \approx f_0 e^{-\alpha r}, \quad g \approx g_0 e^{-\alpha r},$$

$$\frac{f_0}{g_0} = \sqrt{\frac{m_\infty + \omega}{m_\infty - \omega}}, \quad \alpha^2 = m_\infty^2 - \omega^2, \tag{23}$$

where m_∞ and $2e_\infty$is are the mass and charge of this solution for an observer at the infinity.

In contrast with the Reissner-Nordström this solution exists for both cases $(|e_\infty|/m_\infty) > 1$ and $(|e_\infty|/m_\infty) < 1$ but for us is essential the first case with $(|e_\infty|/m_\infty) > 1$. In this case the classical Einstein-Maxwell theory leads to the "naked" singularity. The presence of the spacetime foam drastically changes this result: ***the appearance of the virtual wormholes can prevent the formation of the "naked" singularuty in the Reissner-Nordström solution with*** $|e|/m > 1$.

Our interpretation of this solution is presented on the Fig.(4).

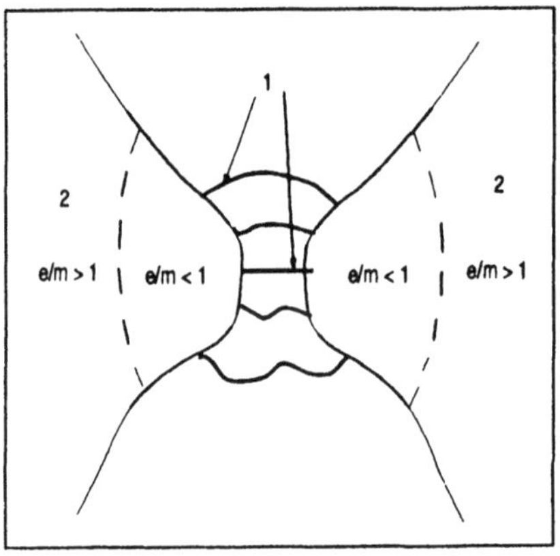

Figure 4: **1** are the quantum (virtual) WHs, **2** are two solutions with $|e_\infty|/m_\infty > 1$. Such object can be named as ***wormhole with quantum throat.***

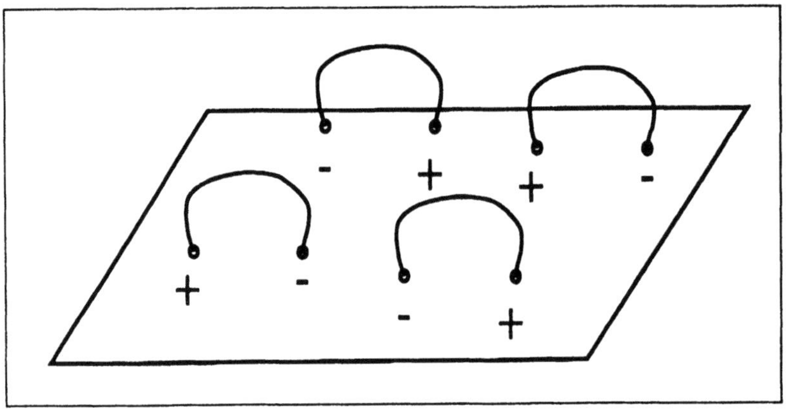

Figure 5: The handles attached to the same spacetime.

3.2 Quantum wormholes with non-separated mouthes

The second possibitiy [12] is presented on the Fig.(5).

We consider the 5D Kaluza-Klein theory : gravity + torsion + spinor field. The Lagrangian in this case is

$$\begin{aligned} \mathcal{L} \;=\; & \sqrt{-G}\left\{-\frac{1}{2k}\left(R^{(5)} - S_{ABC}S^{ABC}\right) + \right. \\ & \left. \frac{\hbar c}{2}\left[i\bar{\psi}\left(\gamma^C \nabla_C - \frac{mc}{i\hbar}\right)\psi + h.c.\right]\right\} \end{aligned} \tag{24}$$

where $\nabla_C = \partial_C - \frac{1}{4}(\omega_{\bar{A}\bar{B}C} + S_{\bar{A}\bar{B}C})\gamma^{[\bar{A}}\gamma^{\bar{B}]}$ is the covariant derivative, G is the determinant of the 5D metric, $R^{(5)}$ is the 5D scalar curvature, S_{ABC} is the antisymmetrical torsion tensor, A, B, C are the 5D world indexes, $\bar{A}, \bar{B}, \bar{C}$ are the 5-bein indexes, $\gamma^B = h^B_{\bar{A}}\gamma^{\bar{A}}$, $h^B_{\bar{A}}$ is the 5-bein, $\gamma^{\bar{A}}$ are the 5D γ matrixes with usual definitions $\gamma^{\bar{A}}\gamma^{\bar{B}} + \gamma^{\bar{B}}\gamma^{\bar{A}} = 2\eta^{\bar{A}\bar{B}}$, $\eta^{\bar{A}\bar{B}} = (+,-,-,-,-)$ is the signature of the 5D metric, ψ is the spinor field which effectively and approximately describes the spacetime foam, $[]$ means the antisymmetrization, $\hbar$, c and m are the usual constants. After dimensional reduction we have

$$\begin{aligned} \mathcal{L} = & \sqrt{-g}\left\{-\frac{1}{2k}\left(R + \frac{1}{4}F_{\alpha\beta}F^{\alpha\beta}\right) + \right. \\ & \frac{\hbar c}{2}\left[i\bar{\psi}\left(\gamma^\mu \tilde{\nabla}_\mu - \frac{1}{8}F_{\bar{\alpha}\bar{\beta}}\gamma^{\bar{5}}\gamma^{[\bar{\alpha}}\gamma^{\bar{\beta}]} - \right.\right. \\ & \left.\left.\left. \frac{1}{4}l^2_{Pl}\left(\gamma^{[\bar{A}}\gamma^{\bar{B}}\gamma^{\bar{C}]}\right)\left(i\bar{\psi}\gamma_{[\bar{A}}\gamma_{\bar{B}}\gamma_{\bar{C}]}\psi\right) - \frac{mc}{i\hbar}\right)\psi + h.c.\right]\right\} \end{aligned} \tag{25}$$

$$S^{\bar{A}\bar{B}\bar{C}} = 2l^2_{Pl}\left(i\bar{\psi}\gamma^{[\bar{A}}\gamma^{\bar{B}}\gamma^{\bar{C}]}\psi\right) \tag{26}$$

where g is the determinant of the 4D metric, $\tilde{\nabla}_\mu = \partial_\mu - \frac{1}{4}\omega_{\bar{a}\bar{b}\mu}\gamma^{[\bar{a}}\gamma^{\bar{b}]}$ is the 4D covariant derivative of the spinor field without torsion, R is the 4D scalar curvature, $F_{\alpha\beta} = \partial_\alpha A_\beta - \partial_\beta A_\alpha$ is the Maxwell tensor, $A_\mu = h^{\bar{5}}_\mu$ is the electromagnetic potential, α, β, μ are the 4D world indexes, $\bar{\alpha}, \bar{\beta}, \bar{\mu}$ are the 4D vier-bein indexes, $h^{\bar{\mu}}_\nu$ is the vier-bein, $\gamma^{\bar{\mu}}$ are the 4D γ matrixes with usual definitions $\gamma^{\bar{\mu}}\gamma^{\bar{\nu}} + \gamma^{\bar{\nu}}\gamma^{\bar{\mu}} = 2\eta^{\bar{\mu}\bar{\nu}}$, $\eta^{\bar{\mu}\bar{\nu}} = (+,-,-,-)$ is the signature of the 4D metric. Varying with respect to $g_{\mu\nu}$, $\bar{\psi}$ and A_μ leads to the following equations

$$\begin{aligned} R_{\mu\nu} - \frac{1}{2}g_{\mu\nu}R = & \frac{1}{2}\left(-F_{\mu\alpha}F^\alpha_\nu + \frac{1}{4}g_{\mu\nu}F_{\alpha\beta}F^{\alpha\beta}\right) \quad + \\ & 4l^2_{Pl}\left[\left(i\bar{\psi}\gamma_\mu\tilde{\nabla}_\nu\psi + i\bar{\psi}\gamma_\nu\tilde{\nabla}_\mu\psi\right) + h.c.\right] \quad - \\ & 2l^2_{Pl}\left[F_{\mu\alpha}\left(i\bar{\psi}\gamma^{\bar{5}}\gamma_{[\nu}\gamma^{\alpha]}\psi\right) + F_{\nu\alpha}\left(i\bar{\psi}\gamma^{\bar{5}}\gamma_{[\mu}\gamma^{\alpha]}\psi\right)\right] \quad - \\ & 2g_{\mu\nu}l^4_{Pl}\left(i\bar{\psi}\gamma^{[\bar{A}}\gamma^{\bar{B}}\gamma^{\bar{C}]}\psi\right)\left(i\bar{\psi}\gamma_{[\bar{A}}\gamma_{\bar{B}}\gamma_{\bar{C}]}\psi\right) \quad , \end{aligned} \tag{27}$$

$$\begin{aligned} & D_\nu H^{\mu\nu} = 0, \; H^{\mu\nu} = F^{\mu\nu} + \tilde{F}^{\mu\nu} \quad , \\ & \tilde{F}^{\mu\nu} = 4l^2_{Pl}\left(i\bar{\psi}\gamma^{\bar{5}}\gamma^{[\mu}\gamma^{\nu]}\psi\right) = 4l^2_{Pl}E^{\mu\nu\alpha\beta}\left(i\bar{\psi}\gamma_{[\alpha}\gamma_{\beta]}\psi\right) \quad , \end{aligned} \tag{28}$$

$$i\gamma^\mu\tilde{\nabla}_\mu\psi - \frac{1}{8}F_{\bar{\alpha}\bar{\beta}}\left(i\gamma^{\bar{5}}\gamma^{[\bar{\alpha}}\gamma^{\bar{\beta}]}\psi\right) \quad -$$

$$\frac{1}{2}l_{Pl}^2\left(i\gamma^{[\bar{A}}\gamma^{\bar{B}}\gamma^{\bar{C}]}\psi\right)\left(i\bar{\psi}\gamma_{[\bar{A}}\gamma_{\bar{B}}\gamma_{\bar{C}]}\psi\right) = 0, \tag{29}$$

where $\omega_{\bar{a}\bar{b}\mu}$ is the 4D Ricci coefficients without torsion, $E^{\mu\nu\alpha\beta}$ is the 4D absolutely antisymmetric tensor. The most interesting is the Maxwell equation (28) which permits us to discuss the physical meaning of the spinor field. We would like to show that this equation in the given form is similar to the electrodynamic in the continuous media. Let we remind that for the electrodynamic in the continuous media two tensors $\bar{F}^{\mu\nu}$ and $\bar{H}^{\mu\nu}$ are introduced [13] for which we have the following equations system (in the Minkowski spacetime)

$$\bar{F}_{\alpha\beta,\gamma} + \bar{F}_{\gamma\alpha,\beta} + \bar{F}_{\beta\gamma,\alpha} = 0, \tag{30}$$

$$\bar{H}^{\alpha\beta}_{\ \ ,\beta} = 0 \tag{31}$$

and the following relations between these tensors

$$\bar{H}_{\alpha\beta}u^\beta = \varepsilon\bar{F}_{\alpha\beta}u^\beta, \tag{32}$$

$$\bar{F}_{\alpha\beta}u_\gamma + \bar{F}_{\gamma\alpha}u_\beta + \bar{F}_{\beta\gamma}u_\alpha = \mu\left(\bar{H}_{\alpha\beta}u_\gamma + \bar{H}_{\gamma\alpha}u_\beta + \bar{H}_{\beta\gamma}u_\alpha\right) \tag{33}$$

where ε and μ are the dielectric and magnetic permeability respectively, u^α is the 4-vector of the matter. For the rest media and in the 3D designation we have

$$\varepsilon\bar{E}_i = \bar{E}_i + 4\pi\bar{P}_i = \bar{D}_i, \quad \text{where} \quad \bar{E}_i = \bar{F}_{0i}, \quad \bar{D}_{0i} = \bar{H}_{0i}, \tag{34}$$

$$\mu\bar{H}_i = \bar{H}_i + 4\pi\bar{M}_i = \bar{B}_i, \quad \text{where} \quad \bar{B}_i = \epsilon_{ijk}\bar{F}^{jk}, \quad \bar{H}_i = \epsilon_{ijk}\bar{H}^{jk}, \tag{35}$$

where P_i is the dielectric polarization and M_i is the magnetization vectors, ϵ_{ijk} is the 3D absolutely antisymmetric tensor. Comparing with the (28) Maxwell equation for the spacetime foam in the 3D form

$$E_i + \tilde{E}_i = D_i \quad \text{where} \quad E_i = F_{0i}, \quad \tilde{E}_i = \tilde{F}_{0i}, \quad D_i = H_{0i} \tag{36}$$

$$B_i + \tilde{B}_i = H_i \quad \text{where } B_i = \epsilon_{ijk}F^{jk}, \ \tilde{B}_i = \epsilon_{ijk}\tilde{F}^{jk}, \ H_i = \epsilon_{ijk}H^{jk} \tag{37}$$

we see that the following notations can be introduced.

$$\tilde{E}_i = 4l_{Pl}^2\epsilon_{ijk}\left(i\bar{\psi}\gamma^{[j}\gamma^{k]}\psi\right) \tag{38}$$

is the polarization vector of the spacetime foam and

$$\tilde{B}_i = -4l_{Pl}^2\epsilon_{ijk}\left(i\bar{\psi}\gamma^5\gamma^{[j}\gamma^{k]}\psi\right) \tag{39}$$

is the magnetization vector of the spacetime foam.

The physical reason for this is evidently: each quantum WH is like to a moving dipole (see Fig.(6) which produces microscopical electric and magnetic fields.

4 Supergravity as a possible model of the spacetime foam

The above-mentioned reasonings tells us that the most important for such kind models of the spacetime foam is the presence of a nonminimal interaction term

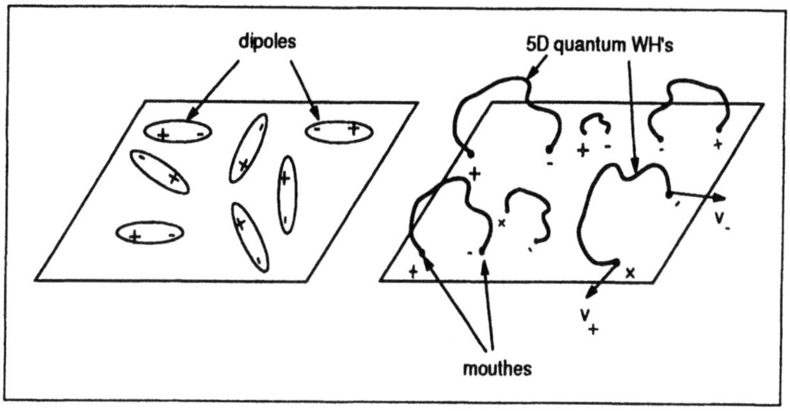

Figure 6: For the 4D observer each mouth looks as a moving electric charge. This allows us in some approximation imagine the spacetime foam as a continuous media with a polarization. $v_\pm$ are velocities of (+/-) moving charges.

between spinor and electromagnetic fields. For example, the N=2 supergravity [14] which contains the vier-bein e^a_μ, Majorana-Rarita-Schwinger field ψ_μ, photon A_μ and a second Majorana spin-$\frac{3}{2}$ field φ_μ has the following term in Lagrangian

$$\begin{aligned} \mathcal{L}_{se} &= \frac{\kappa}{\sqrt{2}}\bar{\psi}_\mu \left(eF^{\mu\nu} + \frac{1}{2}\gamma_5 \tilde{F}^{\mu\nu} \right) \varphi_\nu + \cdots, \\ \tilde{F}_{\mu\nu} &= e_{\mu\nu\alpha\beta}F^{\alpha\beta} \end{aligned} \tag{40}$$

Such term usually occurs in supergravities which have some gauge multiplet of supergravity and some matter multiplet. Taking into account the previous reasonings we can suppose that ***supergravity theories can be considered as approximate models of the spacetime foam.***

5 Conclusions

Thus, here we have suggested the approximate model for the description of the spacetime foam. This model is based on the assumption that the whole spacetime is 5 dimensional but G_{55} is the dynamical variable only inside quantum topological handles (wormholes). In this case 5D gravity has the solution which we have used as a model of the single quantum wormhole. The properties of this solution is such that we can assume that the quantum topological handles (after reducing the cross section to zero) can be approximately described by some effective spinor field.

The topological handles of the spacetime foam either can be attached to one space or connect two different spaces. In the first case we have something like to strings between two D-branes and such object can demonstrate a model of preventing the formation the naked singularity with relation $e > m$ (such object can be named as ***wormhole with quantum throat***). In the second case the spacetime foam looks as a dielectric with quantum handles as dipoles.

Such model leads to the very interesting experimental consequences. We see that the spacetime foam van have a 5D structure which one is connected with the electric/magnetic field. This observation allows us to presuppose that the very strong electric/magnetic field can open a door into 5 dimension! The question is: as is great should be this field ? The electric field E_i in the CGSE units and e_i in the "geometrized" units can be connected by formula

$$e_i = \frac{G^{1/2}}{c^2} E_i = \left(2.874 \times 10^{-25}\, cm^{-1}/gauss\right) E_i, \tag{41}$$

$$[e_i] = cm^{-1}, \quad [E_i] = V/cm \tag{42}$$

As we see the value of e_i is defined by some characteristic length l_0. It is possible that l_0 is a length of the 5th dimension. If $l_0 = l_{Pl}$ then $E_i \approx 10^{57} V/cm$ and this field strength is in the Planck region, and is will beyond experimental capabilities to create. But if l_0 has a different value it can lead to much more realistic scenario for the experimental capability to open door into 5th dimension.

Another interesting conclusion of this paper is that supergravity theories having nonminimal interaction between spinor and electromagnetic fileds can be considered as approximate and effective models of the spacetime foam.

6 Acknowledgment

I would like to acknowledze grant KR-154 for financial support of this project.

References

[1] C. Misner and J. Wheeler, Ann. of Phys., **2**, 525, 1957; J. Wheeler, Ann. of Phys., **2**, 604, 1957.

[2] J. Wheeler, "Neutrinos, Gravitation and Geometry", Princeton Univ. Press, 1960.

[3] A. Chodos and S. Detweiler, Gen. Rel. Grav. **14**, 1982, 879-890.

[4] G. Clément, Gen. Rel. Grav. **16**, 1984, 477-489; G. Clément, Gen. Rel. Grav. **16**, 1984, 131-138.

[5] V. Dzhunushaliev, Grav. Cosmol., **3**, 240,1997.

[6] Bronnikov K., Int.J.Mod.Phys. **D4**, 491, 1995, Grav. Cosmol., **1**, 67,1995.

[7] V. Dzhunushaliev, Mod. Phys. Lett. A **13**, 2179, 1998.

[8] V. Dzhunushaliev, "Matching condition on the EH and the holography principle", gr-qc/9907086, to be published in Int. J. Mod. Phys. D.

[9] V. Dzhunushaliev, H.-J.Schmidt, Grav. Cosmol. **5**, 187, 1999.

[10] V. Dzhunushaliev, "Wormhole with Quantum Throat", gr-qc/0005008, to be published in Grav. Cosmol.

[11] F. Finster, J. Smoller, S.-T. Yau, Phys. Lett. **A259**, 431, 1999.

[12] V. Dzhunushaliev, "An Approximate Model of the Spacetime Foam", gr-qc/0006016.

[13] L.D. Landau and E.M. Lifshitz, "Electrodynamics of Continuous Media", Pergamon Press, Oxford - London - New Jork - Paris, 1960.

[14] S. Ferrara and P. V. Nieuwenhuizen, Phys. Rev. Lett. **37**, 1669, 1976.

Anti-Grand Unification and the Phase Transitions at the Planck Scale in Gauge Theories

L.V.Laperashvili
Institute of Theoretical and Experimental Physics,
B.Cheremushkinskaya 25, 117218 Moscow, Russia

The way of the Grand unification of all interactions and the role of supersymmetry in GUTs are the problems of paramount importance in the contemporary elementary particle physics. However, at present time experiment doesn't indicate any manifestation of the Supersymmetry (see reviews [1-3]).

In this connection, the Anti–Grand Unification Theory (AGUT) was developed as a realistic alternative to SUSY GUTs by H.B.Nielsen (Niels Bohr Insitute, Denmark) and his collaborators: D.L.Bennett, C.D.Froggatt, L.V.Laperashvili, I.Picek and Y.Takanishi [4]-[17]. According to the AGUT the supersymmetry doesn't come into existence up to the Planck energy scale:

$$\mu_{Pl} = 1.2 \cdot 10^{19}\ GeV. \tag{1}$$

The SM is based on the group:

$$SMG = SU(3)_c \otimes SU(2)_L \otimes U(1)_Y, \tag{2}$$

The AGUT suggests that at the scale $\mu_G \lesssim \mu_{Pl}$ there exists the more fundamental group G containing N_{gen} copies of the Standard Model Group SMG:

$$G = SMG_1 \otimes SMG_2 \otimes ... \otimes SMG_{N_{gen}} \equiv (SMG)^{N_{gen}} \tag{3}$$

where the integer N_{gen} designates the number of quark and lepton generations. If $N_{gen} = 3$, then the fundamental gauge group G is:

$$G = (SMG)^3 = SMG_1 \otimes SMG_2 \otimes SMG_3. \tag{4}$$

or the generalized ones:

$$G_f = (SMG)^3 \otimes U(1)_f, \tag{5}$$

which is suggested by the fitting of fermion masses existing in the SM (see Refs. [9]-[11]) and

$$G_{ext} = (SMG \otimes U(1)_{B-L})^3, \tag{6}$$

which takes into account the see–saw mechanism with the right–handed neutrino [13],[14] and also gives the reasonable fitting of the SM fermion masses.

[0]E-mail:laper@heron.itep.ru, larisa@vxitep.itep.ru

Frontiers of Fundamental Physics 4, Edited by Sidharth and Altaisky,
Kluwer Academic/Plenum Publishers, New York, 2001

The group G_f contains the following gauge fields: $3 \times 8 = 24$ gluons, $3 \times 3 = 9$ W-bosons and $3 \times 1 + 1 = 4$ abelian gauge bosons.

At first sight, this $(SMG)^3 \times U(1)_f$ group with its 37 generators seems to be just one among many possible SM gauge group extensions. However, it is not such an arbitrary choice. There are at least reasonable requirements (postulates) on the gauge group G (or G_f, or G_{ext}) which have uniquely to specify this group. It should obey the following postulates (the first two are also valid for SU(5) GUT):

1. G or G_f should only contain transformation transforming the known 45 Weyl fermions (= 3 generations of 15 Weyl particles each) – counted as left handed, say – into each other unitarily, so that G (or G_f) must be a subgroup of U(45): $G \subseteq U(45)$.

2. No anomalies, neither gauge nor mixed. AGUT assumes that only straightforward anomaly cancellation takes place and forbids the Green-Schwarz type anomaly cancellation [18].

3. AGUT should NOT UNIFY the irreducible representations under the SM gauge group, called here SMG (see Eq.(2)).

4. G is the maximal group satisfying the above-mentioned postulates.

There are five Higgs fields named ϕ_{WS}, S, W, T, ξ in the extended G_f–AGUT by Froggatt and Nielsen [11],[12]. These fields break the AGUT to the SM what means that their vacuum expectation values (VEV) are active. The field ϕ_{WS} corresponds to the Weinberg—Salam theory, $<S>= 1$, so that we have only three free parameters — three VEVs $<W>, <T>$ and $<\xi>$ to fit the experiment in the framework of this model.

The authors of Refs.[11],[12] used them with aim to find the best fit to conventional experimental data for all fermion masses and mixing angles in the SM (see Table I):

Table I: Best fit to conventional experimental data. All masses are running masses at 1 GeV except the top qurk mass M_t which is the pole mass.

	Fitted	Experimental
m_u	3.6 MeV	4 MeV
m_d	7.0 MeV	9 MeV
m_e	0.87 MeV	0.5 MeV
m_c	1.02 GeV	1.4 GeV
m_s	400 MeV	200 MeV
m_μ	88 MeV	105 MeV
M_t	192 GeV	180 GeV
m_b	8.3 GeV	6.3 GeV
m_τ	1.27 GeV	1.78 GeV
V_{us}	0.18	0.22
V_{cb}	0.018	0.041
V_{ub}	0.0039	0.0035

The result is encouraging. The fit is given by the χ^2 function (called here $\tilde{\chi}^2$). The lowest value of $\tilde{\chi}^2 (\approx 1.87)$ gives the following VEVs:

$$<S>= 1; \qquad <W>= 0.179; \qquad <T>= 0.071; \qquad <\xi>= 0.099. \qquad (7)$$

The extended Anti–GUT theory by Nielsen and Takanishi [13],[14], which is described by the group of symmetry G_{ext} (see Eq.(6)), was suggested with aim to explain the neutrino oscillations.

Introducing the right–handed neutrino in the model, the authors replaced the assumption 1 and considered U(48) group instead of U(45), so that G_{ext} is a subgroup of U(48): $G_{ext} \subseteq U(48)$. This group ends up having 7 Higgs fields falling into 4 classes according to the order of magnitude of the expectation values:

1) The smallest VEV Higgs field plays role of the SM Weinberg–Salam Higgs field ϕ_{WS} having the weak scale value $<\phi_{WS}>= 46\ GeV/\sqrt{2}$.

2)The next smallest VEV Higgs field breaks all families $U(1)_{(B-L)}$ group, which is broken at the see–saw scale. This VEV is $<\phi_{(B-L)}>\sim 10^{12}$ GeV. Such a field is absent in the "old" extended AGUT.

3) The next 4 Higgs fields are W, T, ξ and χ, which have VEVs of the order of a factor 10 to 50 under the Planck unit. That means that if intermediate propagators have scales given by the Planck scale, as it is assumed in the AGUT in general, then they will give rise to suppression factors of the order 1/10 each time they are needed to cause a transition. The field χ is absent in the "old" G_f–AGUT. It was introduced in Refs.[13],[14] for the purpose of the study neutrinos.

4) The last one, with VEV of the same order as the Planck scale, is the Higgs field S. It had VEV $<S>= 1$ in the "old" extended AGUT by Froggatt and Nielsen (with G_f group of symmetry), but this VEV is not equal to unity in the "new" extended AGUT. Therefore there is a possibility to observe phenomenological consequences of the field S in the Nielsen–Takanishi model.

Typical fit to the masses and mixing angles for the SM leptons and quarks in the framework of the G_{ext}–AGUT is given by Table II.

Table II: Best fit to conventional experimental data in the "new" AGUT.

	Fitted	Experimental
m_u	3.1 MeV	4 MeV
m_d	6.6 MeV	9 MeV
m_e	0.76 MeV	0.5 MeV
m_c	1.29 GeV	1.4 GeV
m_s	390 MeV	200 MeV
m_μ	85 MeV	105 MeV
M_t	179 GeV	180 GeV
m_b	7.8 GeV	6.3 GeV
m_τ	1.29 GeV	1.78 GeV
V_{us}	0.21	0.22
V_{cb}	0.023	0.041
V_{ub}	0.0050	0.0035

All masses are running masses at 1 GeV except the top qurk mass M_t which is the pole mass.

The lowest value of $\tilde{\chi}^2$ is ≈ 1.46. In contrast to the "old" extended AGUT, the new results are more encouraging.

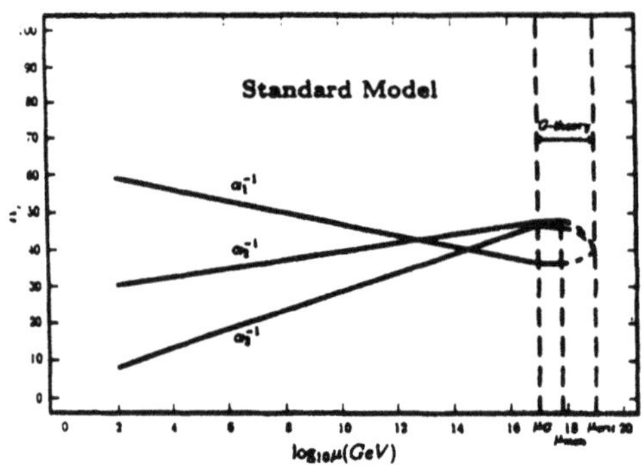

Figure 1

The AGUT approach is used in conjuction with the Multiple Point Principle (MPP) proposed by D.L.Bennett and H.B.Nielsen [6]. According to this principle Nature seeks a special point — the Multiple Critical Point (MCP) — which is a point on the phase diagram of the fundamental regulirized gauge theory G (or G_f, or G_{ext}), where the vacua of all fields existing in Nature are degenerate having the same vacuum energy density. This is the Multiple Point Principle.

Such a phase diagram has axes given by all coupling constants considered in theory. Then all (or just many) numbers of phases meet at the MCP.

In the AGUT at some point μ_G the group G (or G_f, or G_{ext}) undergoes spontaneous breakdown to the diagonal subgroup:

$$G \longrightarrow G_{diag.subgr.} = \{g, g, g || g \in SMG\}, \tag{8}$$

which is identified with the usual (lowenergy) group SMG.

Multiple Point Model assumes the existence of MCP at the Planck scale, insofar as gravity may be "critical" at the Planck scale.

The position of the MCP is shown in Fig.1: this is the point $\mu = \mu_G$.

The idea of the MPP has its origin from the lattice investigations of gauge theories. In particular, Monte Carlo simulations on lattice of U(1)-, SU(2)- and SU(3)- gauge theories indicate the existence of a triple (critical) point, which is a boundary point of three first order phase transitions.

Using the Monte Carlo results on lattice, it is possible to make theoretical calculations of the critical coupling constants and obtain slightly more accurate predictions of the AGUT for the SM fine structure constants.

In the SM the usual definition of coupling constants is used:

$$\alpha_1 = \frac{5}{3}\frac{\alpha}{\cos^2\Theta_{\overline{MS}}}, \qquad \alpha_2 = \frac{\alpha}{\sin^2\Theta_{\overline{MS}}}, \qquad \alpha_3 \equiv \alpha_s = \frac{g_s^2}{4\pi}, \tag{9}$$

where α and α_s are the electromagnetic and strong fine structure constants, respectively. All values are defined in the Modified minimal substraction scheme ($\overline{MS}$). Using experimentally given parameters and the renormalization group equations (RGE), it is possible to extrapolate the experimental values of three inverse running constants α_i^{-1} (here $i = 1, 2, 3$ corresponds to U(1), SU(2) and SU(3) groups) from Electroweak scale to the Planck scale. The precision of the LEP data allows to make this extrapolation with small errors. Assuming that the RGEs are contingent not encountering new particles up to $\mu \lesssim \mu_{Pl}$ and doing the extrapolation with one Higgs doublet under the assumption of a "desert", the following result for the inverses $\alpha^{-1}_{Y,2,3}$ ($\alpha_Y \equiv \frac{3}{5}\alpha_1$) was obtained in Ref.[6]:

$$\alpha_Y^{-1}(\mu_{Pl}) \approx 55.5; \qquad \alpha_2^{-1}(\mu_{Pl}) \approx 49.5; \qquad \alpha_3^{-1}(\mu_{Pl}) \approx 54. \tag{10}$$

The extrapolation of $\alpha_i^{-1}(\mu)$ up to the point $\mu = \mu_G$ is shown in Fig.1. The AGUT predicts their values at the scale $\mu_G \sim 10^{18}$ in terms of the critical couplings $\alpha_{i,crit}$ taken from the lattice gauge theory [19]-[23]:

$$\alpha_i(\mu_{Pl}) = \frac{\alpha_{i,crit}}{N_{gen}} = \frac{\alpha_{i,crit}}{3} \tag{11}$$

for i=2,3 and

$$\alpha_1(\mu_{Pl}) = \frac{\alpha_{1,crit}}{\frac{1}{2}N_{gen}(N_{gen}+1)} = \frac{\alpha_{1,crit}}{6} \tag{12}$$

for U(1).

According to the AGUT, at the Planck scale the running constants α_1 (or $\alpha_Y \equiv \frac{3}{5}\alpha_1$), α_2 and α_3, as chosen by Nature, are just the ones corresponding to the MCP.

There exists a simple explanation of the relations (11) and (12). As it was mentioned above, the group G breaks down at $\mu = \mu_G$. It should be said that at the very high energies $\mu \geq \mu_G \lesssim \mu_{Pl}$ (see Fig.1) each generation has its own gluons, own W's etc. The breaking makes only linear combination of a certain color combination of gluons which exists below $\mu = \mu_G$ and down to the low energies.

We can say that the phenomenological gluon is a linear combination (with amplitude $1/\sqrt{3}$ for $N_{gen} = 3$) for each of the AGUT–gluons of the same color combination. This means that coupling constant for the phenomenological gluon has a strength that is $\sqrt{3}$ times smaller, if as we effectively assume that three AGUT SU(3) couplings are equal to each other.

Then we have the following formula connecting the fine structure constants of G–theory (e.g. AGUT) and low energy surviving diagonal subgroup $G_{\text{diag.subg.}} \subseteq (SMG)^3$ given by Eq.(8):

$$\alpha^{-1}_{\text{diag},i} = \alpha^{-1}_{\text{1st gen.},i} + \alpha^{-1}_{\text{2nd gen.},i} + \alpha^{-1}_{\text{3rd gen.},i}. \tag{13}$$

Here i = U(1), SU(2), SU(3), and i=3 means that we talk about the gluon couplings.

For non–Abelian theories we immediately obtain Eq.(11) from Eq.(13) at the critical point (MCP).

In contrast to non-Abelian theories, in which the gauge invariance forbids the mixed (in generations) terms in the Lagrangian of G–theory, the U(1)–sector of the AGUT contains such mixed terms what explains the difference between the expressions (11) and (12).

Using Monte Carlo results on lattice the AGUT predicts [6] :

$$\alpha_Y^{-1}(\mu_{MCP}) = 55 \pm 6, \quad \alpha_2^{-1}(\mu_{MCP}) = 49.5 \pm 3, \quad \alpha_3^{-1}(\mu_{MCP}) = 57 \pm 3, \tag{14}$$

in correspondence with the result (10).

According to Eq.(12), the first values of Eqs.(10) and (13) gives the following estimation for the U(1) fine structure constant at $\mu = \mu_{MCP}$:

$$\alpha^{-1}_{crit} \sim 9. \tag{15}$$

The Monte Carlo simulations of the lattice U(1) gauge theory gives [20]-[22] :

$$\alpha_{crit} \approx 0.20 \pm 0.015, \quad \text{or} \quad \alpha_{crit}^{-1} \approx 5. \tag{16}$$

However, it is possible to show (see [6]) that quantum fluctuations encrease the value of α_{crit}^{-1} giving the value (15).

The following hypothesis was stated in Refs.[16],[17]: it is possible that the existence of monopoles at superhigh energies — at a short distance from the Planck energy scale (1) — plays an essential role in the phase transitions at the Planck scale when all fields existing in the SM turn into the new phase, say, (super)string phase.

In the previous works [6],[7],[15] the investigations of the phase transition phenomena and, in particular the calculation of the U(1) critical coupling constant, were connected with the existence of artifact monopoles in the lattice gauge theory and also in the Wilson loop action model, which we proposed in Ref.[15].

Now, instead of using the lattice or Wilson loop cut-off, we are going to introduce physically existing monopoles into the theory as fundamental fields.

Developing a version of the local field theory of the Higgs scalar monopoles and electrically charged particles, we consider an Abelian gauge theory in the Zwanziger formalism [24]-[28] and look for a/or rather several phase transitions connected with the monopoles forming a condensate in the vacuum.

The Zwanziger formalism [24],[25] (see also [26],[27] and review [28]) considers two potentials $A_\mu(x)$ and $B_\mu(x)$ describing one physical photon with two physical degrees of freedom.

Now and below we call this theory QEMD ("Quantum ElectroMagnetoDynamics"). In QEMD the total field system of the gauge, electrically (Ψ) and magnetically (Φ) charged fields (with charges e and g, respectively) is described by the partition function which has the following form in the Euclidean space:

$$Z = \int [DA][DB][D\Phi][D\Phi^+][D\Psi][D\Psi^+]e^{-S}, \tag{17}$$

where

$$S = \int d^4x L(x) = S_{Zw}(A, B) + S_{gf} + S_{(matter)}. \tag{18}$$

The Zwanziger action $S_{Zw}(A, B)$ is given by:

$$S_{Zw}(A,B) = \int d^4x[\frac{1}{2}(n\cdot[\partial\wedge A])^2 + \frac{1}{2}(n\cdot[\partial\wedge B])^2 +$$

$$\frac{i}{2}(n\cdot[\partial\wedge A])(n\cdot[\partial\wedge B]^*) - \frac{i}{2}(n\cdot[\partial\wedge B])(n\cdot[\partial\wedge A]^*)], \tag{19}$$

where we have used the following designations:

$$[A\wedge B]_{\mu\nu} = A_\mu B_\nu - A_\nu B_\mu, \quad (n\cdot[A\wedge B])_\mu = n_\nu(A\wedge B)_{\nu\mu},$$

$$G^*_{\mu\nu} = \frac{1}{2}\epsilon_{\mu\nu\lambda\rho}G_{\lambda\rho}. \tag{20}$$

In Eqs.(19) and(20) the unit vector n_μ represents the fixed direction of the Dirac string in the 4–space.

The action $S_{(matter)} = \int d^4x L_{(matter)}(x)$ describes the electrically and magnetically charged matter fields. S_{gf} is the gauge–fixing action (see [26]).

Let us consider now the Lagrangian $L_{(matter)}$ describing the Higgs scalar fields $\Psi(x)$ and $\Phi(x)$ interacting with gauge fields $A_\mu(x)$ and $B_\mu(x)$, respectively:

$$L_{(matter)}(x) = \frac{1}{2}|D_\mu\Psi|^2 + \frac{1}{2}|\tilde{D}_\mu\Phi|^2 - U(\Psi,\Phi), \tag{21}$$

where

$$D_\mu = \partial_\mu - ieA_\mu, \quad \text{and} \quad \tilde{D}_\mu = \partial_\mu - igB_\mu \tag{22}$$

are covariant derivatives;

$$U(\Psi,\Phi) = \frac{1}{2}\mu_e^2|\Psi|^2 + \frac{\lambda_e}{4}|\Psi|^4 + \frac{1}{2}\mu_m^2|\Phi|^2 + \frac{\lambda_m}{4}|\Phi|^4 + \lambda_1|\Psi|^2|\Phi|^2 \tag{23}$$

is the Higgs potential for the electrically and magnetically charged fields Ψ and Φ.

The complex scalar fields: $\Psi = \psi + i\zeta$ and $\Phi = \phi + i\chi$ contain the Higgs (ψ, ϕ) and Goldstone (ζ, χ) boson fields.

The Lorentz invariance is lost in the Zwanziger Lagrangian (19) because it depends on a fixed vector n_μ, but this invariance regained for the quantized values of coupling constants e and g obeying the Dirac relation:

$$e_i g_j = 2\pi n_{ij}, \qquad n_{ij} \in Z. \tag{24}$$

Considering the electric and magnetic fine structure constants: $\alpha = \frac{e^2}{4\pi}$ and $\tilde{\alpha} = \frac{g^2}{4\pi}$ we have the invariance of the QEMD under the interchange $\alpha \leftrightarrow \tilde{\alpha}$.

For $n_{ij} = 1$ from the Dirac relation (24) we have:

$$\alpha\tilde{\alpha} = \frac{1}{4}. \tag{25}$$

The effective potential in the Higgs model of electrodynamics for a charged scalar field was calculated in the one-loop approximation for the first time by the authors of Ref.[29] (see also review [30]). Using this method we can construct the effective potential (also in the one–loop approximation) for the theory described by the partition function (17) with the action S.

Let us consider now shifts: $\Psi = \Psi_B + \hat{\Psi}(x)$, $\Phi(x) = \Phi_B + \hat{\Phi}(x)$ with Ψ_B and Φ_B as background fields and calculate the following expression for the partition function in the one-loop approximation:

$$Z = \int [DA][DB][D\hat{\Phi}][D\hat{\Phi}^+][D\hat{\Psi}]D\hat{\Psi}^+] \times$$

$$\exp\{-S(A,B,\Phi_B + \hat{\Phi}, \Psi_B + \hat{\Psi}) - \int d^4x[\frac{\delta S(\Phi)}{\delta\Phi(x)}|_{\Phi=\Phi_B}\hat{\Phi}(x)$$

$$+\frac{\delta S(\Psi)}{\delta\Psi(x)}|_{\Psi=\Psi_B}\hat{\Psi}(x) + h.c.]\} = \exp\{-F(\Psi_B,\Phi_B,e^2,g^2,\mu_e^2,\mu_m^2,\lambda_e,\lambda_m)\}. \tag{26}$$

Using the representations $\psi = Re\Psi$, $\phi = Re\Phi$, we obtain the effective potential:

$$V_{eff} = F(\psi_B, \phi_B, e^2, g^2, \mu_e^2, \mu_m^2, \lambda_e, \lambda_m) \tag{27}$$

given by the function F of Eq.(26) for the constant background fields: $\Psi_B = \psi_B = \text{const}$, $\Phi_B = \phi_B = \text{const}$.

The effective potential (27) has several minima. Their position depends on $e^2, g^2, \mu_{e,m}^2$ and $\lambda_{e,m}$. If the first local minimum occurs at $\psi_B = 0$ and $\phi_B = 0$, it corresponds to the Coulomb-like phase in our description.

We are interested in the phase transition from the Coulomb-like phase "$\psi_B = \phi_B = 0$" to the confinement phase "$\psi_B = 0,\ \phi_B = \phi_0 \neq 0$". In this case the one-loop effective potential for monopoles coincides with the expression of the effective potential calculated by authors of Ref.[29] for scalar electrodynamics and extended to the massive theory in Ref.[31].

Assuming the existence of the first vacuum at $\phi_B = 0$ and using from now the designations: $\mu = \mu_m$, $\lambda = \lambda_m$, we have the effective potential in the Higgs monopole model described by the following expression equivalent to that considered in Ref.[29]:

$$V_{eff}(\phi^2) = \frac{\mu_{run}^2}{2}\phi_B^2 + \frac{\lambda_{run}}{4}\phi_B^4 + \frac{\mu^4}{64\pi^2}\log\frac{(\mu^2 + 3\lambda\phi_B^2)(\mu^2 + \lambda\phi_B^2)}{\mu^4}, \tag{28}$$

where

$$\lambda_{run}(\phi_B^2) = \lambda + \frac{1}{16\pi^2}[3g^4\log\frac{\phi_B^2}{M^2} + 9\lambda^2\log\frac{\mu^2 + 3\lambda\phi_B^2}{M^2} + \lambda^2\log\frac{\mu^2 + \lambda\phi_B^2}{M^2}], \tag{29}$$

$$\mu_{run}^2(\phi_B^2) = \mu^2 + \frac{\lambda\mu^2}{16\pi^2}[3\log\frac{\mu^2 + 3\lambda\phi_B^2}{M^2} + \log\frac{\mu^2 + \lambda\phi_B^2}{M^2}]. \tag{30}$$

Here M is the cut–off scale.

As it was shown in Ref.[29], the one–loop effective potential (28) can be improved by the consideration of the renormalization group equation (RGE).

According to Refs.[29]-[31], RGE for the improved one–loop effective potential is given by the following expression:

$$(M\frac{\partial}{\partial M} + \beta_\lambda\frac{\partial}{\partial\lambda} + \beta_g\frac{\partial}{\partial g} + \beta_{(\mu^2)}\mu^2\frac{\partial}{\partial\mu^2} - \gamma\phi\frac{\partial}{\partial\phi})V_{eff}(\phi^2) = 0, \tag{31}$$

where the function γ is the anomalous dimension: $\gamma(\frac{\phi}{M}) = -\frac{\partial\phi}{\partial M}$. The γ–expression for monopoles is given by Ref.[29] with replacement $e \to g$:

$$\gamma = -\frac{3g_{run}^2}{16\pi^2}. \tag{32}$$

RGE (31) leads to a new improved effective potential:

$$V_{eff}(\phi^2) = \frac{1}{2}\mu_{run}^2(t)G^2(t)\phi^2 + \frac{1}{4}\lambda_{run}(t)G^4(t)\phi^4, \tag{33}$$

where

$$G(t) \equiv \exp[-\frac{1}{2}\int_0^t dt'\gamma\Big(g_{run}(t'), \lambda_{run}(t')\Big)] \quad \text{with} \quad t = \log(\phi^2/M^2). \tag{34}$$

Let us write now the one-loop potential (28) as

$$V_{eff} = V_0 + V_1, \qquad \text{where} \qquad V_0 = \frac{\mu^2}{2}\phi^2 + \frac{\lambda}{4}\phi^4, \tag{35}$$

$$V_1 = \frac{1}{64\pi^2}[3g^4\phi^4 \log\frac{\phi^2}{M^2} + (\mu^2 + 3\lambda\phi^2)^2 \log\frac{\mu^2 + 3\lambda\phi^2}{M^2}$$

$$+(\mu^2 + \lambda\phi^2)^2 \log\frac{\mu^2 + \lambda\phi^2}{M^2} - 2\mu^4 \log\frac{\mu^2}{M^2}]. \tag{36}$$

We can plug this V_{eff} into RGE (31) and obtain the following RG-equations (see [30]):

$$\frac{d\lambda_{run}}{dt} = \frac{1}{16\pi^2}(3g^4_{run} + 10\lambda^2_{run} - 6\lambda_{run}g^2_{run}), \tag{37}$$

$$\frac{d\mu^2_{run}}{dt} = \frac{\mu^2_{run}}{16\pi^2}(4\lambda^2_{run} - 3g^2_{run}). \tag{38}$$

The Dirac relation and the RGE for electric and magnetic fine structure constants α and $\tilde{\alpha}$ were investigated in detail in the recent paper [27]. The following result was obtained.

If we have the electrically and magnetically charged particles existing simultaneously for $\mu > \mu_{(threshold)}$ (here μ is the energy scale) and if in some region of μ their β-functions are computable perturbatively as a power series in e^2 and g^2, then the Dirac relation is valid not only for the "bare" elementary charges e_0 and g_0, but also for the renormalized effective charges e and g (see [32] and review [28]), and the following RGEs (obtained in Ref.[27]) take place:

$$\frac{d\log\alpha(p)}{dt} = -\frac{d\log\tilde{\alpha}(p)}{dt} = \beta^{(e)}(\alpha) - \beta^{(m)}(\tilde{\alpha}) \quad \text{where} \quad t = \log\frac{\mu^2}{M^2}. \tag{39}$$

These RGEs are in accordance with the Dirac relation (25) and the dual symmetry considered above.

By restricting ourselves to the two-loop approximation for β- functions, we have the following equations (39) for scalar particles:

$$\frac{d\log\alpha(p)}{dt} = -\frac{d\log\tilde{\alpha}(p)}{dt} = \frac{\alpha - \tilde{\alpha}}{12\pi}(1 + 3\frac{\alpha + \tilde{\alpha}}{4\pi} + ...). \tag{40}$$

According to Eq.(40), the two-loop contribution is not more than 30% if both α and $\tilde{\alpha}$ obey the following requirement:

$$0.25 \lesssim \alpha, \tilde{\alpha} \lesssim 1. \tag{41}$$

The lattice simulations of compact QED give the behavior of the effective fine structure constant $\alpha(\beta)$ ($\beta = 1/e_0^2$, and e_0 is the bare electric charge) in the vicinity of the phase transition point (see Refs.[20],[22]).

The following critical values of the fine structure constant α and $\tilde{\alpha}$ was obtained in Ref.[22]:

$$\alpha^{lat}_{crit} \approx 0.20 \pm 0.015, \qquad \tilde{\alpha}^{lat}_{crit} \approx 1.25 \pm 0.10 \qquad \text{at} \qquad \beta_{crit} \approx 1.011. \tag{42}$$

These values almost coincide with the borders of the requirement (41) given by the perturbation theory for β–functions.

In the one–loop approximation, we have:

$$\frac{dg^2_{run}}{dt} = \frac{g^4_{run}}{48\pi^2} - \frac{1}{12}. \tag{43}$$

Here the second term describes the influence of the electrically charged fields on the behavior of the monopole charge.

Investigating the phase transition from the Coulomb–like phase "$\psi_B = \phi_B = 0$" to the phase with "$\psi_B = 0$, $\phi_B = \phi_0 \neq 0$", we see that the effective potential (33) has the first and the second minima appearing at $\phi = 0$ and $\phi = \phi_0$, respectively. They are shown in Fig.2 by the curve "1". These minima of $V_{eff}(\phi^2)$ correspond to the different vacua arising in the model.

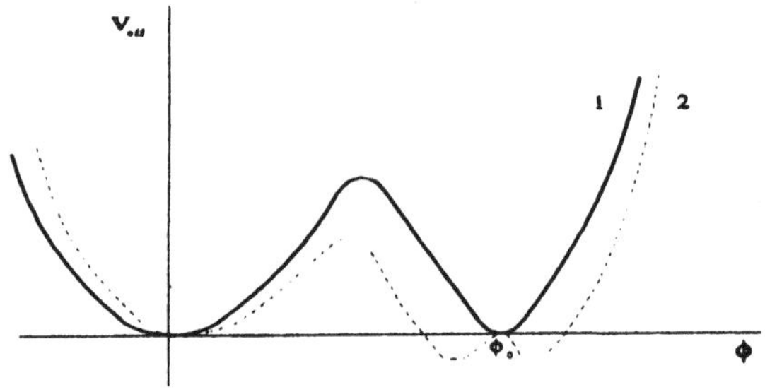

Figure 2: The effective potential V_{eff}: curve 1 corresponds to the "Coulomb-confinement" phase transition; curve 2 describes the existence of two minima corresponding to the confinement phases.

The conditions for the existence of degenerate vacua are given by the following requirements:

$$V_{eff}(0) = V_{eff}(\phi_0^2) = 0, \tag{44}$$

$$V'_{eff}(\phi_0^2) \equiv \frac{\partial V_{eff}}{\partial \phi^2}|_{\phi=\phi_0} = 0, \tag{45}$$

$$V''_{eff}(\phi_0^2) \equiv \frac{\partial^2 V_{eff}}{\partial (\phi^2)^2}|_{\phi=\phi_0} > 0. \tag{46}$$

From the first equation (44) applied to Eq.(33) we have:

$$\mu^2_{run} = -\frac{1}{2}\lambda_{run}(t_0)\phi_0^2 G^2(t_0), \quad \text{where} \quad t_0 = \log(\phi_0^2/M^2). \tag{47}$$

The joint solution of equations $V_{eff}(\phi_0^2) = V'_{eff}(\phi_0^2) = 0$ gives:

$$g^4_{crit} = -2\lambda_{run}(\frac{8\pi^2}{3} + \lambda_{run}). \tag{48}$$

The curve (48) is represented on the phase diagram $(\lambda_{run}; g^4_{run})$ of Fig.3 by the curve "1" which describes a border between the "Coulomb–like" phase with $V_{eff} \geq 0$ and the confinement ones having $V^{min}_{eff} < 0$.

Figure 3: The phase diagram (λ_{run}; $g^4 \equiv g^4_{run}$) corresponding to the Higgs monopole model shows the existence of a triple point A ($\lambda_{(A)} \approx -13.4$; $g^2_{(A)} \approx 18.6$).

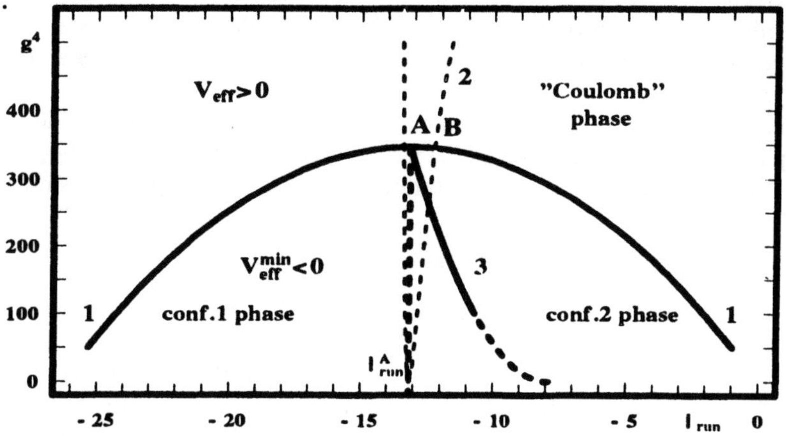

g^4_{run}) corresponding to the Higgs monopole model shows the existence of a triple point A ($\lambda_{(A)} \approx -13.4$; $g^2_{(A)} \approx 18.6$).

The next step is the calculation of the second derivative of the effective potential.

Let us consider now the case when this second derivative changes its sign giving a maximum of V_{eff} instead of the minimum at $\phi^2 = \phi_0^2$. Such a possibility is shown in Fig.2 by the dashed curve "2".

Now two additional minima at $\phi^2 = \phi_1^2$ and $\phi^2 = \phi_2^2$ appear in our theory. They correspond to two different confinement phases related with the confinement of the electrically charged particles. If these two minima are degenerate, then we have the following requirements:

$$V_{eff}(\phi_1^2) = V_{eff}(\phi_2^2) < 0, \quad V'_{eff}(\phi_1^2) = V'_{eff}(\phi_2^2) = 0, \tag{49}$$

which describe the border between the confinement phases "conf.1" and "conf.2" presented in Fig.3 by curve "3". This curve "3" meets the curve "1" at the triple point A.

According to the illustration shown in Fig.3, it is obvious that this triple point A is given by the following requirements:

$$V_{eff}(\phi_0^2) = V'_{eff}(\phi_0^2) = V''_{eff}(\phi_0^2) = 0. \tag{50}$$

In contrast to the requirements:

$$V_{eff}(\phi_0^2) = V'_{eff}(\phi_0^2) = 0, \tag{51}$$

producing the curve "1", let us consider now the joint solution of the following equations:

$$V_{eff}(\phi_0^2) = V''_{eff}(\phi_0^2) = 0. \tag{52}$$

The dashed curve "2" of Fig.3 represents the solution of Eq.(52). This curve is going very close to the maximum of the curve "1". It is natural to assume that the position of the triple point A coincides with this maximum and the corresponding deviation can be explained by our approximate calculations. Taking into account such an assumption, let us consider the border between the phase "conf.1", having the first minimum at nonzero ϕ_0 with $V_{eff}^{min} = c_1 < 0$, and the phase "conf.2 " which reveals two minima with the second minimum being the deeper one and having $V_{eff}^{min} = c_2 < 0$.

This border (described by the curve "3" of Fig.3) was calculated in the vicinity of the triple point A by means of Eq.(49) with ϕ_1 and ϕ_2 represented as $\phi_{1,2} = \phi_0 \pm \epsilon$ with $\epsilon << \phi_0$. The result of such calculations gives the following expression for the curve "3":

$$g_{run}^4 = \frac{5}{2}(5\lambda_{run} + 8\pi^2)\lambda_{run} + 8\pi^4. \tag{53}$$

The piece of the curve "1" to the left of the point A describes the border between the "Coulomb–like" phase and the phase "conf.1".

The right piece of the curve "1" along to the right of the point B separates the "Coulomb" phase and the phase "conf.2". But between the points A and B the phase transition border is going slightly upper the curve "1". This deviation is very small and can't be distinguished on Fig.3.

The numerical solution demonstrates that the triple point A exists in the very neighborhood of the maximum of the curve (48) and its position is approximately given by the following values:

$$\lambda_{(A)} \approx -\frac{4\pi^2}{3} \approx -13.4, \tag{54}$$

$$g_{(A)}^2 = g_{crit}^2|_{\text{for } \lambda_{run}=\lambda_{(A)}} \approx \frac{4\sqrt{2}}{3}\pi^2 \approx 18.6. \tag{55}$$

The triple point values of the electric and magnetic fine structure constant follow from Eqs.(25) and (55):

$$\alpha_{(A)} = \frac{\pi}{g_{(A)}^2} \approx 0.17, \qquad \tilde{\alpha}_{(A)} = \frac{g_{(A)}^2}{4\pi} \approx 1.48. \tag{56}$$

The obtained result is very close to the Monte Carlo lattice result (42). The phase diagram drawn in Fig.3 corresponds to the validity of one–loop approximation (with accuracy of deviations not more than 30%, see Ref.[27]) in the region of parameters:

$$0.17 \lesssim \alpha, \tilde{\alpha} \lesssim 1.5. \tag{57}$$

It is necessary to note that the RGE for λ_{run} indicates a slow convergence of the series over λ (see Ref.[33]) and the one–loop approximation is valid for λ_{run} up to $|\lambda| \lesssim 30$ with accuracy of deviations $< 10\%$.

It is obvious that in our case both phases, "conf.1" and "conf.2", have nonzero monopole condensate in the minima of the effective potential, when $V_{eff}^{min}(\phi_{1,2} \neq 0) < 0$. By this reason, the Abrikosov–Nielsen–Olesen (ANO) electric vortices [34],[35] may exist in these both phases, which are equivalent in the sense of the "string" formation. If electric charges are present in a model, they are placed at the ends of the vortices–"strings" and therefore are confined.

The phase diagram of Fig.3 demonstrates the existence of the confinement phase for $\alpha \geq \alpha_{(A)} \approx 0.17$.

The lattice investigations [20],[22] show that in the confinement phase $\alpha(\beta)$ increases when $\beta = 1/e_0^2 \to 0$ (here e_0 is the bare electric charge) and very slowly approaches to its maximal value: $\alpha_{max} = \frac{\pi}{12} \approx 0.26$ predicted in Ref.[36] (see also Ref.[15]).

It is worthwhile mentioning that the confinement of monopoles can be described by using duality. The Higgs field Ψ, having the electric charge, is responsible for this confinement. The corresponding confinement phases for monopoles are absent on the phase diagram of Fig.3. They can be described by the phase diagram $(\lambda_{run}^e; e_{run}^2)$. The overall phase diagram is three-dimensional and is given by $(\lambda_{run}^m; \lambda_{run}^e; g_{run}^2)$ (e_{run}^2 and g_{run}^2 are related by the Dirac relation).

The result (56) obtained in the framework of the Higgs scalar monopole model gives the following prediction:

$$\alpha_{crit}^{-1} = \alpha_{(A)}{}^{-1} \approx 6, \tag{58}$$

which is comparable with the MPM result (15).

Recent investigations (L.V.Laperashvili, H.B.Nielsen, Bled Workshop, Slovenia, 2000) show that monopoles are confined in the SM up to the Planck scale. But they can exist in the AGUT. Let us assume now the existence of monopoles at superhigh energies $\mu \geq \mu_{mon} \lesssim \mu_{Pl}$. Then RGEs (39) describing also monopoles can lead to the Unification of all interactions at the Planck scale giving the coincidence of all $\alpha_{i,crit}$ at the point $\mu \sim \mu_{Pl}$.

Such a situation is shown in Fig.1. These investigations are in progress.

References

[1] K.A.Olive, "Introduction to Supersymmetry: Astrophysical and Phenomenological Constraints", hep-ph/9911307.

[2] Gi-Chol Cho, Kaoru Hagiwara, "Supersymmetry versus precision experiments revisted", hep-ph/9912260.

[3] P.Igo-Kamenez, "Search for New Particles and New Phenomena Results from e^+e^- Colliders", XXXth Intern. Conf. on High Energy Physics, July 27 - August 2, 2000, Osaka, Japan.

[4] D.L.Bennett, H.B.Nielsen, I.Picek, Phys.Lett.**B208**, 275, 1988.

[5] C.D.Froggatt, H.B.Nielsen, "Origin of Symmetries", World Scientific, 1991.

[6] D.L.Bennett, H.B.Nielsen, Int.J.Mod.Phys. **A9**, 5155, 1994.

[7] L.V.Laperashvili, Yad.Fiz. **57**, 501, 1994 (Review in Russian);
Phys.of Atom.Nucl. **57**, 471, 1994; ibid **59**, 162, 1996.

[8] C.D.Froggatt, L.V.Laperashvili, H.B.Nielsen, "SUSY or NOT SUSY: AntiGUTs, Critical Coupling Universality and Higgs—Top Masses", "SUSY98", Oxford, 10-17 July, 1998, hep-ntst1.rl.ac.uk/susy98/.

[9] H.B.Nielsen, C.D.Froggatt, "Masses and Mixing Angles and Going beyond the Standard Model", in Proceedings of the International Workshop on "What Comes Beyond the Standard Model", Bled, Slovenia, 29 June - 9 July 1998: Ljubljana 1999, p.29.

[10] C.D.Froggatt, G.Lowe, H.B.Nielsen, Phys.Lett. **B311**, 163, 1993; Nucl.Phys. **B414**, 579, 1994; ibid **B420**, 3, 1994.

[11] C.D.Froggatt, H.B.Nielsen, D.J.Smith, Phys.Lett. **B385**, 150, 1996.

[12] C.D.Froggatt, M.Gibson, H.B.Nielsen, D.J.Smith, Int.J.Mod.Phys. **A13**, 5037, 1998.

[13] H.B.Nielsen, Y.Takanishi, Nucl.Phys. **B588**, 281, 2000.

[14] H.B.Nielsen, Y.Takanishi, "Neutrino Mass Matrix in Anti-GUT with See-Saw Mechanism", hep-ph/0011062; to appear in Nucl.Phys.

[15] L.V.Laperashvili, H.B.Nielsen, Mod.Phys.Lett. **A12**, 73, 1997.

[16] L.V.Laperashvili, H.B.Nielsen, "Multiple Point Principle and Phase Transition in Gauge Theories", in: Proceedings of the International Workshop on "What Comes Beyond the Standard Model", Bled, Slovenia, 29 June - 9 July, 1998; Ljubljana 1999, p.15.

[17] L.V.Laperashvili, H.B.Nielsen, "Phase Transition in the Higgs Model of Scalar Fields with Electric and Magnetic Charges", hep-th/0010260; ; to appear in Int.J.Mod.Phys., 2001.

[18] M.B.Green, J.Schwarz, Phys.Lett. B149, 117, 1984.

[19] G.Bhanot, Nucl.Phys.**B205**, 168, 1982; Phys.Rev. **D24**, 461, 1981; Nucl.Phys. **B378** 633, 1992.

[20] J.Jersak, T.Neuhaus and P.M.Zerwas, Phys.Lett. **B133** 103, 1983; Nucl.Phys. **B251**, 299, 1985.

[21] H.G.Everetz, T.Jersak, T.Neuhaus, P.M.Zervas, Nucl.Phys. **B251**, 279, 1985.

[22] J.Jersak, T.Neuhaus, H.Pfeiffer, Nucl.Phys.Proc.Suppl. **83-84**, 491, 2000; Phys.Rev. **D60**, 054502, 1999.

[23] C.P.Bachas and R.F.Dashen, Nucl.Phys. **B210**, 583, 1982.

[24] D.Zwanziger, Phys.Rev. **D3**, 343,880, 1971.

[25] R.A.Brandt, F.Neri, D.Zwanziger, Phys.Rev. **D19**, 1153, 1979.

[26] F.V.Gubarev, M.I.Polikarpov, V.I.Zakharov, Phys.Lett. **B438**, 147, 1998.

[27] L.V.Laperashvili, H.B.Nielsen, Mod.Phys.Lett., **A14**, 2797, 1999.

[28] M.Blagojevic, P.Senjanovic, Phys.Rept. **157**, 234, 1988.

[29] S.Coleman, E.Weinberg, Phys.ReV. **D7**, 1888, 1973.

[30] M.Sher, Phys.Rept. **179**, 274, 1989.

[31] D.Gross, in: Methods of Field Theory, Proc. 1975 Les Houches Summer School, eds R.Balian and J.Zinn-Justin, North Holland, Amsterdam, 1975.

[32] J.Shwinger, Phys.Rev. **44**, 1087, 1996;ibid **151**, 1048, 1055, 1966; ibid **173**, 1536, 1968; Science **165**, 757, 1969; ibid **166**, 690, 1969.

[33] H.Alhendi, Phys.Rev. **D37**, 3749, 1988.

[34] H.B.Nielsen, P.Olesen, Nucl.Phys., **B61**, 45, 1973.

[35] A.A.Abrikosov, Soviet JETP, **32**, 1442, 1957.

[36] M.Lüscher, K.Symanzik, P.Weisz, Nucl.Phys. **B173**, 365, 1980.

The Structure of the Yang–Mills Vacuum Seen by Distant Observers

G. Etesi
Yukawa Institute for Theoretical Physics
Kyoto University, 606-8502 Kyoto, Japan
Email:etesi@yukawa.kyoto-u.ac.jp

1 Introduction

The famous solution of the long-standing $U(1)$-problem in the Standard Model via instanton effects was presented by 't Hooft about two decades ago [1]. This solution demonstrated that ***instantons***, i.e. finite-action solutions of the ***Euclidean*** Yang–Mills-equations discovered by Belavin et al. [2] should be taken seriously in gauge theories. Another problem arose in these models over ***flat*** space-times, however: if instantons really exist, they induce a P- hence CP-violating so-called θ-term in the effective Yang–Mills action. But according to accurate experimental results, such a CP-violation does not exists in QCD, for instance. The most accepted solution to this problem is the so-called ***Peccei–Quinn mechanism*** [3]. A consequence of this mechanism is the existence of a light particle, the so-called ***axion***. This particle has not been observed yet, however.

The question naturally arises whether or not such problematic θ-term must be introduced over more general space-times. The aim of our paper is to claim that the answer is yes.

First, let us summarize the vacuum structure of a gauge theory over Minkowski space-time. We are going to use a Yang–Mills theory framework over $(\mathbb{R}^4, g)$ where g is a Lorentzian metric. Let E be a (trivial) complex vector bundle over $\mathbb{R}^4$ belonging to a finite dimensional complex representation of G. Without loss of generality we choose the gauge group G to be a compact Lie group. Consider a G-connection on this bundle; choosing a particular frame on E, this connection can be identified globally with a $\mathfrak{g}$-valued 1-form A with curvature F_A. We choose the usual Yang–Mills action (by fixing the coupling to be 1):

$$S(A,g) = -\frac{1}{8\pi^2}\int\limits_M \mathrm{tr}\,(F_A \wedge *F_A)\,, \tag{1}$$

in this case $M = \mathbb{R}^4$ and $*$ denotes the Hodge-operator induced by the metric g on $\mathbb{R}^4$. Usually the metric g is fixed and supposed to be the Minkowski metric on $\mathbb{R}^4$. In this case the Euler–Lagrange equations of (1) are the Yang–Mills equations and read as follows:

$$\mathrm{d}_A F_A = 0, \quad \mathrm{d}_A * F_A = 0.$$

Frontiers of Fundamental Physics 4, Edited by Sidharth and Altaisky,
Kluwer Academic/Plenum Publishers, New York, 2001

The simplest solution is the vacuum. The gauge field A is a vacuum field if $F_A = 0$ i.e. its field strength or curvature is equal to zero. By simply connectedness of $\mathbb{R}^4$ such gauge fields can be written in the form $A = f^{-1}\mathrm{d}f$, where $f : \mathbb{R}^4 \to G$ is a smooth function.

But by the existence of a global temporal gauge and the stationarity of the flat metric on $\mathbb{R}^4$ it is enough to consider the restriction of f to a space-like submanifold of Minkowski space-time, i.e. $f : \mathbb{R}^3 \to G$. Minkowski space-time is asymptotically flat as well, so there is a point i^0 called space-like infinity. This point represents the "infinity of space" hence can be added to $\mathbb{R}^3$ completing it to the three-sphere $\mathbb{R}^3 \cup \{i^0\} = S^3$. It is well-known that vacuum fields (possibly after a null-homotopic gauge-transformation around i^0) extend to the whole S^3 consequently classical vacua are classified by maps $f : S^3 \to G$. These maps up to homotopy are given by elements of $\pi_3(G)$. For typical compact Lie groups $\pi_3(G) \cong \mathbf{Z}$. This fact can be interpreted as classical vacua are separated from each other by barriers of finite height i.e. it is impossible to develop two vacua of different winding numbers into each other only through vacuum states. Hence homotopy equivalence reflects the dynamical structure of the theory.

On the other hand if f_1, f_2 are vacua of winding numbers n_1, n_2 respectively, there is a gauge transformation $g : S^3 \to G$ of winding number $n_2 - n_1$ satisfying $gf_1 = f_2$. Consequently we can see that the concept of *dynamical equivalence* of vacua given by the *dynamics* of the theory (i.e. the *homotopy equivalence* of maps $f : S^3 \to G$) is different from that of *symmetry-equivalence* of vacua provided by the *symmetry* of the gauge theory (i.e. the *gauge equivalence* of the above maps).

To avoid this discrepancy, we proceed as follows. Suppose we have constructed the Hilbert space $\mathcal{H}_{\mathbf{R}^4}$ of the corresponding quantum gauge theory. If $|n\rangle \in \mathcal{H}_{\mathbf{R}^4}$ denotes the quantum vacuum state belonging to a classical vacuum f of winding number n, the simplest way to construct a state which is invariant (up to phase) under both dynamical (i.e. homotopy) and symmetry- (i.e. gauge) equivalence is to introduce the state

$$|\theta\rangle := \sum_{n=-\infty}^{\infty} e^{in\theta}|n\rangle \in \mathcal{H}_{\mathbf{R}^4}, \quad \theta \in U(1). \tag{2}$$

These quantum vacuum states are referred to as "θ-vacua".

From the physical point of view, the introduction of θ-vacua is also necessary because the vacuum states of different winding numbers can be joined semi-classically i.e. by a tunneling process induced by non-trivial instantons of the corresponding *Euclidean* gauge theory. Indeed, as it is well known [4][5], in the case of $G = SU(2)$, the instanton number of an instanton is an element $k \in H^4(S^4, \mathbf{Z}) \simeq \mathbf{Z}$ (here S^4 is the one-point conformal compactification of the Euclidean flat $\mathbb{R}^4$. Note that the notion of "instanton number" comes from a very different compactification process comparing to the derivation of "vacuum winding number"). If two vacua, $|n_1\rangle$, $|n_2\rangle$ ($n_1, n_2 \in \pi_3(SU(2)) \simeq \mathbb{Z}$) are given then there is an instanton of instanton number $n_2 - n_1 \in H^4(S^4, \mathbb{Z}) \simeq \mathbb{Z}$ tunneling between them in temporal gauge [4][5]. In other words the true vacuum states are linear combinations of the vacuum states of unique winding numbers yielding again (2).

But the value of θ cannot be changed in any order of perturbation, i.e. it should be treated as a physical parameter of the theory; this implies that tunnelings induce the effective term

$$\frac{\theta}{8\pi^2}\int_{\mathbf{R}^4} \mathrm{tr}\,(F_A \wedge F_A)$$

in addition to action (1). But it is not difficult to see that such a term violates the parity symmetry P hence the CP-symmetry of the theory.

In summary, we have seen that there are at least three different ways to introduce θ-parameters in Yang–Mills theories *over Minkowskian space-time*:

(i) θ is introduced to fill in the gap between the notions of dynamical (i.e. homotopy) and symmetry- (i.e. gauge) equivalence of Yang–Mills vacua. This approach is pure mathematical in its nature;

(ii) θ must be introduced because by instanton effects vacua of definite winding numbers are superposed in the underlying semi-classical Yang–Mills theory;

(iii) θ must be introduced by "naturality arguments", i.e. nothing prevents us to extend the Yang–Mills action at the full quantum level by a P-violating term $\mathrm{tr}\,(F_A \wedge F_A)$ with coupling constant θ.

There is a correspondence between the above three characterizations of the θ in *flat Minkowskian space-time* but in the case of general space-times, clear and careful distinction must be made until a relation or correspondence between the three notions is established. Clearly, (i) is related to the *topology* of the space-time and the gauge group hence it is relatively easy to check whether or not it remains valid in the general case. Concept (ii) is related to the semi-classical structure of the general Yang–Mills theory especially to the existence of instanton solutions in the Wick-rotated theory and their relationship with vacuum tunneling. The validity of concept (iii) is the most subtle one: we need lot of information on global non-perturbative aspects of the general quantum Yang–Mills theory to check if any θ-term survives quantum corrections. In the present state of affairs, having no adequate general theory of Wick rotation, instantons and their physical interpretation, non-perturbative aspects of general Yang–Mills theories, we can examine only the validity of concept (i) in the general case. Its validity or invalidity may serve as a good indicator for the existence and role of θ-terms in general Yang–Mills theories.

The analysis of the vacuum structure of general Yang–Mills theories from the point of view of (i) was carried out by Isham et al. [6][7][8][9][10]. In these papers Isham et al. argue that in the general case concept (i) for introducing θ-terms still continue to hold due to the complicated topology of the spatial surface S and the gauge group G [6]. The classical vacuum structure of these theories becomes more complicated and we cannot avoid the introduction of various CP-violating terms into the effective Lagrangian [10].

We have to emphasize that the approach of Isham et al. to the problem is pure topological in its nature, however. By a result of Witt [11] every oriented, connected three-manifold S appears as Cauchy-surface of a physically reasonable initial data set. It is well-known that the complicated topology of the space-like submanifold S leads to appearance of singularities in space-time if it arises as the Cauchy development of S. Indeed, an early result of Gannon [12] shows that the Cauchy development of a non-simply connected Cauchy surface is geodesically incomplete i.e. singularities occur. If we accept the Cosmic Censorship Hypothesis, these singularities are hidden behind event horizons resulting a non-trivial causal structure for these space-times, too. A theorem of Chruściel–Wald [13] show that distant observers can observe only simply-connected portions of asymptotically flat space-times: all topological property are hidden behind event horizons, eventually resulting again in a topologically simple *effective* space-time. Hence one may doubt if Isham's conclusions remain valid.

In Section 2 we formulate Yang–Mills theories with an arbitrary compact gauge group over general asymptotically flat, stationary space-times. This model provides a good framework for studying classical Yang–Mills vacua over causally non-trivial space-times. In this setup we simply mimic the above analysis concerning classical Yang–Mills vacua and find that although all vacua are topologically equivalent on the causally connected regime of the space-time, the appearance of a natural boundary condition on the event horizons (also a consequence of the causal structure) finally makes concept (i) still remain valid. A modification appears compared with Isham and other's pure topological considerations, however in the sense that generally the vacuum structure in our case has the same complexity as in the flat Minkowskian case. This demonstrates the "stability" of the θ-problem.

Finally, in Section 3, we present a few properties of $SU(2)$-Schwarzschild instantons which make the relationship between vacuum tunneling and instantons not clear.

The idea of studying relationship between micro- or virtual black holes, wormholes and θ-vacua is not new. For example, see Hawking [14] and Preskil et al. [15].

2 Vacua: A General Space-Time Model

Let $\widetilde{S}$ be a connected, oriented, closed three-manifold and let $S := \widetilde{S} \setminus \{i^0\}$ where $i^0 \in \widetilde{S}$ is a point. Using the result of Witt [11] we can choose an asymptotically flat initial data set (S, h, k), where h is a smooth Riemannian metric while k is a symmetric $(0,2)$-type tensor field on S both satisfying suitable fall-off conditions in a neighbourhood of i^0 (this point is called ***spatial infinity***). We suppose these initial data are given by some matter field represented by a stress-energy tensor $T|_S$ obeying the dominant energy condition. Consider the maximal Cauchy development of (S, h, k) denoted by (M, g). This space-time is ***globally hyperbolic*** with Cauchy surface S by construction and $M \cong S \times \mathbb{R}$.

Choose a complex vector bundle E over M associated to the gauge group G via a complex representation and a G-connection A on it. Consider an Yang–Mills theory with action (1). In this paper we will focus on ***vacuum solutions on a gravitational background*** of this theory i.e. pairs (A, g) where A is a smooth flat G-connection on the bundle E while g is a smooth Lorentzian metric on M (which is a solution of the Einstein's equations with a matter field given by a stress-energy tensor T. For technical reasons we suppose T satisfies the strong energy condition). We will refer the collection (E, A, M, g) to as an ***Yang–Mills vacuum setup***.

We take two more restrictions. First, we will assume that (M, g) is ***asymptotically flat***. At a first look this means that there is a conformal embedding $i : (M, g) \to (\widetilde{M}, \widetilde{g})$ such that the image of the Cauchy surface can be completed to a maximal space-like submanifold $\widetilde{S}$ by adding the space-like infinity $i^0 \in \widetilde{M}$ to it: $i(S) \cup \{i^0\} = \widetilde{S}$. Moreover the infinitely distant points of M, represented by $\partial i(M)$ are divided naturally into three classes: the future and past null infinities $\mathcal{I}^{\pm}$ and the already mentioned spatial infinity i^0. The ***asymptotically flat outer region*** of M is defined to be the set $N := J^-(\mathcal{I}^+) \cap i(M)$ which is a manifold (here $J^\pm(K)$ denotes the causal past or future of a set $K \subset \widetilde{M}$). The metric $\widetilde{g}$ is related by a conformal factor Ω to the original metric; although the details are unrelevant in our considerations, we remark that $\widetilde{g}$ is not necessarily smooth in i^0. Note that if $M \setminus N =: B$ is not empty then (M, g) contains a ***black hole region*** B. We denote by $H \subset M$ its *event horizon*. Clearly, $\partial N = \partial B = H$. For details, see [17].

Secondly, we assume that (M, g) is ***stationary***.

In summary, we focus our attention to each stationary, asymptotically flat, globally hyperbolic Yang–Mills vacuum setups (E, A, M, g) with an arbitrary compact gauge group G. We address the problem of describing the topology of Yang–Mills vacua ***seen by an observer in the asymptotically flat region*** of the space-time (M, g). Clearly, at least classically, only this part of the space-time can be relevant for ordinary macroscopic observers. To achieve this, we refer to a general result of Chruściel and Wald on asymptotically flat outer regions [13].

Theorem (Chruściel–Wald). *Let (M, g) be a globally hyperbolic, asymptotically flat, stationary space-time with a matter field represented by a stress-energy tensor T satisfying the strong energy condition. Then the outer asymptotically flat region N of (M, g) is simply connected, i.e. $\pi_1(N) = 1$.*

Moreover, if M contains a black hole region, then all connected components of the event horizon $H \subset M$ are homeomorphic to $S^2 \times \mathbb{R}$. $\diamond$

By global hyperbolicity, there is a global time function $T : M \to \mathbb{R}$. Let $S_t := T^{-1}(t)$ $(t \in \mathbb{R})$ be a Cauchy surface and $\widetilde{S}_t = i(S_t) \cup \{i^0\}$ its conformal completion. Of course $\widetilde{S}_t \cong \widetilde{S}$ for all $t \in \mathbb{R}$.

In light of the above theorem, $V_t := N \cap S_t$, a Cauchy surface for the outer region, is an oriented simply connected three-manifold. If M contains black hole domains then $\partial V_t \neq \emptyset$ and all boundary components are homeomorphic to a two-sphere S^2 ("the event horizon of a stationary black hole has no handles"). By conformal completion we may consider rather the three-manifold $\widetilde{V}_t := \widetilde{N} \cap \widetilde{S}_t$ which is moreover compact.

Now we are ready to describe the Yang–Mills-vacuum structure over $(N, g|_N)$. The simply connectedness of N implies that a restricted flat Yang–Mills bundle $E|_N$ is trivial whatever G is hence a vacuum Yang–Mills connection $A|_N$ may be regarded as a $\mathfrak{g}$-valued 1-form on $E|_N$. In an appropriate trivialization of the bundle $E|_N$, we may regard $A|_N$ as a $\mathfrak{g}$-valued 1-form over N instead of $E|_N$. Moreover, also by simply connectedness, a smooth Yang–Mills field $A|_N$ is vacuum if and only if there is a smooth function $f : N \to G$ obeying $A|_N = f^{-1}\mathrm{d}f$. Using again the simply connectedness of N, there is a gauge-transformation over N such that all vacuum fields can be transformed into temporal gauge, i.e. $A_0|_N = 0$ where the A_0 component is defined by the time function T. Consequently in temporal gauge, by exploiting the stationarity of the metric g, all flat connections are characterized by smooth functions $f_t : V_t \to G$. It is easily seen that $\pi_2(G) = 0$ implies that always exists a *null-homotopic* gauge transformation in the neighbourhood of the space-like infinity $i^0 \in \widetilde{V}_t$ such that the gauge-transformed function $\widetilde{f}_t$ extends as the identity to i^0 hence we are dealing with smooth functions $\widetilde{f} : \widetilde{V} \to G$ (we denote $\widetilde{V}_t$ and $\widetilde{f}_t$ simply by $\widetilde{V}$ and $\widetilde{f}$ if there is no danger of confusion).

A pure Yang–Mills theory being conformally invariant, we may consider our original Einstein–Yang–Mills theory over $(\widetilde{M}, \widetilde{g})$ instead of the original space-time. The restriction of the extended flat Yang–Mills bundle $\widetilde{E}|_{\widetilde{N}}$ is trivial even in this case. Certain physical quantities of the extended theory may suffer from singularities on the boundary $\partial i(M)$ but classical Yang–Mills vacua extend smoothly to the whole $(\widetilde{M}, \widetilde{g})$ as we have seen. In other words the studying of the vacuum sector of the extended Yang–Mills theory is correct.

Summing up, we can see that dynamically (i.e. homotopically) inequivalent vacua of the Yang–Mills theory are classified by the homotopy classes of smooth maps $\tilde{f} : \widetilde{V} \to G$ satisfying $\tilde{f}(i^0) = e \in G$, usually written as

$$\left[(\widetilde{V}, i^0), (G, e)\right]. \tag{3}$$

Now suppose that (M, g) contains black hole(s). In this case $\widetilde{V}$ is a simply connected manifold ***with boundary***. Such manifolds, considered as CW-complexes, have only cells of dimension less than three. Hence by the Cellular Approximation Theorem, every map $\tilde{f} : \widetilde{V} \to G$ descends to a homotopic map with values only on the cells of G having dimension less than three. Consequently, being $\pi_2(G) = 0$, G can be replaced by the simple Postnikov-tower $P_2 = K(\pi_1(G), 1)$ where $K(\pi_1(G), 1)$ is an Eilenberg–Mac Lane space yielding

$$\left[(\widetilde{V}, i^0), (G, e)\right] \cong \left[\widetilde{V}, K(\pi_1(G), 1)\right] \cong H^1(\widetilde{V}, \pi_1(G)) = 0. \tag{4}$$

The result is zero because $\widetilde{V}$ is simply connected. For details, see for instance [16]. Consequently all vacuum states are homotopy equivalent i.e. can be deformed into each other only through vacuum states ***over an outer, asymptotically flat portion N of the space-time*** (M, g). Clearly, classically only this part is relevant for a distant observer.

This result can be explained from a different point of view as well. Since the outer part N of M is also globally hyperbolic, the space-like submanifold V forms a Cauchy surface for N. Consequently if we know the initial values of two gauge fields, A and A' say, on $V \subset S$, we can determine their values over the whole *outer* space-time $N \subset M$ by using the field equations. This implies that the values of the fields A and A' "beyond" the event horizon in a moment are irrelevant for an observer outside the black hole. But we just proved that every vacuum fields restricted to V are homotopic. Roughly speaking, homotopical differences between Yang–Mills vacua "can be swept" into a stationary black hole.

Via (4) for arbitrary smooth functions $\tilde{f}_0 : \widetilde{V} \to G$ and $\tilde{f}_1 : \widetilde{V} \to G$ there is a homotopy

$$\widetilde{F}_T : \widetilde{V} \times [0, 1] \to G \tag{5}$$

satisfying $\widetilde{F}_T(x, 0) = \tilde{f}_0(x)$ and $\widetilde{F}_T(x, 1) = \tilde{f}_1(x)$ for all $x \in \widetilde{V}$. Taking two Cauchy-surfaces $V_0 \subset S_0 := T^{-1}(0)$ and $V_1 \subset S_1 := T^{-1}(1)$ we can regard the two functions as vacua $\tilde{f}_0 : \widetilde{V}_0 \to G$ and $\tilde{f}_1 : \widetilde{V}_1 \to G$. In the homotopy $\widetilde{F}_T$ the subscript T shows that the "time" required for the homotopy is measured by the time function T naturally associated to the globally hyperbolic space-time (M, g). We call this homotopy as ***T-homotopy***.

But on physical grounds, such a deformation or homotopy is effective only if it can be carried out *in finite proper time according to a distant observer's clock*. Such homotopies will be referred to as *finite γ-homotopies* or *effective homotopies*.

Let $\gamma : \mathbb{R}^+ \to N$ be a smooth, time-like, future directed curve in the outer region $N \subset M$ with $\gamma(0) \in V_0$ representing an observer moving in N. We denote by τ its proper time i.e. the natural affine parameter of the curve γ obeying $g(\dot\gamma(\tau), \dot\gamma(\tau)) = -1$. Moreover, let $\beta_1 : [0,1] \to V_1$ be a continuous space-like curve in the fixed Cauchy-surface V_1 approaching one connected component of the horizon i.e. $\beta_1(0) \notin H_1 = H \cap V_1$ while $\beta_1(1) \in H_1$. Define

$$\tau_{\beta_1(s)} := \inf_{\tau \in \mathbb{R}} \{\gamma(\tau) \in J^+(\beta_1(s))\}.$$

We prove the following simple lemma:

Lemma. *Let (M, g) be an asymptotically flat, stationary space-time with outer asymptotically flat region N and black hole region B and event horizon $H = \partial N = \partial B$. Consider the curves $\beta_1 : [0,1] \to V_1$ and $\gamma : \mathbb{R}^+ \to N$ defined above. Then*

$$\lim_{s \to 1} \tau_{\beta_1(s)} = \infty.$$

Proof. The proof is very simple. Clearly, if for a non-space-like curve $\alpha : \mathbb{R} \to M$ the condition $\mathrm{im}\alpha \cap B \neq \emptyset$ holds then $\mathrm{im}\alpha \subset B$ by the definition of the black hole $B \subset M$. Consequently if by assumption $\beta_1(1) \in H_1 \subset B$ then all non-space-like curves α with the property $\alpha(\tau') = \beta_1(1)$ ($\tau' \in \mathbb{R}$) never enter N hence never meet γ showing $\gamma(\tau) \notin J^+(\beta_1(1))$ for all $\tau \in \mathbb{R}^+$ i.e. $\tau_{\beta_1(1)} = \infty$. By continuity we get the result. ◇

Now we are ready to understand the above homotopy from the point of view of a distant observer. Fix an observer γ. Clearly, the T-homotopy (5) can be written as a γ-homotopy as

$$\widetilde{F}_T(x, t) = \tilde{f}_t(x) = \widetilde{F}_\gamma(x, \tau).$$

Take $x := \beta_1(s)$ then we get

$$\widetilde{F}_T(\beta_1(s), 1) = \tilde{f}_1(\beta_1(s)) = \widetilde{F}_\gamma(\beta_1(s), \tau_{\beta_1(s)}).$$

But by the Lemma we can see that as we approach the horizon, i.e. $s \to 1$, $\tau_{\beta_1(s)}$ diverges hence typically the T-homotopy cannot be a finite γ-homotopy in other words it cannot be "finished" in finite proper time measured by γ.

From here we can see that given a T-homotopy (5), it gives rise to an effective homotopy if and only if there is a neighbourhood $H \subset U \subset N$ of the horizon in N with the property $\tilde{f}_0|_{U_0} = \tilde{f}_1|_{U_1}$. This implies

$$\tilde{f}_0|_{H_0} = \tilde{f}_1|_{H_1}.$$

This result can be interpreted as a natural boundary condition on each connected component of the horizon for effectively deformable vacua. Since each boundary component is homeomorphic to the two-sphere S^2 and $\pi_2(G) = 0$ we may select a function in each homotopy class obeying $\tilde{f}_0|_{H_0} = e$. We just remark that exactly this is the physical reason for keeping the functions as identity in the space-like infinity i^0 when we discuss homotopy classes of vacua over Minkowskian space-time.

Taking into account that $\partial\widetilde{V}_t = H_t$, ***the classes of effectively deformable vacua*** are given by the homotopy classes of functions $\tilde{f} : \widetilde{V} \to G$ with the property $\tilde{f}(\partial\widetilde{V}) = \tilde{f}(i^0) = e \in G$. The homotopy is restricted to obey these boundary conditions. This set is denoted by

$$\left[(\widetilde{V}, \partial\widetilde{V}, i^0), (G, e)\right]$$

and replaces (3).

This set is typically non-trivial. It is possible to show that for a typical classical, compact Lie group G (i.e. for $U(n)$ with $n \geq 2$ and $SO(n)$, Spin(n) with $n \neq 4$ and $SU(n)$, $Sp(n)$ for all n and G_2, F_4, E_6, E_7, E_8) we have

$$\left[(\widetilde{V}, \partial\widetilde{V}, i^0), (G, e)\right] \cong \mathbf{Z}$$

while

$$\left[(\widetilde{V}, \partial\widetilde{V}, i^0), (U(1), e)\right] \cong 0,$$

and

$$\left[(\widetilde{V}, \partial\widetilde{V}, i^0), (SO(4), e)\right] \cong \left[(\widetilde{V}, \partial\widetilde{V}, i^0), (\mathrm{Spin}(4), e)\right] \cong \mathbf{Z} \oplus \mathbf{Z}.$$

The proof is based on the calculations of Isham [6] and is presented in [18].

Being these vacua of definite winding number non gauge invariant, we have to introduce again linear combinations as (2) in this more general situation. Consequently we can see that approach (i) to the θ-parameter, mentioned in the Introduction, still makes sense in the general case.

The natural question arises: are there instanton solutions in the corresponding Wick-rotated theory? What is the physical relevance of these solutions? Do they induce semi-classical tunneling between the vacuum states of different effective winding number? If yes, beyond (i) we have another, more physical, reason to introduce θ-vacua because of concept (ii), also mentioned in the Introduction.

3 Instantons: A Special Case

Next we turn our attention to the semi-classical sector of Yang–Mills theories. Unfortunately we are unable to present such general treatment as in the case of vacua because of several reasons: any satisfying formulation of the Wick rotation is yet unknown, the moduli space and analytical properties of instantons over general (mainly non-compact) Riemannian four-manifolds are not available in this moment.

Rather we restrict our attention to the $SU(2)$ gauge theory over the *Euclidean Schwarzschild manifold*. Even this simple example can provide lot of puzzles concerning instantons, vacuum tunneling, etc.

The Wick rotation (M_+, g_+) of Schwarzschild space-time (M_-, g_-) is given by the replacement $t \mapsto i\tau$ in the original Schwarzschild metric and locally takes the form [17]

$$\mathrm{d}s^2 = \left(1 - \frac{2m}{r}\right)\mathrm{d}\tau^2 + \left(1 - \frac{2m}{r}\right)^{-1}\mathrm{d}r^2 + r^2(\mathrm{d}\Theta^2 + \sin^2\Theta\mathrm{d}\phi^2). \quad (6)$$

Here

$$r \in (2m, \infty), \quad \tau \in [0, 8\pi m), \quad \Theta \in [0, \pi], \quad \phi \in [0, 2\pi).$$

It is well-known [17] that this metric can be realized as a complete, analytical metric on $M_+ = S^2 \times \mathbb{R}^2$ if and only if τ is periodic with period $8\pi m$. In this picture τ plays the role of the angular variable while r is the radial coordinate of the polar coordinate system of $\mathbb{R}^2$. Note that although M_- and M_+ are actually homeomorphic they are endowed with metrics in a very different way. The periodicity of imaginary time (i.e. the existence of a finite temperature, namely the Hawking-temperature, in the original theory) causes many difficulties. First, this implies if $m \to 0$ in (6), we recover the flat metric on $\mathbb{R}^3$ only, not on $\mathbb{R}^4$ since $\tau \in [0, 8\pi m)$.

Surprisingly even in the Euclidean Schwarzschild case, the full moduli space of $SU(2)$-instantons is unknown; only sporadic explicit solutions exist (for a survey see [19]). We can present a special instanton in this model found by Charap and Duff [20]. Since M_+ is a four dimensional spin manifold equipped with an Einstein metric g_+, the projection of its Levi–Cività connection onto a suitable $\mathfrak{su}(2)$ component of $\mathfrak{so}(4) \cong \mathfrak{su}(2)_+ \oplus \mathfrak{su}(2)_-$ produces a self-dual connection on (M_+, g_+):

$$A_+(\tau, r, \Theta, \phi) = \frac{1}{2}\sqrt{1 - \frac{2m}{r}}\mathrm{d}\Theta\,\mathbf{i} + \frac{1}{2}\sqrt{1 - \frac{2m}{r}}\sin\Theta\,\mathrm{d}\phi\,\mathbf{j} + \frac{1}{2}\left(\cos\Theta\mathrm{d}\phi - \frac{m}{r^2}\mathrm{d}\tau\right)\mathbf{k}.$$

We have used the identification $\mathfrak{su}(2) \cong \mathrm{Im}\,\mathbb{H}$ given by $\{\sigma_1, \sigma_2, \sigma_3\} \mapsto \{\mathbf{i}, \mathbf{j}, \mathbf{k}\}$. It is not difficult to calculate the action of this connection that is equal to 1 consequently this is an instanton field in the Wick-rotated theory. We would like to transform it into τ-temporal gauge in order to understand its relation to vacua. This trial is obstructed by the fact that the embedding $M_+ \subset \mathbb{R}^4$ is not compatible with the metric g_+. More explicitly, we can see that a gauge transformation which transforms the above instanton into temporal gauge ought to take the form

$$h(\tau, r, \Theta, \phi) = \exp \mathbf{k}\left(f(r, \Theta, \phi) - \frac{2m\tau}{r^2}\right),$$

where f is an arbitrary smooth function. The map h should be periodic in τ with period $8\pi m$ but it is impossible. We get the same result for the other known Euclidean Schwarzschild-instantons listed in [19].

These observations suggest that these instantons *may not* related to any tunneling-phenomenon in the Lorentzian theory. From the Lorentzian point of view, however, we have seen that effectively homotopically different vacua can be introduced. Is it possible that these vacua cannot be connected semi-classically by instantons? A similar problem was studied by Bitar et al. [21] in two dimensional σ-models; the authors proved that the instanton structure of this model is more complicated than the vacuum sector. In spite of these properties the authors were able to interpret instantons as tunnelings in their σ-model.

4 Concluding Remarks

In this letter we have studied the concept of θ-vacua in general Yang–Mills theories.

In light of our results, we can see that for outer observers in stationary, asymptotically flat space-times θ-vacua do occur. On the other side, in the Euclidean

Schwarzschild theory at least, instantons also exist, but apparently they cannot connect together homotopically different vacuum states essentially due to the Hawking-temperature of the Schwarzschild black hole.

In summary we can see that through topological and symmetry-considerations θ-vacua seem to be very general and stable objects but the understanding of their role in the semi-classical and the true quantum dynamics of the theory requires lot of future work.

References

[1] G. 't Hooft, Phys. Rev. Lett. **37**,1976, 8.

[2] A.A. Belavin, A.M. Polyakov, A.S. Schwarz, Yu.S. Tyupkin, Phys. Lett. **B59**, 1975, 85.

[3] R.P. Peccei, H.R. Quinn, Phys. Rev. **D16**, 1977, 1791.

[4] L-P. Cheng, L-F.Li, "Gauge Theory of Elementary Particle Physics", Clarendon Press, 1984.

[5] M. Kaku, "Quantum Field Theory", Oxford University Press, Oxford, 1993.

[6] C.J. Isham, in "Old and new Questions in Physics, Cosmology, Philosophy and Theoretical Biology", Ed.: A. Van Der Merwe, Plenum Press, New York, 1983, 189.

[7] C.J. Isham, Trieste Diff. Geom. Meth., 1981, 171.

[8] C.J. Isham, G. Kunstatter, Journ. Math. Phys. **23**, 1982, 1668.

[9] C.J. Isham, G. Kunstatter, Phys. Lett. **B102**, 1981, 417.

[10] S. Deser, M.J. Duff, C.J. Isham, Phys. Lett. **B93**, 1980, 419.

[11] D.M. Witt, Phys. Rev. Lett. **57**, 1986, 1386.

[12] D. Gannon, J. Math. Phys. **16** , 1975, 2364.

[13] P.T. Chruściel, R.M. Wald, Class. Quant. Grav. **11**, 1994, L147.

[14] S.W. Hawking, Phys. Rev. **D53**, 1996, 3099.

[15] J. Preskill, S.P. Trivedi, M.B. Wise, Phys. Lett. **B223**, no.1. 1989, 26.

[16] E.H. Spanier " Algebraic Topology", Springer–Verlag, Berlin, 1966.

[17] R.M. Wald "General Relativity", University of Chicago Press, Chicago, 1984.

[18] G. Etesi, submitted to Class. Quant. Grav., hep-th/0011157;

[19] G. Etesi, T. Hausel, Journ. Geom. Phys. **37**, 2001, 126, E-print: hep-th/0003239;

[20] J.M. Charap, M.J. Duff, Phys. Lett. **B71**, 1977, 219.

[21] K.M. Bitar, S. Chang, G. Grammer Jr., J.D. Stack, Phys. Rev. **D19**, 1979, 1214.

Scale relativity and non-differentiable fractal space-time

Laurent Nottale
CNRS UMR 8631, DAEC, Observatoire de Paris-Meudon
F-92195 Meudon Cedex, France
E-mail: laurent.nottale@obspm.fr

21 November 2000

1 Introduction

The theory of scale relativity [14] is an attempt to study the consequences of giving up the hypothesis of space-time differentiability. One can show [14] [15] that a continuous but nondifferentiable space-time is necessarily *fractal*. Here the word fractal [12] is taken in a general meaning, as defining a set, object or space that shows structures at all scales, or on a wide range of scales. More precisely, one can demonstrate [17] that a continuous but nondifferentiable function is explicitly resolution-dependent, and that its length $\mathcal{L}$ tends to infinity when the resolution interval tends to zero, i.e. $\mathcal{L} = \mathcal{L}(\varepsilon)_{\varepsilon \to 0} \to \infty$. This theorem and other properties of non-differentiable curves have been recently analysed in detail by Ben Adda and Cresson [4]. It naturally leads to the proposal that the concept of *fractal space-time* [21] [25] [14] [7] is the geometric tool adapted to the research of such a new description. In such a generalized framework including all continuous functions, the usual differentiable functions remain included, but as very particular and rare cases.

Since a nondifferentiable, fractal space-time is explicitly resolution-dependent, the same is a priori true of all physical quantities that one can define in its framework. We thus need to complete the standard laws of physics (which are essentially laws of motion in classical physics) by laws of scale, intended to describe the new resolution dependence. We have suggested [13] that the principle of relativity can be extended to constrain also these new scale laws.

Namely, we generalize Einstein's formulation of the principle of relativity, by requiring *that the laws of nature be valid in any reference system, whatever its state.* Up to now, this principle has been applied to changes of state of the coordinate system that concerned the origin, the axes orientation, and the motion (measured in terms of velocity and acceleration). In scale relativity, we assume that the space-time resolutions are not only a characteristic of the measurement apparatus, but acquire a universal status. They are considered as essential variables, inherent to the physical description. We define them as characterizing the "state of scale" of the reference system, in the same way as the velocity characterizes its state of motion. The principle of scale relativity consists of applying the principle of relativity to

Frontiers of Fundamental Physics 4, Edited by Sidharth and Altaisky,
Kluwer Academic/Plenum Publishers, New York, 2001

such a scale-state. Then we set a principle of ***scale-covariance***, requiring that the equations of physics keep their form under resolution transformations.

In the present paper, we shall review various levels of development of the theory, then consider some of its consequences in the domains of elementary particles, cosmology and gravitational structure formation.

2 Galilean scale relativity

2.1 Standard fractal laws

As we shall first see, simple fractal scale-invariant laws can be identified with a "Galilean" version of scale-relativistic laws. Indeed, let us consider a non-differentiable coordinate $\mathcal{L}$. Our basic theorem that links non-differentiability to fractality implies that $\mathcal{L}$ is an explicit function $\mathcal{L}(\varepsilon)$ of the resolution interval ε. As a first step, one can assume that $\mathcal{L}(\varepsilon)$ satisfies the simplest possible scale differential equation one may write, namely, the first order equation:

$$\frac{d\ln\mathcal{L}}{d\ln(\lambda/\varepsilon)} = \delta, \tag{1}$$

where δ is a constant. The solution is a fractal, power-law dependence:

$$\mathcal{L} = \mathcal{L}_0(\lambda/\varepsilon)^{\delta}, \tag{2}$$

where δ is the scale dimension, i.e., $\delta = D - D_T$, the fractal dimension minus the topological dimension. The Galilean structure of the group of scale transformation that corresponds to this law can be verified in a straightforward manner from the fact that it transforms in a scale transformation $\varepsilon \rightarrow \varepsilon'$ as

$$\ln\frac{\mathcal{L}(\varepsilon')}{\mathcal{L}_0} = \ln\frac{\mathcal{L}(\varepsilon)}{\mathcal{L}_0} + \delta(\varepsilon)\ln\frac{\varepsilon}{\varepsilon'} \quad ; \quad \delta(\varepsilon') = \delta(\varepsilon). \tag{3}$$

This transformation has exactly the structure of the Galileo group, as confirmed by the law of composition of dilations $\varepsilon \rightarrow \varepsilon' \rightarrow \varepsilon''$, which writes $\ln\rho'' = \ln\rho + \ln\rho'$, with $\rho = \varepsilon'/\varepsilon$, $\rho' = \varepsilon''/\varepsilon'$ and $\rho'' = \varepsilon''/\varepsilon$.

2.2 Breaking of the scale symmetry

More generally, one can write a first order equation where the scale variation of $\mathcal{L}$ depends on $\mathcal{L}$ only, $d\mathcal{L}/d\ln\varepsilon = \beta(\mathcal{L})$. The function $\beta(\mathcal{L})$ is a priori unknown but, always taking the simplest case, we may consider a perturbative approach and take its Taylor expansion. We obtain the equation:

$$\frac{d\mathcal{L}}{d\ln\varepsilon} = a + b\mathcal{L} + \ldots \tag{4}$$

This equation is solved in terms of a standard power law of power $\delta = -b$, broken at some relative scale λ (which is a constant of integration):

$$\mathcal{L} = \mathcal{L}_0\,[1 + \left(\frac{\lambda}{\varepsilon}\right)^{\delta}]. \tag{5}$$

Depending on the sign of δ, this solution represents either a small-scale fractal behavior (in which the scale variable is a resolution), broken at larger scales, or a large-scale fractal behavior (in which the scale variable ε would now represent a changing window for a fixed resolution λ), broken at smaller scales.

2.3 Euler-Lagrange scale equations

In the previous approach, we have considered as primary variables the position $\mathcal{L}$ and the resolution ε. However, we are naturally led, in the scale-relativistic approach, to reverse the definition and the meaning of variables. The scale dimension δ can be generalized in terms of an essential, fundamental *variable*, that would remain constant only in very particular situations (namely, in the case of scale invariance, that corresponds to "scale-freedom"). It plays for scale laws the same role as played by time in motion laws. We have proposed to call "djinn" this varying scale dimension. The new approach amounts to work in a "space-time-djinn" rather than only in space-time, thus including motion and scale behaviour in the same 5-dimensional description. The resolution can now be defined as a derived quantity in terms of the fractal coordinate and of the djinn:

$$\bar{V} = \ln(\lambda/\varepsilon) = \frac{d\ln\mathcal{L}}{d\delta}. \tag{6}$$

Our identification of standard fractal behavior as Galilean scale laws can now be fully proven. We assume that, as in the case of motion laws, scale laws can be constructed from a Lagrangian approach. A scale Lagrange function $\bar{L}(\ln\mathcal{L}, \bar{V}, \delta)$ is introduced, from which a scale-action is constructed:

$$\bar{S} = \int_{\delta_1}^{\delta_2} \bar{L}(\ln\mathcal{L}, \bar{V}, \delta)d\delta. \tag{7}$$

The action principle, applied on this action, yields a scale-Euler-Lagrange equation

$$\frac{d}{d\delta}\frac{\partial\bar{L}}{\partial\bar{V}} = \frac{\partial\bar{L}}{\partial\ln\mathcal{L}}. \tag{8}$$

The simplest possible form for the Lagrange function is the equivalent for scales of what inertia is for motion, i.e., $\bar{L} \propto \bar{V}^2$ and $\partial\bar{L}/\partial\ln\mathcal{L} = 0$ (no scale "force"). The Lagrange equation writes in this case:

$$\frac{d\bar{V}}{d\delta} = 0 \Rightarrow \bar{V} = cst. \tag{9}$$

The constancy of $\bar{V} = \ln(\lambda/\varepsilon)$ means here that it is independent of the scale-time δ. Then Eq. (6) can be integrated in terms of the usual power law behavior, $\mathcal{L} = \mathcal{L}_0(\lambda/\varepsilon)^\delta$. This reversed viewpoint has several advantages which allow a full implementation of the principle of scale relativity:
(i) The scale dimension takes its actual status of "scale-time", and the logarithm of resolution $\bar{V}$ its status of "scale-velocity", $\bar{V} = d\ln\mathcal{L}/d\delta$.
(ii) This leaves open the possibility of generalizing our formalism to the case of four independent space-time resolutions. Indeed, from $\mathcal{L}^\mu, \mu = 1,2,3,4$ and δ one can now build a 4-component resolution vector, $\bar{V}^\mu = \ln(\lambda^\mu/\varepsilon^\mu) = d\ln\mathcal{L}^\mu/d\delta$.

(iii) At an even more profound level, one can jump to a 5-dimensional covariant representation (the "space-time-djinn") in which the djinn δ is given rank 0 and the four fractal space-time coordinates ranks 1 to 4. This leads to a 10 parameter rotation group in which we recover 3 rotations in space (xy, yz and zx), 3 Lorentz boosts (xt,yt and zt) and 4 resolution transformations ($t\delta$,$x\delta$,$y\delta$,$z\delta$). These new symetries lead to the emergence of four new conservative quantities (momenta), which we tentatively identify with the charges (see next section).

3 Special and generalized scale-relativity

3.1 Special scale relativity

It is well known that the Galileo group of motion is only a degeneration of the more general Lorentz group. The same is true for scale laws. Indeed, if one looks for the general linear laws of scale that come under the principle of scale relativity, one finds that, once they are expressed in logarithm form, they have the structure of the Lorentz group [13]. Therefore, in special scale relativity, we have suggested to substitute to the Galilean law of composition of dilations $\ln(\varepsilon'/\lambda) = \ln\rho + \ln(\varepsilon/\lambda)$ the more general log-Lorentzian law:

$$\ln\frac{\varepsilon'}{\lambda} = \frac{\ln\rho + \ln(\varepsilon/\lambda)}{1 + \ln\rho \ln(\varepsilon/\lambda)/\ln^2(\lambda_P/\lambda)}. \tag{10}$$

The scale dimension (or "djinn", δ) becomes itself a variable. More precisely, the couple it makes with the fractal coordinate $[\delta, \ln(L/\lambda)]$ becomes a scale vector. In the simplified case when $\delta(\lambda) = 1$ (i.e., fractal dimension 2 at transition scale) and $L(\lambda) = \lambda$, it writes:

$$\delta(\varepsilon) = \frac{1}{\sqrt{1 - \ln^2(\varepsilon/\lambda)/\ln^2(\lambda_P/\lambda)}}, \tag{11}$$

where λ is the fractal-nonfractal transition scale (e.g., the Compton length of a particle). In such a law, there appears a minimal (or maximal) scale of space-time resolution which is invariant under dilations and contractions, and plays the same role for scales as that played by the velocity of light for motion.

Toward the small scales, this invariant length-scale is naturally identified with the Planck scale, $\lambda_P = (\hbar G/c^3)^{1/2}$, that now becomes impassable and plays the physical role that was previously devoted to the zero point. The same is true in the cosmological domain, with an inversion of the scale laws: there appears a maximal, impassable scale of resolution that plays the physical role of the infinite, that we have identified with the length-scale $\mathbb{L} = \Lambda^{-1/2}$, where Λ is the cosmological constant .

Some consequences of this new interpretation of the Planck length-time-scale have been considered elsewhere [13][15], concerning in particular the unification of fundamental fields. Let us briefly discuss here its consequence as concerns the status of units. We already know that special motion-relativity has changed the status of space and time units. Indeed, the very existence of space-time implies to use the same units for length and time intervals. This is achieved since 1985, the unit of length being derived from the unit of time, and c fixed. Therefore, the genuine nature of velocities is adimensional pure numbers always smaller than one.

A new step can be made in this direction using special scale relativity. Indeed, in its framework, every length (or time) interval is written in terms of its ratio over the Planck length (time)-scale. The Planck length and time scales thus appear as natural units of length and time intervals, whose genuine nature is found to be pure, adimensional numbers larger than one. In the end, this implies that space and time units do not really exist, since, in the same way as the limitation of 3-velocity is a pure effect of projection from 4-space-time to 3-space, the Planck limit is the simple result of projection from 5 dimensional to 4 dimensional space. More generally, if one replaces the three fundamental constants G, $\hbar$ and c by their expressions in terms of the Planck time, length and mass in any equation of physics, all quantities appear in these equations in terms of their ratio over the corresponding Planck units, which, ultimately, vanish from physics. As a consequence, the time is certainly come to re-found our system of units, in particular concerning masses, which should be referred to the Planck mass ($\approx 2 \times 10^{-5}$ g).

3.2 Scale-motion coupling and mass-charge relations

The theory of scale relativity also allows to get new insights about the physical meaning of gauge invariance [15]. In the scale laws recalled hereabove, only scale transformations at a given point were considered. But we may also wonder about what happens to the structures in scale-space of a scale-dependent object such as an electron or another charged particle, when it is displaced. Consider anyone of these structures, lying at some (relative) resolution ε (such that $\varepsilon < \lambda$, where λ is the Compton length of the particle) for a given position of the particle. In a displacement, the relativity of scales implies that the resolution at which this given structure appears in the new position will a priori be different from the initial one. In other words, $\varepsilon = \varepsilon(x, t)$ is now a function of the space-time coordinates, and we expect the occurrence of ***dilations of resolutions induced by translations***, so that we are led to introduce a covariant derivative:

$$e\,\frac{D\varepsilon}{\varepsilon} = e\,\frac{d\varepsilon}{\varepsilon} - A_\mu dx^\mu, \tag{12}$$

where a four-vector A_μ must be introduced since dx^μ is itself a four-vector and $d\ln\varepsilon$ a scalar (in the case of a global dilation).

However, if one wants such a "field" A_μ to be physical, it should be defined whatever the initial scale from which we started. Starting from another scale $\varepsilon' = \rho\varepsilon$, we get the same expression as in Eq.(12), but with A_μ replaced by A'_μ. Therefore, we obtain the relation:

$$A'_\mu = A_\mu + e\,\partial_\mu \ln\rho, \tag{13}$$

which depends on the relative "state of scale", $\bar{V} = \ln\rho = \ln(\varepsilon/\varepsilon')$, that is now a function of the coordinates.

One may therefore identify A_μ with the electromagnetic potential, and Eq.(13) with the property of gauge invariance. Now we know that applying a gauge transformation to the electromagnetic field implies to change also the wave function of the electron, that becomes:

$$\psi' = \psi\, e^{i4\pi\alpha \ln\rho} \tag{14}$$

where α is a coupling constant. While in Galilean scale relativity, the scale ratio ρ is unlimited, in the more general framework of special scale relativity it is limited by the Planck-scale/Compton-scale ratio. This limitation implies the quantization of charge, following a general mass-charge relation [15]: $\alpha \ln(m_P/m) = k/2$, where k is integer. In order to compare such a relation with experimental data, one should account for the electroweak theory, according to which the electromagnetic coupling is only 3/8 of its high energy value (plus radiative corrections). We get:

$$\frac{8}{3}\alpha_{em} \ln\left(\frac{m_P}{m_e}\right) = 1 \tag{15}$$

where $\alpha_{em} = 1/137.036$ is the low energy fine structure constant and m_e is the electron mass. This relation is implemented with a relative precision of 2×10^{-3}, becoming 10^{-4} when accounting for threshhold effects [15].

This approach can be generalized, since, as recalled hereabove, we can define four different and independent dilations along the four space-time resolutions instead of only one global dilation. The above U(1) field is then expected to be embedded into a larger field, in agreement with the electroweak and grand unification theories, and the charge e to be one element of a more complicated, "vectorial" charge. Some hints about such a generalization will be given in what follows.

More generally, we shall be led to look for the general non-linear scale laws that satisfy the principle of scale relativity. Such a generalized framework implies working in a five-dimensional space-time-djinn. The development of such a "general scale-relativity" lies outside the scope of the present paper and will be considered in forthcoming works.

4 Theoretical predictions of masses and couplings

In the new framework, theoretical predictions of some of the free parameters of the standard model become possible. We have presented and checked such predictions in previous works [13] [14] [15]. But in the recent years, there has been an improvement of several experimental measurements [28], so that it may now be interesting to check them again with these new values. They are, respectively for the top quark mass, Higgs boson mass, W and Z boson masses, strong coupling constant at Z scale, fine structure constant at Z scale, and $\sin^2\theta$ of weak mixing angle at Z scale in the modified minimal substraction scheme (where it is defined through the SU(2) charge g and the U(1) charge g'):

$m_t = 174.3 \pm 5.1\,\mathrm{GeV}$; $m_H = 108 - 220\,\mathrm{GeV}$
$m_W = 80.42 \pm 0.04\,\mathrm{GeV}$; $m_Z = 91.1872 \pm 0.0021$
$\alpha_S(m_Z)^{-1} = 0.118 \pm 0.002$; $\alpha(m_Z)^{-1} = 128.92 \pm 0.03$
$\hat{s}_Z^2 = \frac{g'^2}{g^2+g'^2} = 0.23117 \pm 0.00016$

4.1 Fine structure constant

In [15], we derived a prediction of the fine structure constant (i.e. the electromagnetic coupling). It was based on the suggestion that the bare (infinite energy) value of the electroweak coupling (which becomes finite in special scale-relativity) is $4\pi^2$. The fact that 3 among the 4 gauge bosons acquire mass through the Higgs mechanisms leads to a multiplying factor 8/3, so that one expects that $\alpha_\infty^{-1} = 32\pi^2/3$. The difference between the infinite energy and Z or low energy values was computed using the solutions to the renormalization group equation for the running coupling. The prediction at the Z value for 1 Higgs doublet is (see more detail in [15] [20]):

$$\alpha(m_Z)^{-1} = \frac{32\pi^2}{3} + \frac{22\pi}{3} + \frac{6}{\pi^2} = 128.922\,. \tag{16}$$

This result compares very well with the experimental value, 128.92 ± 0.03.

4.2 Strong coupling

From the conjecture that the strong coupling value reaches the critical value $1/4\pi^2$ at unification scale (i.e. $m_P/2\pi$ in the special scale-relativistic modified standard model), we obtained a predicted value $\alpha_S(m_Z)^{-1} = 0.1155 \pm 0.0002$ from the solution to the renormalization group equation of the running coupling [13] [15]. This expectation remains in agreement (within about one σ) with the recently improved experimental value 0.118 ± 0.002.

4.3 SU(2) coupling

In [15], we also attempted to apply the mass-charge relation to the SU(2) coupling α_2. We found that the relation

$$3\,\alpha_{2Z}\,C_Z = 4 \tag{17}$$

was precisely achieved at the Z scale. However the factor 3 was not accounted for in that work. The solution to this problem relies on the generalization of scale (i.e. gauge) transformations to dilations which are no longer global, but instead may be different on the internal resolutions corresponding to the various coordinates. The group SU(2) corresponds to rotations in a 3-dimensional scale space. Therefore the phase term in a fermion field will write:

$$\alpha_2 \ln(\frac{\varepsilon_x}{\lambda}) + \alpha_2 \ln(\frac{\varepsilon_y}{\lambda}) + \alpha_2 \ln(\frac{\varepsilon_z}{\lambda}) < 3\,\alpha_2 \ln(\frac{\lambda_P}{\lambda}), \tag{18}$$

since the same coupling applies to the three variables, and since all three resolutions are limited at small scales by the Planck scale. From Eq. (17) we expect a value $\alpha_{2Z}^{-1} = 29.8169 \pm 0.0002$. The present precise experimental value is:

$$\alpha_{2Z}^{-1} = \alpha_Z^{-1} \times \hat{s}_Z^2 = 29.802 \pm 0.027, \tag{19}$$

which lies within 1σ of the theoretical prediction.

4.4 Vacuum expectation value of the Higgs field

As recalled hereabove, there are fundamental arguments for introducing a bare inverse coupling at infinite energy (i.e., in special scale relativity, at Planck length-scale) given by the critical value $4\pi^2$. Moreover, our re-interpretation of gauge invariance as scale-invariance on space-time resolution led us to construct general relations between couplings and scale ratios. Therefore one expects the emergence of a new fundamental scale given by:

$$\ln\left(\frac{\lambda}{\lambda_P}\right) = 4\pi^2, \tag{20}$$

where λ_P is the Planck length-scale. This relation may provide a solution to the hierarchy problem, according to which there is a misunderstood factor $\approx 10^{17}$ between the electroweak scale and the Planck scale (expected to be the full unification scale). Indeed the length-scale λ defined above is $e^{4\pi^2} = 1.397 \times 10^{17}$ larger than the Planck scale. As a first approximation, we can apply this relation to mass ratios. This gives a mass scale of 87.39 GeV, intermediate between the Z and W masses. However, mass-scales and length-scales are no longer directly inverse in the scale-relativity framework. There is a "log-Lorentz" factor between them (when they are referred to low energies). Namely, by taking as reference the electron Compton scale, the new mass-scale is more precisely given by:

$$\ln\left(\frac{m}{m_e}\right) = \frac{\ln(\lambda_e/\lambda)}{\sqrt{1 - \ln^2(\lambda_e/\lambda)/C_e^2}}. \tag{21}$$

With the currently accepted value of the gravitational constant (for which the error is now thought to be 12 times larger than previously given, see [28]), we obtain for the fundamental constant $C_e = \ln(\lambda_e/\lambda_P) = 51.52797(70)$. Then the new theoretically predicted mass scale is

$$m_v = 123.23 \pm 0.09\,\text{GeV}, \tag{22}$$

which is closely linked to the vacuum expectation value v of the Higgs field, since the present experimental value of $v/\sqrt{2} = m_W/g$ (where g is the SU(2) weak charge) is $123.11 \pm 0.03\,\text{GeV}$. Now some work remains to be done to really understand why the new mass-scale should have precisely this interpretation.

4.5 Mass of the Higgs boson

The framework of generalized scale-relativity provides one with possibilities to make theoretical predictions of the value of the Higgs boson mass. The (summarized) argument is as follows.

In today's electroweak scheme, the Higgs boson is considered to be separated from the electroweak field. Moreover, a more complete unification is mainly seeked in terms of attempts of "grand" unifications with the strong field. However, one may wonder whether, maybe in terms of an effective theory at intermediate energy, one could achieve a more tightly unified purely electroweak theory. Recall indeed that in the present standard model, the weak and electromagnetic fields are mixed, but there remains four free parameters, which can e.g. be taken to be the Higgs boson mass, the vacuum expectation value of the Higgs field and the Z and W masses. In the attempt sketched out hereafter, the Higgs field is assumed to be a part of the total field, so that only two free parameters would be left. As a consequence, the Higgs boson mass and the W/Z mass ratio could be derived in such a model.

Recall that the structure of the present electroweak boson content is as follows. There is a SU(2) gauge field, then involving three fields of null mass (i.e. $2 \times 3 = 6$ degrees of freedom), a U(1) null mass field (2 d.f.) and a Higgs boson complex doublet (4 d.f.), which makes 12 degrees of freedom in all. Through the Glashow-Salam-Weinberg mechanism, 3 of the 4 components of the Higgs doublet become longitudinal components of the weak field which therefore acquires mass ($3 \times 3 = 9$

d.f.), while the photon remains massless (2 d.f.), so that there remains a Higgs scalar which is nowadays experimentally searched (1 d.f.).

Now, we have suggested a new interpretation of gauge invariance as being scale invariance on the internal resolutions, considered as intrinsic to the description of the particle-fields (at scale smaller than their Compton length in restframe). As a first step we considered only global dilations, which led us to a U(1) invariance and to the relations between mass scale and coupling constant recalled above. But more generally one may consider four independant scale transformations on the four space-time resolutions, i.e., $(\ln \varepsilon_x, \ln \varepsilon_y, \ln \varepsilon_z, \ln \varepsilon_t)$. This means that the scale space (i.e., here the gauge space) is at least four-dimensional (but note that this is not the final word on the subject, since this does not yet include the fifth "djinn" dimension δ). Moreover, the mixing relation between the B [U(1)] and W_3 [SU(2)] fields may also be interpreted as a rotation in the full gauge space. Therefore we expect the appearance of a 6 component antisymmetric tensor field (linked to the rotations in this space), corresponding in the simplest case to a SO(4) group. Such a zero mass field would yield 12 degrees of freedom by itself alone, so that it is able to include the electromagnetic and weak fields, but also the residual Higgs field.

What about the Higgs boson in such a unified framework ? We shall tentatively explore the possibility that it appears as a separated scalar only as a low energy approximation, while in the new framework it would be one of the components of the unified field (in analogy with energy appearing as scalar at low velocity, while it is ultimately a component of the energy-momentum four-vector).

Such an attempt is supported by the form of the electroweak Lagrangian (we adopt Aitchison's [3] notations). Its Higgs scalar boson part writes:

$$L_H = \frac{1}{2}\,\partial_\mu \sigma \partial^\mu \sigma - \frac{1}{2}\,m_H^2 \sigma^2 - \frac{1}{2}\lambda^2 \sigma^4. \tag{23}$$

The vacuum expectation value v of the Higgs field is computed from the square (mass term) and quartic term, so that the Higgs mass is related to v and λ as:

$$m_H = \sqrt{2}\, v\, \lambda. \tag{24}$$

A prediction of the constant λ would therefore lead to a prediction of the Higgs mass. Now, a non-Abelian field writes in terms of its potential :

$$F^{\alpha\mu\nu} = \partial^\mu W^{\alpha\nu} - \partial^\nu W^{\alpha\mu} - g\, c^\alpha_{\beta\gamma} W^{\beta\mu} W^{\gamma\nu}, \tag{25}$$

where g is the (now unique) charge and $c^\alpha_{\beta\gamma}$ the structure coefficients of the Lie algebra associated to the gauge group. Its Lagrangian writes:

$$L_W = -\frac{1}{4} F^{\mu\nu} F_{\mu\nu}. \tag{26}$$

Therefore, it includes W^4 terms coming from the W^2 terms in the field. Now our ansatz consists of identifying some of these W^4 terms, of coefficient $-\frac{1}{4}\, g^2\, (c^\alpha_{\beta\gamma})^2$, with the Higgs boson σ^4 term of coefficient $-\frac{1}{2}\,\lambda^2$.

Namely, let us first separate the six components of the total field in two sub-systems, $[W_1, W_2, W_3]$ and $[B_1, B_2, B_3]$. The three W's can be identified with the standard SU(2) field and, say, B_1 with the $\mathrm{U(1)_Y}$ field.Two vectorial fields remain,

B_2^μ and B_3^μ. They will contribute in a non-vanishing way to the quartic term in the Lagrangian by their cross product. At the approximation (considered here) where their space components are negligible, we find:

$$-B_{2\mu} B_{3\nu} B_3^\mu B_2^\nu = -[B_2^0 B_3^0]^2. \tag{27}$$

Finally, we make the identification of these time components with the residual Higgs boson, $B_2^0 = B_3^0 = \sigma$. This allows a determination of the constant λ according to the relation:

$$\lambda^2 = \frac{g^2 c^2}{2}, \tag{28}$$

where the squared Lie coefficient $c^2 = 1$ in the case of an SO(4) group. Provided the global charge is identical to the SU(2) charge, and since the W mass is given by $m_W = gv/\sqrt{2}$, one finally obtains a Higgs boson mass :

$$m_H = \sqrt{2c^2}\, m_W. \tag{29}$$

More generally, one must make the sum of all the terms that contribute to the final Higgs boson, and since the c's take the values 0, ± 1 for a large class of groups, we expect $m_H = \sqrt{2k}\, m_W$ with k integer. In particular, the simplest case $k = 1$ yields a theoretical prediction [20]:

$$m_H = \sqrt{2}\, m_W = 113.73 \pm 0.06\,\text{GeV}, \tag{30}$$

which is in agreement with current constraints and with a possible recent detection at CERN. Although this calculation is still incomplete and although the self-consistency of this model remains to be established, we hope that at least some of its ingredients could reveal to be useful in more complete attempts [Lehner and Nottale, in preparation].

4.6 Cosmological constant and vacuum energy density

In [15], we were able to make a theoretical prediction of the value of the cosmological constant. Recall that, in the special scale-relativistic framework, new dilation laws having a log-Lorentz form have been introduced [13], that lead to re-interpret the length-scale $\mathbb{L} = \Lambda^{-1/2}$, where Λ is the cosmological constant, as an impassable, maximal resolution scale, and the Planck length-scale λ_P as a minimal length-scale, invariant under dilations of resolutions.

One of the most difficult open questions in present cosmology is the problem of the vacuum energy density and of its manifestation as an effective cosmological constant [33][6]. The scale relativity approach generalizes Zeldovich's [34] approach. It allows one to suggest a solution to this problem and to connect it with Dirac's large number hypothesis (see also Sidharth [32]). The first step toward our solution consists in considering the vacuum as fractal, (i.e., explicitly scale dependent). As a consequence, the Planck value of the vacuum energy density (that gave rise to the 10^{120} discrepancy with observational limits) is relevant only at the Planck scale, and becomes irrelevant at the cosmological scale. We expect the vacuum energy density ρ to be solution of a (renormalisation group-like) scale differential equation:

$$\frac{d\rho}{d\ln r} = \Gamma(\rho) = a + b\rho + O(\rho^2), \tag{31}$$

where ρ has been normalized to its Planck value, so that it is always < 1, allowing us to perform a Taylor expansion of $\Gamma(\rho)$. This equation is solved as:

$$\rho = \rho_c \, [1 + \left(\frac{r_0}{r}\right)^{-b}], \tag{32}$$

where $\rho_c = -a/b$ can be identified with the cosmological energy density. We recover the well-known combination of a power law behavior at small scales and of scale-independence at large scale, with a fractal/non-fractal transition about some scale r_0 that comes out as an integration constant.

The second step toward a solution is to realize that, when considering the various field contributions to the vacuum density, we may always chose $< E >= 0$ (i.e., renormalize the energy density of the vacuum). But consider now the gravitational self-energy of vacuum fluctuations. It writes:

$$E_g = \frac{G}{c^4} \frac{< E^2 >}{r}. \tag{33}$$

The Heisenberg relation prevent from making $< E^2 >= 0$, so that this gravitational self-energy cannot vanish. This relation writes $< E^2 >^{1/2} = \hbar c/r$, so that we obtain the asymptotic high energy behavior:

$$\rho_g = \rho_P \left(\frac{\lambda_P}{r}\right)^6, \tag{34}$$

where λ_P is the Planck length and ρ_P is the Planck energy density. From this equation we can make the identification $b = -6$. Therefore we obtain the complete behavior of the vacuum energy density:

$$\rho = \rho_c \, [1 + \left(\frac{r_0}{r}\right)^6]. \tag{35}$$

For $r >> r_0$, it becomes invariant and is equal to the cosmological energy density ρ_c which manifest itself as a cosmological constant, while for $r << r_0$ it varies very rapidly with scale. We have recovered here, in another way, the statement according to which the vacuum energy density may be the sum of a constant, geometrical, large scale term and of a quantum term (here considered as scale dependent). Moreover, the value of the large scale energy density is given (up to a factor of 2) by the value it reaches at the transition scale r_0, so that the problem of determining the cosmological constant now amounts to determining this length-scale.

We are now able to demonstrate one of Eddington-Dirac's large number relations, and to write it in terms of invariant quantities (i.e., we do not need varying constants to implement it in this form).

Indeed, introducing the maximal scale-relativistic length scale $\mathbb{L} = \Lambda^{-1/2}$, we get the relation:

$$\mathbb{K} = \frac{\mathbb{L}}{\lambda_P} = \left(\frac{r_0}{\lambda_P}\right)^3 = \left(\frac{m_P}{m_0}\right)^3, \tag{36}$$

where r_0 is the Compton length associated with the mass scale m_0. Then the power 3 in Dirac's relation is understood as coming from the power 6 of the gravitational self-energy of vacuum fluctuations and of the power 2 that connects the invariant impassable scale $\mathbb{L}$ to the cosmological constant, following the relation $\Lambda = 1/\mathbb{L}^2$.

Now a complete solution to the problem would be reached only provided the transition scale r_0 be identified. We have suggested [15] that this scale be nothing but the QCD scale, i.e., that the final value of the cosmological constant is fixed at the quark-hadron transition during the Big-Bang. Indeed, before the epoch of this quark-hadron transition, the expansion of the Universe is concerned with free quarks. When the quark distance reaches the size of hadrons (about one Fermi), confinement begins to occur, quarks can no longer be subjected to the expansion, which now applies between hadrons. But, considering the vacuum, one may make the conjecture that the vacuum energy density outside hadrons has been "frozen" at the value it keeps inside hadrons.

Now the QCD scale for 6 quark flavours is found to be $\lambda_{QCD} = 66 \pm 10$ MeV. Note that several fundamental scales of physics fall very close to this energy: the classical radius of the electron, that yields the e^+e annihilation cross section at the energy of the electron mass and corresponds to an energy 70.02 MeV; the effective mass of quarks in the lightest meson, $m_\pi/2 = 69.78$ MeV; the diameter of nucleons, that corresponds to an energy 2×64 MeV. We have suggested to identify the transition scale r_0 with this particular scale and we have obtained [14] [15]:

$$\mathbb{K} = (5.3 \pm 2.0) \times 10^{60}, \tag{37}$$

allowing us to predict a cosmological constant $\Lambda = 1.36 \times 10^{56} cm^2$, i.e. a reduced cosmological constant

$$\Omega_\Lambda = 0.36\, h^{-2}, \tag{38}$$

where $h = H_0/100$ km/s.Mpc. Now the Hubble constant has been recently determined with an improved precision to be $H_0 = 70 \pm 10$ km/s.Mpc. Therefore our theoretical prediction yields a reduced cosmological constant $\Omega_\Lambda = 0.70 \pm 0.25$. Recent measurements using the Hubble diagram of SNe I [9] [29] [30] and the angular power spectrum of the cosmic microwave radiation [5] point precisely toward the same value, 0.7 ± 0.2.

5 Generalized Schrodinger equation

One can demonstrate [14] [15] [17] that, when giving up the hypothesis of differentiability of space-time coordinates, Newton's fundamental equation of dynamics can be integrated in the form of a Schrödinger-like equation. Indeed, as recalled in the introduction, a non-differentiable continuum is necessarily fractal, and trajectories in such a space (or space-time) own (at least) the following three properties:

(i) The test-particles can follow an infinity of potential trajectories: this leads one to use a fluid-like description, $v = v(x(t), t)$.

(ii) The geometry of each trajectory is fractal (of dimension 2). Each elementary displacement is then described in terms of the sum, $dX = dx + d\xi$, of a mean, classical displacement $dx = v\,dt$ and of a fractal fluctuation $d\xi$ whose behavior satisfies the principle of scale relativity (in its simplest "Galilean" version). It is such that $< d\xi >= 0$ and $< d\xi^2 >= 2\mathcal{D}dt$. The existence of this fluctuation implies introducing new second order terms in the differential equations of motion.

(iii) The motion is locally irreversible, i.e., the ($dt \leftrightarrow -dt$) reflection invariance is broken, leading to a two-valuedness of the velocity vector that we represent in terms of a complex velocity, $\mathcal{V} = (v_+ + v_-)/2 - i(v_+ - v_-)/2$.

These three effects can be combined to construct a complex time-derivative operator which writes

$$\frac{\mathrm{d}'}{dt} = \frac{\partial}{\partial t} + \mathcal{V} \cdot \nabla - i \mathcal{D} \Delta \tag{39}$$

where the mean velocity $\mathcal{V} = \mathrm{d}' x/dt$ is now complex and $\mathcal{D}$ is a parameter characterizing the fractal behavior of trajectories (namely, it defines the fractal-nonfractal transition in scale space).

Since the mean velocity is complex, the same is true of the Lagrange function, then of the generalized action $\mathcal{S}$. Setting $\psi = e^{i\mathcal{S}/2m\mathcal{D}}$, Newton's equation of dynamics becomes $m\, \mathrm{d}'\mathcal{V}/dt = -\nabla\phi$, and can be integrated in terms of a generalized Schrödinger equation [14]:

$$\mathcal{D}^2 \,\Delta\, \psi + i\mathcal{D}\frac{\partial}{\partial t}\psi = \frac{\phi}{2m}\,\psi. \tag{40}$$

Since the imaginary part of this equation is the equation of continuity, $\rho = \psi\psi^{\dagger}$ can be interpreted as giving the probability density of the particle position.

Such an approach can be applied to standard quantum mechanics in the microphysical domain, but also, as an approximation, to macroscopic problems of gravitational structuration [17]. In this case, even though it takes this Schrödinger-like form, this equation is still an equation of gravitation, so that it must keep the fundamental properties it owns in Newton's and Einstein's theories. Namely, it must agree with the equivalence principle [16] [10] [1], i.e., it must be independent of the mass of the test-particle and GM must provide the natural length-unit of the system under consideration. As a consequence, the parameter $\mathcal{D}$ takes the form $\mathcal{D} = \frac{GM}{2w}$, where w is a fundamental constant that has the dimension of a velocity. Actually, the ratio $\alpha_g = \frac{w}{c}$ stands out as a macroscopic gravitational coupling constant [1] [2] [24].

It has been shown that this approach accounts for several structures observed in the Solar System [14], including planet distances, eccentricities, and mass distribution [23], obliquities and inclinations of planets and satellites [18]), giant planet satellite distances [11], parabolic comet perihelions [Nottale & Schumacher, in preparation]. Moreover, it also allows one to predict and understand structures observed on a large range of scales, from binary stars [22], to binary galaxies [15] [Tricottet & Nottale, in preparation], and the distribution of galaxies at the scale of the local supercluster [22]. It has been also demonstrated that the first newly discovered extra-solar planetary systems come under the same structures, in terms of the same universal constant as in our own Solar System [16] [24] [19].

Let us finally remark that this theory could also help solving the problem of dark "matter". Recall that, by dynamical and gravitational lensing arguments, one observes the effects of a potential energy additional to that coming from visible matter, from the scale of galaxies to cosmological scales. In current attempts, this potential is tentatively attributed to missing matter. However, one can now suggest a different, more direct explanation. Indeed, by separating the real and imaginary parts of the generalized Schrödinger equation we get respectively a generalized Euler-Newton equation and the continuity equation (which is therefore now part of the dynamics):

$$m\,(\frac{\partial}{\partial t} + V \cdot \nabla)V = -\nabla(\phi + Q), \tag{41}$$

$$\frac{\partial \rho}{\partial t} + \mathrm{div}(\rho V) = 0. \tag{42}$$

Then this system of equations is equivalent to the classical one, except for the introduction of an extra potential energy term Q that writes:

$$Q = -2m\mathcal{D}^2 \frac{\Delta\sqrt{\rho}}{\sqrt{\rho}}. \tag{43}$$

This additional potential energy, (the so-called Böhm potential), can be understood in our framework as a manifestation of the fractal and non-differentiable geometry of space itself. We suggest that it may explain the various effects that have been attributed up to now to dark matter, since it contains a non-zero vacuum energy term. For example, for the fundamental level in a Kepler potential (the expected potential exterior to a galaxy), one finds a constant term $-\frac{1}{2}mw_0^2$ indicating a constant rotation velocity w_0 as observed. First trials of comparison with the observed behavior of these effects have given encouraging results [L.N., in preparation].

6 Conclusion

After having summarized the main lines of development of the theory, we have, in the present contribution, updated some of its theoretical predictions, then we have shown that they continue to agree with recently improved experimental results.

Moreover, we have recalled that scale relativity, when combined with the laws of gravitation, provides us with a general theory of the structuring of gravitational systems [15] [17]. This new approach is complementary of the standard one. In situations where we can no longer follow individual trajectories,we may jump to a statistical description in terms of probability amplitudes which are solutions of a generalized Schrödinger equation. This result suggests, in accordance with recent similar conclusions [31] [26] [27] [8] that the Schrödinger equation could be universal, i.e. that it may have a larger domain of application than previously thought, but with an interpretation different from that of standard quantum mechanics.

Acknowlegments. It is a pleasure to thank Dr. Sidharth for his kind invitation to this Symposium.

References

[1] A.G. Agnese, R. Festa, Phys. Lett. A227, 165, 1997.

[2] M. Agop, H. Matsuzawa, I. Oprea, et al., Australian. J. Phys. 53, 217, 1999.

[3] I. Aitchison, "An informal introduction to gauge field theories", Cambridge University Press, 1982.

[4] F. Ben Adda, J. Cresson, C.R.Acad.Sci. Paris, Serie I, 330, 261, 2000.

[5] P. de Bernardis, et al., Nature 404, 955, 2000.

[6] S.M. Carroll, W.H. Press, E.L. Turner, Ann. Rev. A. & A. 30, 499, 1992.

[7] M.S. El Naschie, Chaos, Solitons & Fractals, 2, 211, 1992.

[8] M.S. El Naschie, in "Quantum mechanics, Diffusion and Chaotic Fractals", eds. M.S. El Naschie, O.E. Rossler and I. Prigogine, Pergamon, 1995, pp. 93, 185

[9] P.M. Garnavich, S. Jha, P. Challis, et al., ApJ 509, 74, 1998.

[10] D.M. Greenberger D.M, Found. Phys. 13, 903, 1983.

[11] R. Hermann, G. Schumacher, R. Guyard , A&A 335, 281, 1998.

[12] B. Mandelbrot, "The fractal geometry of nature" Freeman, 1983.

[13] L. Nottale, Int. J. Mod. Phys. A7, 4899, 1992.

[14] L. Nottale, "Fractal Space-Time and Microphysics: Towards a Theory of Scale Relativity", World Scientific, 1993.

[15] L. Nottale, Chaos, Solitons & Fractals 7, 877, 1996.

[16] L. Nottale, A&A Lett. 315, L9, 1996.

[17] L. Nottale, A&A 327, 867, 1997.

[18] L. Nottale, Chaos, Solitons & Fractals 9, 1035, 1998a.

[19] L. Nottale, Chaos, Solitons & Fractals 9, 1043, 1998b.

[20] L. Nottale, in "Sciences of the Interface", International Symposium in honor of O. Rössler, ZKM Karlsruhe, 18-21 May 2000, Ed. H. Diebner, 2000.

[21] L. Nottale, & J. Schneider, J. Math. Phys., 25, 1296, 1984.

[22] L. Nottale, G. Schumacher, in" Fractals and Beyond: complexities in the sciences", ed. M.M. Novak World Scientific, 1998, p. 149

[23] L. Nottale, G. Schumacher, J. Gay, A&A 322, 1018, 1997.

[24] L. Nottale, G. Schumacher, Lefèvre E.T., A&A 361, 379, 2000.

[25] G.N. Ord, J. Phys. A: Math. Gen., 16, 1869, 1983.

[26] G.N. Ord, Chaos, Solitons & Fractals 7, 821, 1996.

[27] G.N. Ord G.N, A.S. Deakin, Phys. Rev. A 54, 3772, 1996.

[28] Particle Data Group, The European Physical Journal, C15, 1, 2000 (http://www-pdg.lbl.gov/)

[29] S. Perlmutter, G. Aldering, M. Della Valle, et al., , Nature 391, 51, 1998.

[30] A.G. Riess, A.V. Filippenko, P. Challis, et al., AJ 116, 1009, 1998.

[31] O. Rössler, in "Quantum mechanics, Diffusion and Chaotic Fractals", eds. M.S. El Naschie, O.E. Rössler and I. Prigogine, Pergamon, 1995, p. 105.

[32] B.G. Sidharth, Int. J. Mod. Phys. A13, 2599, 1998.

[33] S. Weinberg, Rev. Mod. Phys. 61, 1, 1989.

[34] Ya. B., Zeldovich, JETP Lett. 6, 316, 1967.

't Hooft dimensional regularization implies transfinite Heterotic String theory and dimensional transmutation

M.S. El Naschie
DAMTP, Cambridge, England, UK.

1 Introduction

In the present work we aim at showing how 't Hooft celebrated method of dimensional regularization[1] could be used to demonstrate the infinite dimensional topology of spacetime and leads to the transfinite form of the classical heterotic string theory[2]-[4].

It is of course a particular pleasure to know that Prof. 't Hooft is present at this meeting as one of the guests of honour and we will therefore not attempt to give detailed account of his well known method nor in fact of string theory which may be found in many excellent books and published articles[1, 5]-[9]. Nevertheless we need to recall a minimum of background informations in order to motivate our analysis. In 't Hooft dimensional regularization one have to avoid the poles[1, 10]

$$1/(4-D)$$

by taking the dimensionality D to be slightly smaller than the four dimensionality of our spacetime[1, 10]. However at least in principle and taking the preceding interpretation at its face value, there is another option present to go around this problem by taking the dimensionality of spacetime to be slightly larger than 4.

Now suppose there is a theory of which we do not know the exact form and which happened to be free of poles because it has the right everything including the right dimension.

Calling this unknown theory M, then we have a dimension D_M which may be used to replace the pole terms[1, 10]

$$\frac{1}{D-4} \Rightarrow D_M$$

Setting $D = D_M$, this ideal dimension is easily found as the solution of a simple quadratic equation

$$(D_M - 4)D_M = 1$$

with two solutions

$$D_{M1} = 4 + \phi^3$$

and

$$D_{M2} = -\phi^3$$

Frontiers of Fundamental Physics 4, Edited by Sidharth and Altaisky,
Kluwer Academic/Plenum Publishers, New York, 2001

where

$$\phi = (\sqrt{5} - 1)/2$$

Now $D_{M1} = 4 + \phi^3$ happened to be precisely the expectation value of topological and the Hausdorft dimension of $\mathcal{E}^{(\infty)}$ i.e. $\sim \langle n \rangle = D_{M1} = 4 + \phi^3$ while the sum of D_{M1} and D_{M2} gives the exact embedding topological dimension[2].

$$n = (4 + \phi^3) - \phi^3$$
$$= 4$$

This $n = 4$ together with $d_c^{(0)} = \phi$ describes $\mathcal{E}^{(\infty)}$ completely as evident from the bijection formula[2]

$$d_c^{(n)} = (1/d_c^{(0)})^{n-1}$$

For $n = 4$ and $d_c^{(0)} = \phi$ one finds

$$d_c^{(4)} = (1/\phi)^3$$
$$= 4 + \phi^3$$
$$= 4 + (\bar{4}).$$

We note that the same result is obtained when treating spacetime exactly as radiation and apply a Planckian distribution to them as discussed in detail at various previous occasions[2]. The picture which is then transmitted by $\mathcal{E}^{(\infty)}$ theory is that of a "timeless" fully spatialized space with infinite number of compactified small spherical dimensions of radius of about $10^{-33} cm$ which interact together like a Maxwell-Boltzmann gas and then depending on the resolution of our observation, this space will give the impression as if it were a four dimensional smooth space or a space of higher and higher dimensions and increasing nonsmooth and chaotic outlook[2]-[5] as we sharpen our experimental accuracy.
This may be explained as following:
Since

$$4 + \phi^3 = 1/\phi^3$$
$$= 4 + \cfrac{1}{4 + \cfrac{1}{4 + \cfrac{1}{4 + \cdots}}}$$

then, setting ϕ^3 leads to a superficially contradictory statement, namely that

$$4 + 0 = \frac{1}{0} = \infty$$

This pseudo contradiction is easily removed by the correct interpretation of the topology of $\mathcal{E}^{(\infty)}$ because this space has three vital dimensions. The first is the formal (branching polymer-like) dimension $n_f = \infty$. The second is the expectation value of the Hausdorff dimension[2]

$$\langle d_c \rangle =\sim \langle n \rangle = 4 + \phi^3$$

and finally the third is the embedding topological spacetime dimension $n = 4$

Without going into detailed analysis we just have to recall that further considerations lead us to introduce four more dimensions which account first for the spin 1/2 additional "dimension" of the Fermi-Dirac's statistics namely $n_F = 4 + 1 = 5$ and $\sim \langle n \rangle_F = (4 + \phi^3)(1 + \phi^3) = (4 + \phi^3) + 1 = 5 + \phi^3$ as discussed in[3, 4]. The second set of dimension accounts for pure "special" topology namely $n = 4 - 1 = 3$ and $\sim \langle n \rangle_{sp} = (4 + \phi^3) - 1 = 3 + \phi^3$. It should be noted that even $n_{sp} = 3$ could be interpreted as an expectation value for

$$\langle d_c^{(0)} \rangle = 1/\langle d_c^{(1)} \rangle = 2$$

because[2]

$$\sim \langle n \rangle = \frac{1 + d_c^{(0)}}{1 - d_c^{(0)}} = \frac{1 + (1/2)}{1 - (1/2)} = 3$$

The same interpretation applies to $n = 4$ and $n = 5$ for the corresponding values $d_c^{(0)} = 3/5$ and $d_c^{(0)} = 2/3$[2].

2 The Transfinite Heterotic String Theory

The remarkable thing about the topology of $\mathcal{E}^{(\infty)}$ is that once we have found the dimensions 5 and $5 + \phi^3$ then we can derive the largest of the dimension of Heterotic string theory and far more surprisingly we can derive the electromagnetic fine structure constant which is found to be in an amazingly close agreement with the experimental value. Thus the multiplication of (5) with $(5 + \phi^3)$ gives the heterotic dimension[3, 4]

$$\begin{aligned} D_T^{(-)} &= (5)(5 + \phi^3) \\ &= 26 + k(= 0.18033989) \\ &= 26.18033989 \end{aligned}$$

and a second multiplication with $5 + \phi^3$ gives the electromagnetic fine structure constant

$$\bar{\alpha}_0 = (5)(5 + \phi^3)^2 = (D_T^{(-)})(5 + \phi^3) = 137.082039325 = 137 + \phi^5(1 - \phi^5)$$

3 Physics and the Topology of $\mathcal{E}^{(\infty)}$

Introducing the Cooper-pair charge $g = 2e$ and proceeding as discussed in earlier papers[2]-[4] we find the following

$$(\bar{\alpha}_0)(\phi) = 84.72135957; (\frac{\bar{\alpha}_0}{2}(\phi) = 42.36067977$$

$$(\bar{\alpha}_0)(\phi^2) = 52.36067977; (\frac{\bar{\alpha}_0}{2}(\phi^2) = 26.18033989$$

$$(\bar{\alpha}_0)(\phi^3) = 32.36067977; (\frac{\bar{\alpha}_0}{2}(\phi^3) = 16.18033989$$

$$(\bar{\alpha}_0)(\phi^4) = 20 \quad ; (\frac{\bar{\alpha}_0}{2}(\phi^4) = 10$$

$$(\bar{\alpha}_0)(\phi^5) = 12.36067977; (\frac{\bar{\alpha}_0}{2}(\phi^5) = 6.18033989$$

$$(\bar{\alpha}_0)(\phi^6) = 7.639320242; (\frac{\bar{\alpha}_0}{2}(\phi^6) = 3.819660121.$$

The right hand column gives all the relevant dimensionalities of the transfinite heterotic string theory while the first figure is nothing but the nonsupersymmetric "quantum gravity" coupling constant $\bar{\alpha}_g = 42 + 2k$ at the point of unification of all the four fundamental forces. In the case of supersymmetry which seems to be increasingly the more likely theoretical scenario consistent with Heterotic string theory the same coupling constant becomes

$$\bar{\alpha}_{gs} = (\phi)(\bar{\alpha}_g) = 26 + k = 26.18033989$$

Next we demonstrate how the theoretical number of gauge bosons which is identical to the dimensions of $E_8 \times E_8$ namely 496 can be found with the help of the $4 + \phi^3$ based theory. This is simply given by

$$\begin{aligned} \langle n_{GS} \rangle &= \frac{(\alpha_0)}{3 + \phi^3} \\ &= \frac{1604.984471}{3.236067977} \\ &= 495.9674775 = 496 - k^2 \end{aligned}$$

where $k = \phi^3(1 - \phi^3) = 0.18033989$.
There is another meaningful way of interpreting this result as the sum of four groups of guage bosons namely

$$\begin{aligned} (\frac{\bar{\alpha}_0}{2})(1/\phi^3)^3 + (\frac{\bar{\alpha}_0}{2})(1/\phi)^2 + (\frac{\bar{\alpha}_0}{2})(\phi)^2 &= 290.3444 + 179.444 \\ &+ 26.18033989 \\ &= 495.9674775 \end{aligned}$$

This may be also expressed as

$$\langle n_{Gs} \rangle = (\sqrt{\bar{\alpha}_0})(\bar{\alpha}_g) = 495.9674775$$

or as

$$\langle n_{Gs} \rangle = (\bar{\alpha}_0)(3 + \phi) = (137 + k_0)(3 + \phi) = 496 - k^2$$

The number of scalar fields namely 96[9] is given in $\mathcal{E}^{(\infty)}$ theory by

$$(10 - \phi^2)(10) = 96 + k$$

The dimension of the moduli space[9] of Vacua namely $136 - 4 = 132$ is given by

$$\bar{\alpha}_0 - 5 = (137 + k_0) - 5 = 132 + k_0$$

The 56 electric charges connected to the $E_8 \times E_8$ gauge field[9] is given in $\mathcal{E}^{(\infty)}$ theory by

$$[(4 + \phi^3) + 1](10) = 56 + k$$

and the number of the 28 gauge fields is given by

$$\begin{aligned}(3-k)(10) &= (2.81966011)(10)\\ &= 28+k\end{aligned}$$

In conclusion we should mention a recent observation regarding the so called dimensional transmutation in QCD lattice calculations namely[6]

$$\begin{aligned}\left[\sqrt{K}/\Lambda_{Latt}\right]_{a\to 0} | \Rightarrow (2)(D^{(-)}) &= 52+k\\ &= (\frac{\bar{\alpha}_0}{2})(\phi)\\ &= 52.36067977\end{aligned}$$

where k is the QCD string tension and (a) is the lalttice (finite difference) spacing. In other words the considerable success of K.Wilson's method lay in the close connection between lattice spacetime and $\mathcal{E}^{(\infty)}$ and this in turn shows the reality of what has been suspected for a long time namely the quantum spacetime is radically different from the smooth spacetime of classical mechanics and general relativity[2]-[5]. At a minimum viewing spacetime in this way motivates efficient computational methods which give results compatible with experiments.

4 Indirect experimental confirmation of the transfinite heterotic string theory

One of the most persuasive argument for the reality of the transfinite heterotic string theory may be the following which is based on a renormalization method in $\mathcal{E}^{(\infty)}$.

Let us consider first the four terms conformal weight equation for the classical heterotic strings

$$\frac{4}{26}+\frac{6}{26}+\frac{16}{26}+\frac{0}{26}=1$$

The joint eigenvalue is thus given by the Dunkerly theorems as

$$\frac{26}{4}+\frac{26}{6}+\frac{26}{16}+1=13.458332$$

where we have noted the normalized vacuum contribution $0/26 \Rightarrow 1$. Consequently the ten fold value 134.5833 differs considerably from the electromagnetic fine structure constant $\bar{\alpha}_0 = 137 + k_0$. Now we repeat the same analysis however this time using the transfinite heterotic theory.

Proceeding in this way one finds

$$\frac{4-k}{26+k}+\frac{6+k}{26+k}+\frac{16+k}{26+k}+\frac{0}{26+k}=1$$

Consequently we have

$$\begin{aligned}\frac{26+k}{4-k}+\frac{26+k}{6+k}+\frac{26+k}{16+k}+1 &= 137.0820393\\ &= \bar{\alpha}_0\end{aligned}$$

References

[1] G. 't Hooft, "Under the spell of the gauge principle", World Scientific, Singapore, 1994.

[2] M.S. El Naschie, Int. J. Theor. Phys., 1998, 12 pp.2935-51.

[3] M.S. El Naschie, "On a Heterotic string-based algorithm for the determination of the fine structure constant", Chaos, Solitons & Fractals, Vol.12, No.3, 2001, pp617-621.

[4] M.S. El Naschie, "On an indirect experimental confirmation of Heterotic superstrings via the electromagnetic fine structure constant", Chaos, Solitons & Fractals, Vol.12, No.4, 2001 (To appear).

[5] B.G. Sidharth, "The Chaotic Universe", Nova Science Publishers, New York, In press.

[6] I. Montvay and G. Munster, "Quantum field on a lattice", Cambridge, 1994.

[7] D. Gross, Physica Scripta, 1 (15), 1987, 34.

[8] M. Green, "Superstrings", Scientific American, 225 (3), 1986, 48.

[9] J. Polchinski, "String Theory", Cambridge, 1998.

[10] M.S. El Naschie, "On 't Hooft dimensional regularization in $\mathcal{E}^{(\infty)}$ space", Chaos, Solitons & Fractals, 2000 (To appear).

The Cantorian Gravity Coupling Constant is $\alpha_{gs}(max) = 1/26.18033989$

M.S. El Naschie
DAMTP, Cambridge, England, UK.

1 Introduction

Recent research has shown that the fine structure constant α_0 is a global quantity which does not depend on the details of any particular physical theory but may be determined directly from the global topological structure of $\mathcal{E}^{(\infty)}$ which is assumed to be identical to our actual quantum "micro" space time[1].
We may start here by recalling that $\mathcal{E}^{(\infty)}$ is formally an infinite dimensional "Cantor set-like" space which nevertheless possesses finite exact expectation values for both the topological and the Hausdorff dimensions. Apart from the formal dimension $n_F = \infty$, we have two sets of dimensions which completely determined $\mathcal{E}^{(\infty)}$[1, 2]:

$$D_1 = (4; 5) = (D_{11}, D_{12})$$

and

$$D_2 = (4 + \phi^3); (5 + \phi^3) = (D_{21}, D_{22})$$

where $\phi = \frac{\sqrt{5}-1}{2}$ is the Golden Mean. We stress that while $n_F = \infty$ is postulated, D_1 and D_2 are not assumed but derived using classical probability theory and Planck-like distribution. Thus space and time are treated on equal footing with matter and radiation. Thus $4 + \phi^3$ is found from an exact discrete gamma distribution to be $4 + \phi^3 = (1 + \phi)/(1 - \phi)$ where $\phi = (\sqrt{5} - 1)/2$ and $5 + \phi^3$ is simply $(4 + \phi^3)(1 + \phi^3) = (4 + \phi^3) + 1$.
Proceeding in this way, the inverse of the electromagnetic fine structure constant is easily found to be[1]

$$\begin{aligned}\bar{\alpha}_0 &= (5)(5 + \phi^3)^2 \\ &= (10)[2(1 + \phi)^4] \\ &= 137 + k_0 \\ &= 137 + \phi^5(1 - \phi^5) \\ &= 137.082039325\end{aligned}$$

We note that the most accurate experimental value known at present for $\bar{\alpha}_0$ is $\bar{\alpha}_0 = 137.03598$. Using $\bar{\alpha}_0 = 137 + k_0$, it was shown elsewhere that all dimensions of the $\mathcal{E}^{(\infty)}$ version of the heterotic string theory may be readily generated as following:

First, we multiply $\bar{\alpha}_0$ with $c^+h = d_c^{(0)} = \phi$ repeatedly and find that[1, 2]

$$C_1 = (\bar{\alpha}_0)(\phi) = 84.72135957$$

$$C_2 = (C_1)(\phi)^2 = 52.360679.77$$

$$C_3 = (C_2)(\phi)^3 = 32.36.0.67977$$

$$C_4 = (C_3)(\phi)^4 = 20.$$

$$C_5 = (C_4)(\phi)^5 = 12.3606.7977$$

$$C_6 = (C_5)(\phi)^6 = 7.6393\cdots$$

Note that the initial value C_1 has a simple interpretation being the familiar gauge field strength[3]

$$\| F \|^2 = 48 \int_{R^4} \frac{1}{(1+191^2)^4}$$

when transfered to $\mathcal{E}^{(\infty)}$

$$\| F \|^2 |_{\mathcal{E}^{(\infty)}} = 8 D_T^{(6)} U_1^4 (\frac{1}{1+\phi^3})^4$$

$$= 8(6+k)(4)\frac{1}{1+\phi^3)^4}$$

$$= 84.713$$

Here we have made use of the anticipated result $D_T^{(6)} = \frac{1}{2}(12.360679) = 6 + k$ for the dimension of the transfinite orbifold of our theory[1].
Likewise, the second value $C_2 = 52.360679$ has two interesting interpretations. The first interpretation is as the vital ratio of the lattice of quantum cromodynamic's calculations[4]

$$\left[\sqrt{k}/\Lambda_{latt}\right]_{a\to 0} | \Rightarrow 52 + 2k = (2)(D_T^{(-)})$$

$$= (2)(26.18033)$$

where k is the QCD string tension and Λ_{latt} is the physical parameter which combines the lattice spacing (a) with the coupling parameter $g(a)$.[4].
The second interpretation is that as twice the value of the transfinite Heterotic dimension $D_T^{(-)} = \frac{52+2k}{2} = 26.1803$[1]. In fact if we consider that $q = 2e$ may be interpreted as the charge on the cooper pair then all the dimensions of the recently proposed transfinite Heterotic string theory may be generated as following[1]-[2].

$$(\frac{\bar{\alpha}_0}{2}) = C_1/2 = 42.36067977$$

$$D_T^{(-)} = C_2/2 = 26.18033979$$

$$D_T^{(16)} = C_3/2 = 16.18033989$$

$$D_T^{(4)} = C_4/2 = 10$$

$$D_T^{(6)} = C_5/2 = 6.18033989$$

$$D_T^{(4)} = C_6/2 = 4 - k$$

Setting $k = 0$ in the above expressions, we retrieve the dimensions of the classical Heterotic string theory.
The only expression which is not directly a proper dimension of the Heterotic string theory is of course the first expression $C_1/2 = 42.3606$ which is clearly ten times the expectation value of the topological (or Hausdorff) dimension of $\mathcal{E}^{(\infty)}$,

$$\sim \langle n \rangle \langle \dim_H \mathcal{E}^{(\infty)} \rangle = \sim < \dim_T \mathcal{E}^{(\infty)} >$$

$$= 4 + \phi^3$$

In what follows we will be arguing using various numerical and theoretical methods that

$$\left[\left(\frac{\bar{\alpha}_0}{2}\right)(\phi)\right]^{-1} = 1/42.36067977$$

is the exact smallest coupling constant of quantum gravity, in the absence of supersymmetry[7]. This minimal value will increase in the case of supersymmetry to a second minimal valuee which is eqaual to the inverse of $D^{(-)} = 26 + k$ of the transfinite heterotic string theory[1].
Now since we believe that transfinite heterotic string theory is an accurate description for the quantum universe, then it follows that supersymmetry is most probably real and consequently

$$\bar{\alpha}_g \leq 26.18033989$$

In this connection we may make two important observations gained from scale relativity which seems to have been overlooked by Nottale and which brings his contribution much nearer to the present work namely that[5]

$$(ln\frac{\lambda_e}{\lambda_w})(ln\frac{\lambda_e}{\lambda_z}) = (11.65)(11.772) = 137.1438 \approx \bar{\alpha}_0$$

and

$$(log\frac{\lambda_e}{\lambda_w})(log m_w/m_e) = (5.059)(5.194) = 26.276 \approx D^{(-)}$$

where λ_e, λ_w and λ_z are the wave lengths of the electron, the w and the z respectively and m_e and m_w are the masses[5].

2 The coupling constant of quantum gravity

Quantum gravity, the point at which quantum mechanics and general relativity meet have been always presumed to exist at the Planck scale ($l_p = 10^{-33}cm$), or another related scale such as the GUT scale or the Nottale scale[5]. However it has so far proven impossible to quantize the Einstein equations consistently and a final solution along these classical lines is still outstanding[6, 7].
In $\mathcal{E}^{(\infty)}$ theory we do not use proper quantization which is replaced by transfinite "discretization" and we derive without analytical continuation the Schrodinger and the Dirac equations directly in a similar way to that of a generalized diffusion equation as shown by Ord, Nottale, Nagasawa and the present author in several different but related ways[5].

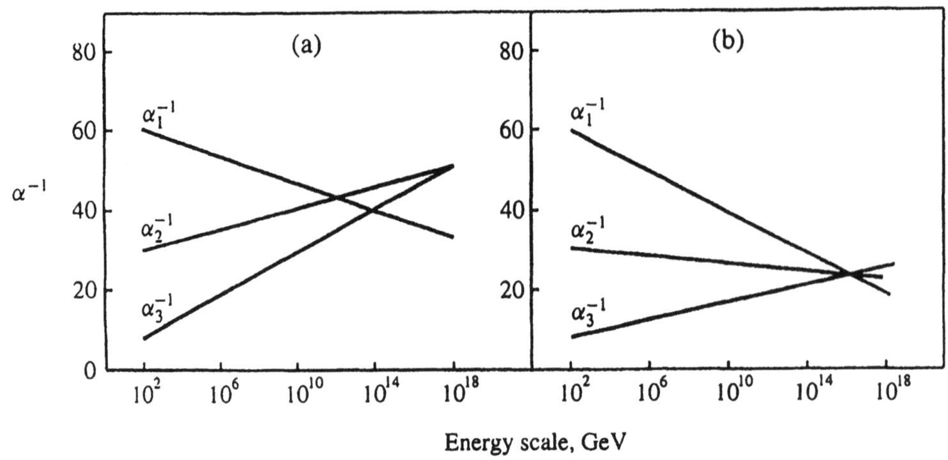

Figure 1: The running coupling constant $\bar{\alpha}_i = \alpha^{-1}{}_i$ as a function of energy for (a) nonsupersymmetric and (b) supersymmetric intersection.

For that reason, it is feasible to attempt a solution for what we may call here "Cantorian" gravity, rather than quantum gravity and find the coupling constants at the point of complete unification of all of the four fundamental forces[6]-[7] without running into troubles with proper quantization which we do not need to use,[2, 9].

A Grand unification without gravity

We may start with the simpler problem of grand unification without gravity. This particular situation is shown graphicly in fig (1a). It is not difficult to estimate the value at which $\alpha_1 = \alpha_2 = \alpha_3 = \alpha_{GUT}$ from Fig 1a and see that[2],[6, 7]

$$\bar{\alpha}_{GUT} \approx 43$$

This value is rather close to our $\mathcal{E}^{(\infty)}$ exact value $\bar{\alpha}_g = 42.3606$. This may be taken to indicate that the grand unification and quantum gravity are both rather close to the Planck length scale. The interesting point however is that when adding supersymmetry, the new minimal value which we can estimate from fig 1b to be

$$\bar{\alpha}_s|_{GUT} \approx 24$$

is also not far off our theoretically exact result namely

$$\bar{\alpha}_{gs} = D^{(-)} = 26.18033988 = (10)(1/\phi)^2$$

In other words, with or without supersymmetry GUT and quantum gravity coupling constants are relatively close. An instructive qualitative estimation may help in understanding this point. In the classical heterotic string theory the unification scale is given in terms of the string coupling g_{st} and the Planck scale by

$$M_{st}/g_{st} \sim (5.27)(10)^{17} Gev$$

while the GUT experimental scale is given by

$$M_{GUT} \sim (2)(10)^{16} GeV$$

Consequently

$$(M_{st})/(M_{GUT} g_{st}) \approx (\frac{5.27}{2})(10) = 26.35 \approx D^{(-)}$$

We could go one step further and make our estimation precise by adding the transfinite correction namely $5.27 \to 5 + \phi^3$ and $g_{st} \to 1 + \phi$. That way we find

$$(1/g_{st})(M_{st})/(M_{GUT}) = 26.18033 = (10)(dc)^{(3)} = \bar{\alpha}_{gs}$$

B Quantum gravity grand unification

In fig (2) (due to S. Weinberg) and fig (3) (due to L. Nottale) gravity is included as shown. In the case of fig (2) one can find a crude estimation by optimizing the error due to three intersections with gravity ($\alpha_g = GE^2$ where G is Newton's constant and E is the energy) instead of only one and find that

$$\bar{\alpha}_g \approx 43 \quad \text{and} \quad \bar{\alpha}_{gs} \approx 26$$

for the nonsupersymmetric and the supersymmetric theory respectively. This result is in a fair quantitative agreement with both the GUT results as well as the exact result of $\mathcal{E}^{(\infty)}$ theory. In the case of fig (3) of the scale relativity we note first that it does not take account of supersymmetry and that the three intersections with gravity are given by $\bar{\alpha}_{g1} = 42.4, \bar{\alpha}_{g2} = 42.8 \text{and} \bar{\alpha}_{g3} = 39.7$.

C Theoretical derivation of the quantum gravity coupling constant

Based on scale relativistic considerations Nottale in[5] conjectured that the quantum gravity coupling constant is given by

$$\bar{\alpha}(\Lambda) = 4\pi^2 = 39.4784176.$$

and that the corresponding fine structure constant is

$$\bar{\alpha}|_{QG} = (8/3)(\bar{\alpha}(\Lambda))$$

$$= 105.27$$

The transfinite corrections to the above results are taken care for by performing the following $\mathcal{E}^{(\infty)}$ transformations.

$$\pi \to \sqrt{10} = \sqrt{D^{(+)}},$$

$$4 \to 4 + \phi^3 =\sim \langle n \rangle$$

and

$$8/3 \to 2 + \phi = \langle d_c^{(2)} \rangle + d_c^{(0)}.$$

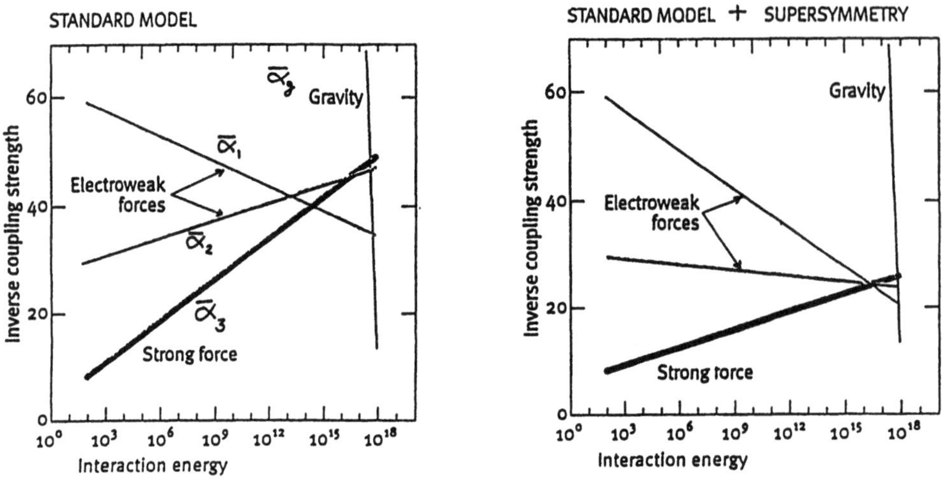

Figure 2

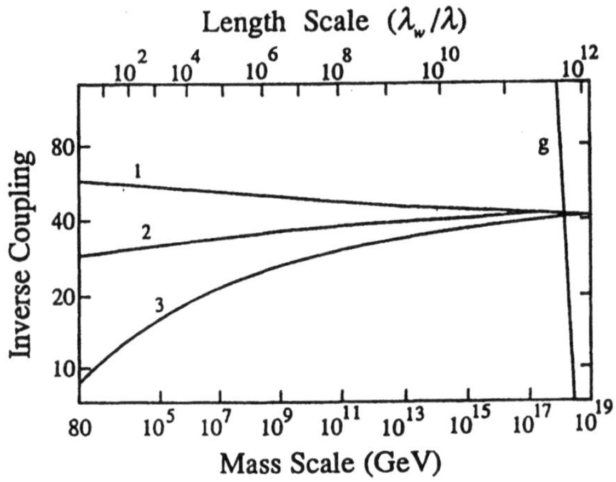

Figure 3: Graphical presentation of Nottale's scale relativistic attempt for unification at $\bar{\alpha}(\Lambda) = 4\pi^2$.

Or alternatively

$$\pi \to 3 + \phi^3 = (\sim \langle n \rangle - 1),$$

$$4 \to (16 + k)/4 = D_T^{(16)}/4$$

and

$$8/3 \to 2 + \phi = \langle d_c^{(2)} \rangle + d_c^{(0)}.$$

Either way one finds

$$\bar{\alpha}(\Lambda) = 4\pi^2 \Rightarrow \bar{\alpha}_g = 42.36067977$$

and

$$\bar{\alpha}|_\Lambda = (8/3)(\bar{\alpha}(\Lambda)) = 105.22 \Rightarrow \bar{\alpha}_0|_{\mathcal{E}^{(\infty)}} = 110.9016$$

It may be instructive to determine $\bar{\alpha}_2 = \bar{\alpha}_1(\Lambda)$ using scale relativity. This is given by

$$\bar{\alpha}_1(\Lambda) \approx (3/8)\bar{\alpha}(\lambda_w) - \frac{5}{8\pi} C_w$$

Again transforming to $\mathcal{E}^{(\infty)}$ by noting that

$$\bar{\alpha}(\lambda_w) \to 127 + k_0,$$

$$5/8 \to \phi$$

and

$$C_w \to (2)(16 + k) = 32 + 2k$$

one finds that

$$\bar{\alpha}_1(\mathcal{E}^{(\infty)})$$

$$= 48.54101969 - (6 + k)$$

$$= 42 + 2k$$

Thus $\bar{\alpha}_1(\mathcal{E}^{(\infty)} = \bar{\alpha}_1$ as should be.
Nontheless we may find different values if we take into accfount different effects of the Higgs doublets and number of generations. In particular we may find that

$$\bar{\alpha}_1(\Lambda) = 48.54101969 - (5)(6 + k)$$

$$= 17.63932021$$

This value is very close indeed to that found by S. Weinberg[7] namely $\bar{\alpha} = 17.5$ for the supersymmetric GUT unification at M_x

$$\frac{g_s^2(M_x)}{4\pi} = 1/17.5$$

and we may mention that the corresponding value for the nonsupersymmetric case was found by Weinberg to be

$$\bar{\alpha}(M_x) = 41$$

Again this seems to be in qualitative agreement with all the preceding conclusions made so far.
Finally we may note the consistency of $\bar{\alpha}_{gs} = 26 + k$ with our renormalization procedure in $\mathcal{E}^{(\infty)}$ and the restriction imposed by the conformal weight unitarity equation

$$[U_R] = \frac{1}{(5+\phi^3)}\left[\phi + \phi^2 + 1/\phi + 1/\phi^2\right] = 1$$

by showing that

$$\begin{aligned}
&(26+k)\left[\phi + \phi^2 + 1/\phi + 1/\phi^2\right] \\
&= (6+k) + (10) + (42.360679) + (68.5410197) \\
&= 137 + k_0 \\
&= \bar{\alpha}_0
\end{aligned}$$

In terms of the exact minimal nonsupersymmetric quantum gravity coupling $\alpha_g = \frac{1}{42.360679}$ this would mean

$$\begin{aligned}
&\bar{\alpha}_g\left[1 + 1/\phi + \phi^2 + \phi^3\right] \\
&= [42.36067977]\left[3 + \phi^3\right] \\
&= 137 + k_0 \\
&= \bar{\alpha}_0
\end{aligned}$$

We note that the contribution $\bar{\alpha} = (42.3606)\phi^3 = 10$ to $\bar{\alpha}_0$ may be regarded as that of the true vacuum given in the original unitarity weight equation by $\phi^\infty = 0$ and renormalized to $1 \rightarrow 10$ in our exact renormalization group procedure.
For further reading on the subject of $\mathcal{E}^{(\infty)}$ theory and various recent results and insights, the reader is referred to the growing body of literature on fractal spacetime physics and scale relativity particularly the recent work of Ord, Nottale and the Indian Physicist Sidharth who has just finished his awaited new book[9].

3 Discussion and conclusions

Coupling constants in quantum spacetime are global quantities with precise expectation value fixed by the topology of spacetime. The definite numerical values which these "constants" have are the result of what is topologically admissible and what is prohibited in an infinite dimensional, transfinite and pointless topological space which has a precise expectation value for its topological as well as it's Hausdorff dimension. In other words this space must be hierarchical. This is thus the reason why we can proceed from general consideration to precise statement about the coupling constants, the number of generations and so forth without establishing the action S in $\mathcal{E}^{(\infty)}$ nor solving a corresponding "differential" equation in a nondifferential disconnected transfinite space. Thus our global analysis used in $\mathcal{E}^{(\infty)}$ follows in a sense Poincare's method and goes even beyond it. In fact it is the traditional praxis of establishing first a governing variational principle or a differential equation

which disguised the fact that the coupling constants of the standard model are not at all dependant on the details of this model.
Thinking in this way one may start regarding $\mathcal{E}^{(\infty)}$ theory as a marriage between infinite dimensional topology and probability theory. The situation is therefore comparable with that of Boltzmanian statistical mechanics as completed by I. Prigogine[8] being a marriage between classical spacetime mechanics and probability theory.
We believe that the results reported in the present and previous work justifies this admittedly radical view that at a very deep level it is very difficult to see where topology is different from what we normally call physics.

Acknowledgement
The author is very indebted to Prof. Martienssen, Prof. Sidharth, Prof. Greiner, Prof. Nottale and Prof. I. Prigogine for valuable discussion and suggestions.

References

[1] M.S. El Naschie, "On a transfinite heterotic string theory", Chaos, Solitons & Fractals, Vol.12, 2001, pp417-421.

[2] M.S. El Naschie, Int. J. Theor. Phys., 1998, 12 pp.2935-51.

[3] M. Nakahara, "Geometry, Topology and Physics", IOP, Bristol, 1998.

[4] I. Montvay and G. Munster, "Quantum field on a lattice", Cambridge, 1994.

[5] L. Nottale, "Fractal spacetime and Microphysics", World Scientific, 1993.

[6] H.M. Georgi, "Grand Unified Theories", in "The New Physics", Ed. P. Davies, Cambridge, 1989, p.425.

[7] S. Weinberg, "The Quantum Theory of Fields", Cambridge, 2000.

[8] G. Nicolis and I. Prigogine, "Self-organization in Nonequilibrium Systems", Wiley, New York, 1977.

[9] B.G. Sidharth, "The Chaotic Universe", Nova Science Publishers, New York, In press.

Fuzzy, Non Commutative SpaceTime: A New Paradigm for A New Century

B.G. Sidharth
B.M. Birla Science Centre, Adarshnagar, Hyderabad - 500 063, India

1 Introduction

At the beginning of the twentieth century several Physicists including Poincare and Abraham amongst others were tinkering unsuccessfully with the problem of the extended electron[1, 2]. The problem was that an extended electron appeared to contradict Special Relativity, while on the other hand, the limit of a point particle lead to inexplicable infinities. These infinities dogged physics for many decades. Infact the Heisenberg Uncertainity Principle straightaway leads to infinities in the limit of spacetime points. It was only through the artifice of renormalization that 't Hooft could finally circumvent this vexing problem, in the 1970s (Cf. paper by 't Hooft in this volume).
Nevertheless it has been realized that the concept of spacetime points is only approximate[3, 4, 5, 6, 7]. We are beginning to realize that it may be more meaningful to speak in terms of spacetime foam, strings, branes, non commutative geometry, fuzzy spacetime and so on[8]. This is what we will now discuss.

2 Two Approaches

We now consider the well known theory of Quantum SuperStrings and also an approach in which an electron is considered to be a Kerr-Newman Black Hole, with the additional input of fuzzy spacetime.
As is well known, String Theory originated from phonenomenological considerations in the late sixties through the pioneering work of Veneziano, Nambu and others to explain features like the s-t channel dual resonance scattering and Regge trajectories[9]. Originally strings were conceived as one dimensional objects with an extension of the order of the Compton wavelength, which would fudge the point vertices of the s-t channel scattering graphs, so that both would effectively correspond to one another (Cf.ref.[9]).
The string itself is governed by the equation[10]

$$\rho\ddot{y} - Ty'' = 0 \tag{1}$$

where the frequency ω is given by

$$\omega = \frac{\pi}{2}\sqrt{\frac{T}{\rho}} \tag{2}$$

$$T = \frac{mc^2}{l}; \rho = \frac{m}{l} \tag{3}$$

$$\sqrt{T/\rho} = c \tag{4}$$

T being the tension of the string, which has to be introduced in the theory, l its length and ρ the line density. The identification (3) gives (4) where c is the velocity of light, and (1) then goes over to the usual d'Alembertian or massless Klein-Gordon equation. It is worth noting that as $l \to 0$ the potential energy which is $\sim \int_0^l T \left(\frac{\partial y}{\partial x}\right)^2 dx$ rapidly oscillates.
Quantization of the states leads to

$$\langle \Delta x^2 \rangle \sim l^2 \tag{5}$$

The string effectively shows up as an infinite collection of Harmonic Oscillators [10]. It follows from the above that the length l of the string turns out to be the Compton wavelength, a circumstance which has been described as one of the miracles of String Theory by Veneziano[11].
The above strings are really Bosonic strings. Raimond[12], Scherk[13] and others laid the foundation for the theory of Fermionic strings. Essentially the relativistic Quantized String is given a rotation, when we get back the equation for Regge trajectories,

$$J \leq (2\pi T)^{-1} M^2 + a_0 \hbar \quad \text{with} \quad a_0 = +1(+2) \text{for the open (closed) string} \tag{6}$$

Attention must be drawn to the additional term a_0 which now appears in (6). It arises from a zero point energy effect. When $a_0 = 1$ we have gauge Bosons while $a_0 = 2$ describes the gravitons. In the full theory of Quantum Super Strings, we are essentially dealing with extended objects rotating with the velocity of light, rather like spinning black holes. The spatial extention is at the Planck scale while features like extra space time dimensions which are curled up in the Kaluza Klein sense and, as we will see, non commutative geometry appear[14, 15].
The above considerations raise the question, can a charged elementary particle be pictured as a Kerr Newman Black Hole, though in a Quantum Mechanical context rather than the General Relativistic case? Indeed it is well known that the Kerr Newman Black Hole itself mimics the electron remarkably well including the purely Quantum Mechanical anomalous $g = 2$ factor[16]. The problem is that there would be a naked singularity, that is the radius would become complex,

$$r_+ = \frac{GM}{c^2} + \imath b, b \equiv \left(\frac{G^2Q^2}{c^8} + a^2 - \frac{G^2M^2}{c^4}\right)^{1/2} \tag{7}$$

where a is the angular momentum per unit mass.
This problem has been studied in detail by the author in recent years[1, 17, 18]. Indeed it is quite remarkable that the position coordinate of an electron in the Dirac theory is non Hermitian and mimics equation (7), being given by

$$x = (c^2 p_1 H^{-1} t + a_1) + \frac{\imath}{2} c\hbar(\alpha_1 - cp_1 H^{-1}) H^{-1}, \tag{8}$$

where the imaginary parts of (7) and (8) are both of the order of the Compton wavelength.

The key to understanding the unacceptable imaginary part was given by Dirac himself[19], in terms of Zitterbewegung. The point is that according to the Heisenberg Uncertainity Principle, space time points themselves are not meaningful- only space time intervals have meaning, and we are really speaking of averages over such intervals, which are atleast of the order of the Compton scale. Once this is kept in mind, the imaginary term disappears on averaging over the Compton scale.
Indeed, from a classical point of view also, in the extreme relativistic case, as is well known there is an extension of the order of the Compton wavelength, within which we encounter meaningless negative energies[20]. With this proviso, it has been shown that we could think of an electron as a spinning Kerr Newman Black Hole. This has received independent support from the work of Nottale[21].

3 Non Commutative Geometry

We are thus lead to the picture where there is a cut off in space time intervals as indicated in the introduction.
In the above two scenarios, the cut off is at the Compton scale (l, τ). Such discrete space time models compatible with Special Relativity have been studied for a long time by Snyder and several other scholars[22, 23, 24]. In this case it is well known that we have the following non commutative geometry

$$\begin{aligned} &[x,y] = (\imath a^2/\hbar)L_z, [t,x] = (\imath a^2/\hbar c)M_x, \\ &[y,z] = (\imath a^2/\hbar)L_x, [t,y] = (\imath a^2/\hbar c)M_y, \\ &[z,x] = (\imath a^2/\hbar)L_y, [t,z] = (\imath a^2/\hbar c)M_z, \end{aligned} \tag{9}$$

where a is the minimum natural unit and L_x, M_x etc. have their usual significance. Moreover in this case there is also a correction to the usual Quantum Mechanical commutation relations, which are now given by

$$\begin{aligned} &[x,p_x] = \imath\hbar[1 + (a/\hbar)^2 p_x^2]; \\ &[t,p_t] = \imath\hbar[1 - (a/\hbar c)^2 p_t^2]; \\ &[x,p_y] = [y,p_x] = \imath\hbar(a/\hbar)^2 p_x p_y; \\ &[x,p_t] = c^2[p_x,t] = \imath\hbar(a/\hbar)^2 p_x p_t; \text{etc.} \end{aligned} \tag{10}$$

where p^μ denotes the four momentum.
In the Kerr Newman model for the electron alluded to above (or generally for a spinning sphere of spin $\sim \hbar$ and of radius l), L_x etc. reduce to the spin $\frac{\hbar}{2}$ of a Fermion and the commutation relations (9) and (10) reduce to

$$[x,y] \approx 0(l^2), [x,p_x] = \imath\hbar[1 + \beta l^2], [t,E] = \imath\hbar[1 + \tau^2] \tag{11}$$

where $\beta = (p_x/\hbar)^2$ and similar equations.
Interestingly the non commutative geometry given in (11) can be shown to lead to the representation of Dirac matrices and the Dirac equation[25]. From here we can get the Klein Gordon equation, as is well known[26, 27], or alternatively we deduce the massless string equation (1), using (4).

This is also the case with superstrings where Dirac spinors are introduced, as indicated in Section 2. Infact in QSS also we have equations mathematically identical to the relations (11) containing momenta. This, which implies (9), can now be seen to be the origin of non-commutativity.

The non commutative geometry and fuzzyness is contained in (11). Infact fuzzy spaces have been investigated in detail by Madore and others[28, 29], and we are lead back to the equation (11). The fuzzyness which is closely tied up with the non commutative feature is symptomatic of the breakdown of the concept of the spacetime points and point particles at small scales or high energies. As has been noted by Snyder, Witten, and several other scholars, the divergences encountered in Quantum Field Theory are symptomatic of precisely such an extrapolation to spacetime points and which necessitates devices like renormalization. As Witten points out[30], "in developing relativity, Einstein assumed that the space time coordinates were Bosonic; Fermions had not yet been discovered!... The structure of space time is enriched by Fermionic as well as Bosonic coordinates."

Interestingly, starting from equation (11), we can deduce that l is the Compton wavelength without however assuming it to be so. Let us write the first equation of (11) as

$$[x, y] = \imath H \tag{12}$$

The relation (12) shows that y plays a role similar to the x component of the momentum, and infact mathematically we have

$$y = \frac{\imath H}{\hbar} p_x \equiv \tilde{h} p_x \tag{13}$$

At the extreme energies and speeds, we would have

$$y = \tilde{h}p = \tilde{h}mc, m\dot{y} = p_y, x = \tilde{h}p_y \tag{14}$$

From (13) it follows that

$$y = H\frac{d}{dx}$$

whence

$$Ty'' = \frac{T}{H}yy' = \frac{T}{H^2}y \cdot y^2 \tag{15}$$

Further from (14) it follows that

$$\rho\ddot{y} = \frac{\rho}{m}\frac{d}{dt}\left(\frac{x}{\tilde{h}}\right) = \frac{\rho}{m^2\tilde{h}} \cdot \frac{y}{\tilde{h}} = \frac{\rho}{m^2\tilde{h}^2}y \tag{16}$$

Fusing (15) and (16) in to one we get

$$\frac{H^2}{\tilde{h}^2}\frac{1}{m^2c^2} \equiv l^2 = \left(\frac{h}{mc}\right)^2 = y^2$$

where l is now the Compton wavelength. This is the explanation for the so called miraculous emergence of the Compton wavelength in string theory, as noted by Veneziano (Cf.ref.[11]).

Finally it may be pointed out that the tension T of String Theory, appears as the energy of the Quantum Mechanical Kerr Newman Black Hole alluded to in Section 2 via the relation (3).

We next have to see how, from the Compton scale above, we arrive at the Planck scale of QSS. For this, we note that from (11), using

$$\Delta p \cdot \Delta x \approx h,$$

we get,

$$\Delta p \cdot \Delta x = \hbar [1 + \frac{l^2}{(\Delta x)^2}]$$

Whence

$$\Delta p (\Delta x)^3 = \hbar [(\Delta x)^2 + l^2] \qquad (17)$$

Witten describes it as an extra correction to the Heisenberg Uncertainity Principle. As long as we are at usual energies, we have the usual Uncertainity Principle, and the usual bosonic or commutative spacetime. At high energies however we encounter the extra term in (17) viz., $\hbar' = \hbar l^2$. With this, the Compton scale goes over from l to l^3, the Planck scale (Cf. also [31]). Equally interesting is the fact that as can be seen from (17), the single x dimension gets trebled. At these Planck scales, therefore, a total of six extra dimensions appear, which are curled up in the Kaluza Klein sense at the Planck scale. This provides an explanation for the puzzling six extra dimensions of QSS.

In any case, the mysterious quantum prescription is really in the non commutative relations, (9) or (11). If we neglect l and higher orders, we recover classical physics. If we neglect l^2 and higher orders we are in the quantum domain including Quantum Field Theory. The effects that we have been seen above result from retaining terms $\sim l^2$.

4 Further Issues

(i) Vortices

As described in detail in [17] the Quantum Mechanical Kerr-Newman Black Hole could also be considered to be a vortex. If we take two parallel spinning vortices separated by a distance d then the angular velocity is given by

$$\omega = \frac{\nu}{\pi d^2},$$

where $\nu = h/m$.

Whence the spin of the system turns out to be h, that is in usual units the spin is one, and the above gives the states ± 1.

There is also the case where the two above vortices are anti parallel. In this case there is no spin, but rather there is the linear velocity given by

$$v = \nu / 2\pi d$$

This corresponds to the state 0 in the spin 1 case.

Together, the two above cases give the three $-1, 0, +1$ states of spin 1 as in the

Quantum Mechanical Theory.
In the case of the Quantum Mechanical Kerr-Newman Black Hole hydrodynamical vortex pictured above, it is interesting that for the bound state, there is really no interaction in the particle physics sense. The interaction comes in because in the above description we really identify a background Zero Point Field with the hydrodynamical flow (Cf.ref.[18] and also[32]). Interestingly in a simulation involving vortices, such an "attraction" was noticed[33].

(ii) Monopoles

It is interesting that the above considerations lead to a characterization of the elusive monopole. Infact a non commutative geometry can be associated with a powerful magnetic field[34], and specialising to the equations (11) we can show that this field B satisfies,

$$Bl^2 \sim \frac{nhc}{2e}$$

which is the celebrated equation of the monopole.

(iii) Duality

A related concept, which one encounters also in String Theory is Duality. Infact the relation (11) leads to (Cf. also equation (17)),

$$\Delta x \sim \frac{\hbar}{\Delta p} + \alpha' \frac{\Delta p}{\hbar} \tag{18}$$

where $\alpha' = l^2$, which in Quantum SuperStrings Theory $\sim 10^{-66}$. Witten has wondered about the basis of (18), but as we have seen, it is a consequence of (11).

In String Theory this is an expression of the duality relation,

$$R \to \alpha'/R$$

This is symptomatic of the fact that we cannot go down to arbitrarily small space-time intervals, below the Planck scale in this case (Cf.ref.[14]).
In the Quantum Mechanical Kerr-Newman Black Hole model of the electron, on the contrary, we are at Compton scale, and the effect of (18) is precisely that seen in point 1 above: We go from the electric charge e to the monopole, as in the Olive-Montonen duality[35], (Cf.also ref.[14]).
There is an interesting meaning to the duality relation arising from (18). While it appears that the ultra small is a gateway to the macro cosmos, we could look at it in the following manner. The first term of the relation (18) which is the usual Heisenberg Uncertainity relation is supplemented by the second term which refers to the macro cosmos.
Let us consider the second term in (18). We write $\Delta p = \Delta N mc$, where ΔN is the Uncertainity in the number of particles, N, in the universe. Also $\Delta x = R$, the radius of the universe which $\sim \sqrt{N} l$, the famous Eddington relationship. It should be stressed that the otherwise emperical Eddington formula, arises quite naturally in a Brownian characterisation of the universe as has been pointed out earlier (Cf.

for example ref.[5]).
We now get,

$$\Delta N = \sqrt{N}$$

Substituting this in the time analogue of the second term of (18), we immediately get, T being the age of the universe,

$$T = \sqrt{N}\tau$$

In the above analysis, including the Eddington formula, l and τ are the Compton wavelength and Compton time of a typical elementary particle, namely the pion. The equation for the age of the universe is also correctly given above. Infact in the closely related model of fluctuational cosmology (Cf. for example ref.[18]) all of the Dirac large number coincidences as also the mysterious Weinberg formula relating the mass of the pion to the Hubble constant, follow as a consequence, and are not emperical. All these relations relating large scale parameters to microphysical constants were shown to be symptomatic of what has been called, stochastic holism (Cf. also ref.[32]), that is a micro-macro connection with a Brownian or stochastic underpinning. Duality, or equivalently, relation (18) is really an expression of this micro-macro link.

(iv) Spin

One could argue that the non commutative relations (11) are an expression of Quantum Mechanical spin. To put it briefly, for a spinning particle the non commutativity arises when we go from canonical to covariant position variables. Zakrzewsk[36] has shown that we have the Poisson bracket relation

$$\{x^j, x^k\} = \frac{1}{m^2}R^{jk}, (c = 1),$$

where R^{jk} is the spin. The passage to Quantum Theory then leads us back to the relation (11).
Conversely it was shown that the relations (11) imply Quantum Mechanical spin[5]. Another way of seeing this is to observe as noted in (13) that (11) implies that $y = \alpha \hat{p}_y$, where α is a dimensional constant viz $[T/M]$ and $\hat{p}_y$ is the analogue of the momentum, but with the Planck constant replaced by l^2. So the spin is given by

$$|\vec{r} \times \vec{p}| \approx 2xp_y \sim 2\alpha^{-1}l^2 = \frac{1}{2}\left(\frac{\hbar}{m^2c^2}\right)^{-1} \times \frac{\hbar^2}{m^2c^2} = \frac{\hbar}{2}$$

as required.

(v) Extremal Black Holes

Going back to the relation (7), we can see that if

$$a = \frac{\hbar}{Mc} \sim \frac{GM}{c^2}$$

then we are at the Planck scale and have a Planck mass Schwarzchild Black Hole. The purely Quantum Mechanical Compton length equals the classical Schwarzchild

radius.
Also if,

$$Q \sim Mc^2, \tag{19}$$

while at the same time the particlee has no spin, so that $a = 0$, we recover a Schwarzchild Black Hole. We observe that if the mass $M \sim$ electron mass then the charge Q from (19) turns out to be $\sim 1000e$, as in the case of the monopole.
Interestingly these parameters also fit a neutrino, whose mass, as recent experiments indicate is given by

$$m \leq 10^{-8} m_e \tag{20}$$

It was further argued[37] that a neutrino could in principle have an electric charge, a millionth that of the electron, while, as the neutrino has no Compton wavelength we can apply in principle equation (19). (20) coupled with this and with the above electric charge shows that indeed equation (19) is satisfied.
Such particles however have a very high Bekenstein temperature

$$\sim 10^{-7} \left(\frac{M_0}{M}\right) K,$$

M_0 being the solar mass and would disintegrate into gamma rays within about $10^{-23} M^3 secs$. So these extremal Black Holes would not be detectable, but this could nevertheless provide a rationale for the puzzling cosmic gamma ray emissions.

(vi) Spacetime

We have seen that the spacetime given by (11) is radically different from its usual description. Infact the usual spacetime is a sort of a stationary spacetime, a low energy approximation, as will be clear by the following argument. We start with the Nelsonian theory in which there is a complex velocity potential $V - \imath U$, due to a double Weiner process. This has been shown to lead to the usual Quantum Mechanical description[38]. Indeed the diffusion equation,

$$\Delta x \cdot \Delta x = \frac{h}{m} \Delta t \equiv \gamma \Delta t$$

can also be written as

$$m \frac{\Delta x}{\Delta t} \cdot \Delta x = h = \Delta p \cdot \Delta x$$

which is the usual Heisenberg description.
Using the WKB approximation, the Nelsonian wave function

$$\psi = \sqrt{\rho} e^{(\imath/\hbar)s}$$

becomes

$$(p_x)^{-\frac{1}{2}} e^{\frac{\imath}{\hbar} \int p(x) dx}$$

whence

$$\rho = \frac{1}{p_x} \tag{21}$$

In this case the condition $U \approx 0$ gives

$$v \cdot \nabla ln(\sqrt{\rho}) \approx 0$$

that is the probability densityh ρ and hence from (21) the momentum varies very slowly with x.
The continuity equation now gives

$$\frac{\partial \rho}{\partial t} + \vec{\nabla}(\rho \vec{v}) = \frac{\partial \rho}{\partial t} = 0$$

which shows that ρ is independent of t also[18]. This is a scenario of, strictly speaking, a single particle universe, without environmental effects, a scenario which is an approximation valid for small incremental changes. It must be pointed out that, as realized by Poincare himself, even a three body problem leads to chaos. (The more physical scenario takes all the particles in the universe into account, leading to what has been called stochastic holism[32]). In this case, we can take limits to vanishing spacetime intervals, as in the usual theory (Cf. 't Hooft loc. cit). Spacetime in this description is a differentiable manifold and instead of the relations (11), spacetime is commutative. Effectively we are neglecting l^2. This has been the backbone of twentieth century physics.
On the other hand according to Witten[39], "String Theory is a part of twenty-first century physics that fell by chance into the twentieth century." It does appear that non commutative fuzzy spacetime is a paradigm for the twenty-first century.

References

[1] B.G. Sidharth, in Instantaneous Action at a Distance in Modern Physics: "Pro and Contra" , Eds., A.E. Chubykalo et. al., Nova Science Publishing, New York, 1999.

[2] F. Rohrlich, "Classical Charged Particles", Addison-Wesley, Reading, Mass., 1965.

[3] A.M. Polyakov, "Gravitation and Quantizations", Eds. B. Julia and J. Zinn-Justin, Les Houches, Session LVII, 1992, Elsevier Science, 1995, p.786-804.

[4] T.D. Lee, Physics Letters, Vol.122B, No.3,4, 10March 1983, p.217-220.

[5] B.G. Sidharth, Chaos, Solitons and Fractals, 12(2001), 173-178.

[6] C. Wolf, Il Nuovo Cimento, Vol.100B, No.3, September 1987, p.431-434.

[7] J.A. Wheeler, "Superspace and the Nature of Quantum Geometrodynamics", Battelles Rencontres, Lectures, Eds., B.S. De Witt and J.A. Wheeler, Benjamin, New York, 1968.

[8] A. Kempf, in "From the Planck Length to the Hubble Radius", Ed. A. Zichichi, World Scientific, Singapore, 2000, p.613-622.

[9] G. Veneziano, Physics Reports, 9, No.4, 1974, p.199-242.

[10] G. Fogleman, Am.J.Phys., 55(4), 1987, pp.330-336.

[11] G. Veneziano, "Quantum Geometric Origin of All Forces in String Theory" in "The Geometric Universe", Eds. S.A. Huggett et al., Oxford University Press, Oxford, 1998, pp.235-243.

[12] P. Ramond, Phys.Rev.D., 3(10), 1971, pp.2415-2418.

[13] J. Scherk, Rev.Mod.Phys., 47 (1), 1975, pp.1-3ff.

[14] W. Witten, Physics Today, April 1996, pp.24-30.

[15] Y. Ne'eman, in Proceedings of the First Internatioinal Symposium, "Frontiers of Fundmental Physics", Eds. B.G. Sidharth and A. Burinskii, Universities Press, Hyderabad, 1999, pp.83ff.

[16] C.W. Misner, K.S. Thorne and J.A. Wheeler, "Gravitation", W.H. Freeman, San Francisco, 1973, pp.819ff.

[17] B.G. Sidharth, Ind.J. Pure & Appld.Phys., Vol.35, 1997, pp.456-471.

[18] B.G. Sidharth, Int.J.Mod.Phys.A, 13 (15), 1998, p.2599ff.

[19] P.A.M. Dirac, "The Principles of Quantum Mechanics", Clarendon Press, Oxford, 1958, pp.4ff, pp.253ff.

[20] C. Moller, "The Theory of Relativity", Clarendon Press, Oxford, 1952, pp.170 ff.

[21] L. Nottale, "Scale relativity and gauge invariance", to appear in Chaos, Solitons and Fractals, special issue in honor of M. Conrad.

[22] H.S. Snyder, Physical Review, Vol.72, No.1, July 1 1947, p.68-71.

[23] V.G. Kadyshevskii, Translated from Doklady Akademii Nauk SSSR, Vol.147, No.6, December 1962, p.1336-1339.

[24] L. Bombelli, J. Lee, D. Meyer and R.D. Sorkin, Physical Review Letters, Vol.59, No.5, August 1987, p.521-524.

[25] B.G. Sidharth, Chaos, Solitons and Fractals, 11 (2000), 1269-1278.

[26] V. Heine, "Group Theory in Quantum Mechanics", Pergamon Press, Oxford, 1960, p.364.

[27] J.R.Klauder, "Bosons Without Bosons" in Quantum Theory and The Structures of Time and Space, Vol.3 Eds by L. Castell, C.F. Van Weiizsecker, Carl Hanser Verlag, Munchen 1979.

[28] J. Madore, Class.Quantum Grav. 9, 1992, p.69-87.

[29] B.L. Cerchiai, J. Madore, S.Schraml, J. Wess, Eur.Phys.J. C 16, 2000, p.169-180.

[30] J. Schwarz, M.B. Green and E. Witten, "SuperString Theory", Vol.I, Cambridge University Press, Cambridge, 1987.

[31] B.G. Sidharth, "Quantum Superstrings and Quantized Fractal Space Time", to appear in Chaos, Solitons and Fractals.

[32] B.G. Sidharth, "The Chaotic Universe", Nova Science Publishers, New York, In press.

[33] J.A. Wheeler, "Geometrodynamics and the Issue of the Final State", Lectures at l"Ecole d'ete de Physique Theorique, Les Houches, Haute-Savole, France, July 1963.

[34] T. Saito, Gravitation and Cosmology, 6 (2000), No.22, pp.130-136.

[35] D.I. Olive, Nuclear Physics B (Proc. Suppl.) 46, 1996, 1-15.

[36] S. Zakrzewski, "Quantization, Coherent States, and Complex Structures", Ed. J.P. Antoine et al., Plenum Press, New York, 1995, p.249ff.

[37] B.G. Sidharth, in Proceedings of the International Symposium on "Frontiers of Fundamental Physics", Vol.2, Universities Press, Hyderabad, 2000, p.138ff. Also xxx.lanl.gov quant-ph/9803048.

[38] L. Nottale, "Fractal Space-Time and Microphysics: Towards a Theory of Scale Relativity", World Scientific, Singapore, 1993, p.312.

[39] B. Greene, "The Elegant Universe", Jonathan Cape, London, 1999, pg.19.

Quarks as Vortices in Vacuum

G. Musulmanbekov
Joint Institute for Nuclear Research,
141980 Dubna Russia
e-mail: genis@cv.jinr.ru

1 Introduction

Despite the many successes of QCD much of low energy hadronic physics cannot be extracted from the first principles of the theory. Quarks were first postulated as constituents of hadrons with masses about hundreds of MeV to describe the spectroscopy of hadrons. Subsequently, it was realized that hard processes are the arena of current quark considerations with masses much lighter than the original constituent ones. Relationship between current and constituent quarks is still s-mooth and unclear. According to the modern picture of hadrons quark–antiquark in mesons and three valence quarks in nucleons are dressed with gluons and "sea" quark–antiquark pairs in such a way that they become much heavier, although the dynamical mechanism of dressing is not understood definitely.

Interplay between hard and soft interactions in high energy experiments strictly relates to interconnection between small and large size quark configurations inside a hadron. We think that this interplay is a manifestation of fluctuations of hadronic matter distributions inside hadrons and these fluctuations in turn are results of correlated motion of valence quarks inside hadrons.

The question arises: is it possible to construct the dynamical system of quarks which can be observed at one instant of time as constituent (dressed) quarks and at the another one - as current (bare) quarks? Proposed by author the semiclassical model of strongly correlated quarks, SCQM, demonstrates how these configurations could be realized inside hadrons(Section 2)[1]. In Section 3 we show that SCQM is identical to breather solution of sine-Gordon equation. Inclusion of spin that is performed using classical considerations leads to representation of quarks as spinning solitons, or vortices in vacuum (Section 4).

2 Strongly Correlated Quark Model

Let us imagine the following hypothetical picture: single, colored quark imbedded into vacuum. Because of vacuum fluctuations one can observe two competing processes: first, polarization of $\bar{q}q$ sea by the color field of valence quark (VQ) and second, the tendency of vacuum fluctuations, due to background Zero Point Radiation Field, to destroy this polarization. As a result, one can say about a vacuum pressure on a single, colored quark. This effect can be interpreted as the "response" of the vacuum on the presence of, say, point defect or dislocation like in solid state

Frontiers of Fundamental Physics 4, Edited by Sidharth and Altaisky,
Kluwer Academic/Plenum Publishers, New York, 2001

physics. What happens if we place in vicinity of this quark the corresponding antiquark? By virtue of opposite color signs their polarization fields interfere in the overlapping region **destructively**. So the pressure of vacuum on quark (antiquark) from outside exceeds that one going from inner space between quark and antiquark. This results in an **attractive** force between quark (dislocation) and antiquark (antidislocation). The density of the remaining part of polarization field around quark (antiquark) is identified with hadronic matter distribution. At maximum displacement in $\bar{q}q-$ system, that corresponds to small overlapping of polarization fields, hadronic matter distributions have maximum values. So quark and antiquark located nearby start moving towards each other. The closer they to one another, the larger destructive interference effect and the smaller hadronic matter distributions are around VQs and the larger their kinetic energies.

For such interacting $\bar{q}q-$ system the total Hamiltonian is

$$H = \frac{m_{\bar{q}}}{(1-\beta^2)^{1/2}} + \frac{m_q}{(1-\beta^2)^{1/2}} + V_{\bar{q}q}(2x), \tag{1}$$

were $m_{\bar{q}}$, m_q- masses of valence antiquark and quark, $\beta = \beta(x)-$ their velocity depending on displacement x and $V_{\bar{q}q}-$ quark–antiquark potential energy with separation $2x$. It can be rewritten as

$$H = \left[\frac{m_{\bar{q}}}{(1-\beta^2)^{1/2}} + U(x)\right] + \left[\frac{m_q}{(1-\beta^2)^{1/2}} + U(x)\right] = H_{\bar{q}} + H_q, \tag{2}$$

were $U(x) = \frac{1}{2}V_{\bar{q}q}(2x)$ is potential energy of quark or antiquark. Therefore, keeping in mind that quark and antiquark are strongly correlated we consider each of them separately as undergoing oscillatory motion in 1+1 dimension. Generalization to three–quark system in baryons is performed according to $SU(3)_{color}$ symmetry: an antiquark is replaced by two correspondingly colored quarks to get color singlet baryon and destructive interference takes place between color fields of three valence quarks. Putting aside the mass and charge differences of valence quarks we may say that inside baryon three quarks oscillate along the bisectors of equilateral triangle. Hereinafter we consider that axis Z is perpendicular to the plane of oscillation XY.

VQ with its polarized surroundings (hadronic matter distribution) form constituent quark. According to our approach potential energy of valence quark, $U(x)$, corresponds to the mass M_Q of constituent quark:

$$U(x) = const \int_{-\infty}^{\infty} dz' \int_{-\infty}^{\infty} dy' \int_{-x}^{\infty} dx' \rho(x, \mathbf{r}') \approx M_Q(x) \tag{3}$$

with

$$\rho(x, \mathbf{r}') = |\varphi(x, \mathbf{r}')|^2 = \left|\varphi_Q(x'+x, y', z') - \varphi_{\overline{Q}}(x'-x, y', z')\right|^2. \tag{4}$$

The knowledge of the mechanism and structure of vacuum polarization around valence quark would give the information about the confining potential. We cannot say at the moment for sure what is the microscopic mechanism of interaction of valence quark with vacuum. It could be instanton induced interactions, excitation of fractal structure of space–time, etc. So we assume, as a first approximation, that the polarization field can be taken in gaussian form:

$$\varphi_Q(\mathbf{r}) = \varphi_Q(x,y,z) = \varphi_Q(x_1,x_2,x_3) = \frac{(\det \hat{A})^{1/2}}{(\pi)^{3/2}} \exp\left(-\mathbf{X}^T \hat{A} \mathbf{X}\right), \quad (5)$$

where exponent is written in quadratic form. The same is for $\varphi_{\overline{Q}}(\mathbf{r})$. We define the mass of constituent quark at maximum displacement

$$M_{Q(\overline{Q})}(x_{\max}) = \frac{1}{3}\left(\frac{m_\Delta + m_N}{2}\right) \approx 360\ MeV, \quad (6)$$

where m_Δ and m_N are masses delta–isobar and nucleon correspondingly. The parameters of the model, namely, masses of VQs, $m_{q(\overline{q})}$, maximum displacement, $x_{\max}$, and parameters of gaussian function, $\sigma_{x,y,z}$, are chosen to be

$$m_{q(\overline{q})} = 5\ MeV, \quad (7)$$

$$x_{\max} = 0.64\ fm, \quad (8)$$

$$\sigma_{x,y} = 0.24\ fm, \quad (9)$$

$$\sigma_z = 0.12\ fm. \quad (10)$$

They are adjusted by comparison of calculated and experimental values of inelastic cross sections, $\sigma_{in}(s)$, and inelastic overlap function $G_{in}(s,b)$ for pp and $\bar{p}p$–collisions[2]. Using (3)–(5) we can calculate the confining potential $U(x)$ and force $F(x) = -\frac{dU}{dx}$. They are shown in Fig. 1. As one can see the confining potential is essentially nonlinear. The behavior of potential evidently demonstrates the relationship between constituent and current quark states inside a hadron. At maximum displacement quark is nonrelativistic, constituent one (VQ surrounded by "polarized sea"), since according to (3) the confining potential corresponds to the mass of constituent quark. At the origin of oscillation, $x = 0$, antiquark–quark in mesons and three quarks in baryons, being close to each other, have maximum kinetic energy and correspondingly minimum potential energy and mass: they are relativistic, current quarks (bare VQs). Intermediate region corresponds to increasing (decreasing) of quark's mass by dressing (undressing) of quarks due to vacuum polarization. This mechanism agrees with local gauge invariance principle. Indeed, phase rotation of wave function of single quark in color space ψ_c on angle θ depending on displacement x of the quark in coordinate space

$$\psi_c(x) \rightarrow e^{i\theta(x)}\psi_c(x) \quad (11)$$

results in it's dressing (undressing) by quark and gluon condensate that corresponds to the transformation of gauge field $A_\mu = (\varphi, 0, 0, 0)$

$$A_\mu(x) \rightarrow A_\mu(x) + \partial_\mu\theta(x). \quad (12)$$

Here we dropped color indices and took into account only scalar component, φ, of gauge field. Thus gauge transformation maps internal (isotopic) space of colored

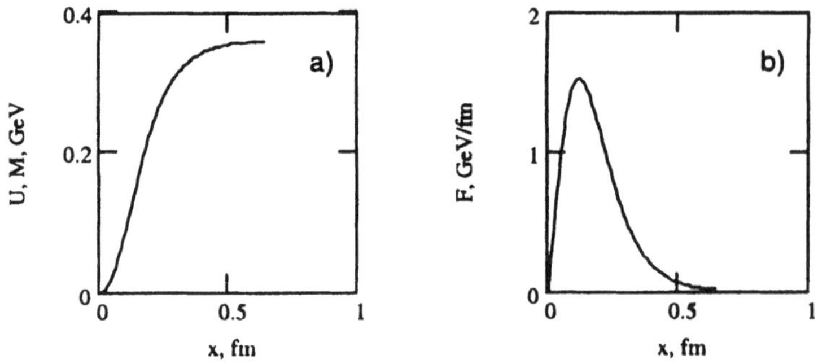

Figure 1: a)Potential energy of valence quark and mass of constituent quark; b)"Confinement" force.

quark onto coordinate space. On the other hand this dynamical picture of VQ dressing (undressing) corresponds to chiral symmetry breaking (restoration). The behaviour of field φ and hadronic matter distribution,ρ, for quark–antiquark system during their oscillations is shown in Fig. 2. Due to this mechanism of VQs oscillations nucleon runs over the states corresponding to the definite terms of the infinite series of Fock space

$$| B\rangle = c_1 \mid q_1 q_2 q_3\rangle + c_2 \mid q_1 q_2 q_3 \bar{q} q\rangle + \ldots \tag{13}$$

Confining force drastically differs from the one given by string models. When VQs are close each other it is very weak and fulfills the "asymptotic freedom" behavior of quarks of QCD. At larger distances between VQs it starts growing rapidly, then reaching maximum value goes down, asymptotically approaching zero. Thus at large distances inside hadrons quarks being in a constituent state are almost free. Hence, it is clear why additive quark model, where quarks are treated as massive, almost unbound and extended objects, works well. We must emphasize that interaction between VQs is not direct but a result of polarization of surrounding vacuum combined with destructive interference. Attractive force between VQs in ground state hadrons does not appear as gluon string but goes from vacuum suppression that predominates from outside. Therefore, our approach reflects the features of bag models, as well. The model is in agreement with the experiments (Fig. 3) for description of VQ structure function inside a nucleon

$$F_2^{ep} - F_2^{en} = \frac{x_F}{3}\left[u_v(x_F) - d_v(x_F)\right]. \tag{14}$$

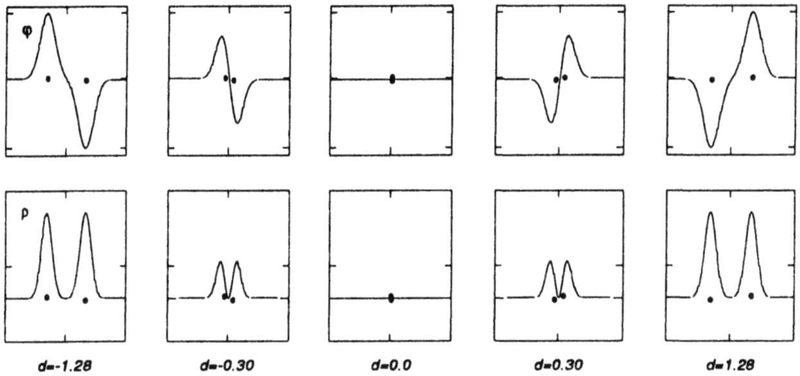

Figure 2: Evolution of field φ, (Eq. (5)), and hadronic matter distribution ρ, (Eq. (4)), in quark–antiquark system during one–half of the period of oscillations; $d = 2x$ – distance in *fermi* between quark and antiquark depicted as dots.

Obviously, our description is classical and we must take into account quantum corrections. Nevertheless, classical consideration of VQ oscillations is justified by E. Schrodinger's approach[3] where he, analyzing the motion of gaussian wave packet for time dependent Schrodinger equation for harmonic oscillator, demonstrated that this wave packet moves in exactly the same way as corresponding classical oscillator. The wave packet does not deform and spread and has a minimum uncertainty. In other words, it looks like a soliton. Moreover, in paper[4] the same method was also applied to solve exactly the problem of simple harmonic oscillator with the force depending on time. In this case, probability densities with the same wave packet structures as that of usual harmonic oscillator can be found to move in such a way that their centroids follow the classical law. In our model VQ with its surroundings can be treated as (nonlinear) wave packet. In the next section we show that these wave packets really possess soliton–like features.

3 Quarks as Solitons

Let us consider sine-Gordon (SG) equation in (1+1) dimension which in scaled form for scalar function $\phi(x,t)$ is given by

$$\Box\phi(x,t) + \sin\phi(x,t) = 0, \tag{15}$$

where $x, t-$ dimensionless[7]. SG system has several application including Josephson–junction transmission, motion of Bloch wall between ferromagnetic domains, motion of dislocation in crystalline structure. Many solutions of the SG system are known. The following solutions are useful to us:

· soliton

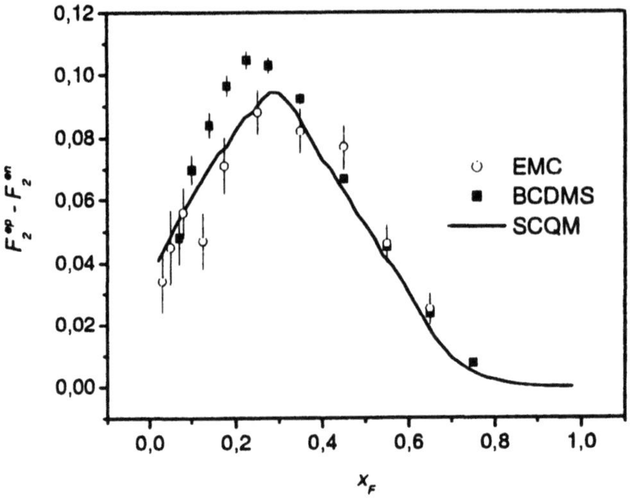

Figure 3: Valence quark structure functions in nucleons; data point are from papers[5, 6].

$$\phi_s(x,t) = -4\tan^{-1}\exp\left[\frac{(x-ut)}{\sqrt{(1-u^2}}\right], \tag{16}$$

· antisoliton

$$\phi_{as}(x,t) = 4\tan^{-1}\exp\left[-\frac{(x-ut)}{\sqrt{(1-u^2}}\right], \tag{17}$$

· breather

$$\phi_{br}(x,t) = 4\tan^{-1}\left[\frac{\sinh\left[ut/\sqrt{(1-u^2}\right]}{v\cosh\left[x/\sqrt{(1-u^2}\right]}\right], \tag{18}$$

were $u-$ dimensionless velocity. Breather is a periodic solution representing bound state of soliton–antisoliton pair which oscillate around their center of mass with the period

$$\tau = (2\pi\sqrt{1+u^2})/u. \tag{19}$$

Profiles of $\phi_{br}(x,t)$ for $t = 0, \tau/4, \tau/2, 3\tau/4$ are shown on Fig. 4a.The existence of such a bound state is understandable from the fact that free soliton and antisoliton approaching each other with velocity $2u$ undergo time advance

$$\Delta = \left(2\sqrt{1-u^2}/u\right) \ln u \tag{20}$$

implying that soliton and antisoliton attract each other and mutually accelerate. Approximating the soliton–antisoliton pair by a system of one single effective degree of freedom with the total energy

$$E = \frac{M_s}{\sqrt{1-\dot{x}}} + \frac{M_{as}}{\sqrt{1-\dot{x}}} + V_{s-as}(2x) = \frac{2M}{\sqrt{1-\dot{x}}} + W(x) \tag{21}$$

and using time shift (16) Vinciarelli[8], Troost[9] and Hsu [10] derived the effective potential for breather solution. Here M is a classical soliton mass. The result of Hsu for potential W is given in the form of infinite series

$$x(W) = \sqrt{W + 2M}\left[b_1 \ln(-W) + \sum_{n=2} b_n W^{n-2}\right]. \tag{22}$$

The shape of the potential $V_{s-as}(x) = W(x/2)$ is very close to our "confining" potential shown on Fig. 1b. Using nonrelativistic approximation Troost and Vinciarelli obtained the analytic formula

$$V_{s-as}(x) = 2M \tanh^2(x/2) - 2M. \tag{23}$$

Their results agree with Hsu's in large $x-$ region and deviate at small x which is the relativistic region.

Similarity of our model (SCQM) to breather solution of SG system is seen from the behavior of breather density profile

$$\varphi_{br}(x,t) = \frac{d\phi_{br}}{dx}, \tag{24}$$

drawn on Fig. 4. During the oscillations soliton–antisoliton pair evolves like our quark–antiquark system (Fig. 2a), i.e. at maximal displacement ($t = \tau/4, 3\tau/4$) soliton and antisoliton profiles, $\varphi_{br}(x,t)$, are maximal and at minimum displacement ($t = 0, \tau/2$) they "annihilate". It is not surprising because our quark–antiquark system was built in close analogy with the model of dislocation–antidislocation[11], which in its continuous limit is described by breather solution of SG equation. So our "confining" potential evidently is close to that one calculated by Hsu (18). In constructing the model we conjectured that vacuum polarization field around valence quark has a gaussian form which differs from SG soliton density profile in proper frame

$$\varphi_s(x,t) = \frac{d\phi_s}{dx} = \frac{const}{\cosh x}. \tag{25}$$

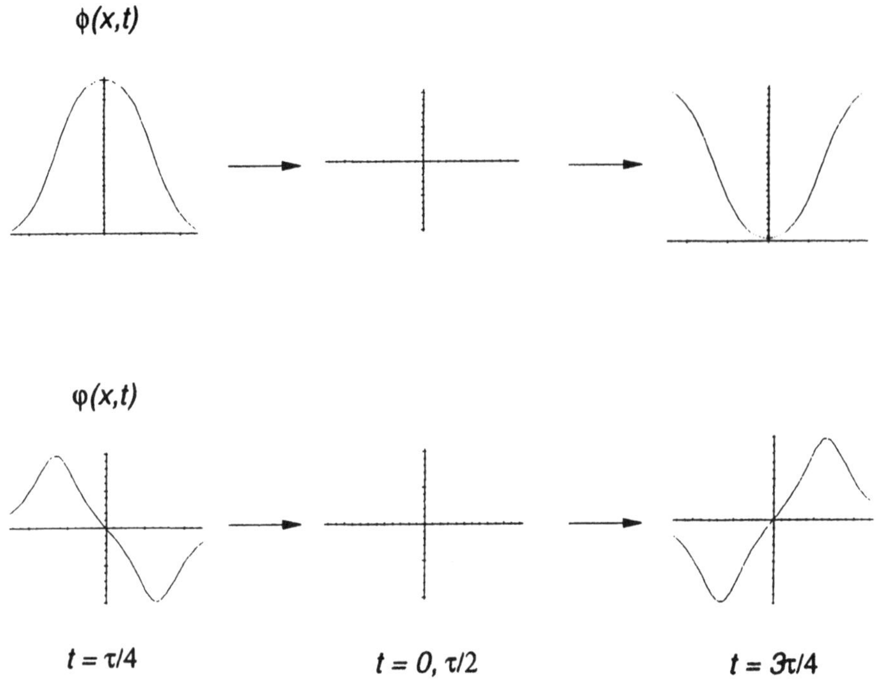

Figure 4: Evolution of breather,ϕ, and its density profile, φ, during the oscillation.

Thus in more realistic model vacuum polarization field around valence quark must correspond to this density profile. Then mesonic systems will be identical to the model of dislocation–antidislocation in crystalline structures.

Since the above consideration of quarks as solitons is purely classical the important problem is to construct quantum states around them. Although soliton solution of SG equation looks like extended (quantum) particle the relation of classical solitons to quantum particles is not so trivial. Technique of quantization of classical solitons with usage of different methods has been developed by many authors. The most known of them is semiclassical method of quantization (WKB) which allows one to relate classical periodic orbits (breather solution of SG) with the quantum energy levels[12].

4 Quarks with Spin

So far we dealt with scalar polarization field around VQ. Since quarks are fermions the question arises: How can one include spin in the frame of classical considerations? This problem has a long history[13, 14, 15]. According to the prevailing belief, the spin is quantum feature of microparticle and has no classical analog. However, Belinfante [13] as early as in 1939 showed that the spin of electron may be regarded as an angular momentum generated by a circulating flow of energy, or a momentum density, in the wave field of electron. Furthermore, a comparison between calculations of angular momentum in the Dirac and electromagnetic fields, performed by Ohanian[14], shows that the spin of the electron is entirely analogous to the angular momentum carried by classical circularly polarized wave.

We follow the classical picture of electron spin considered by Hiqbie[15]. Let us imagine point electric charge, e, and magnetic bar, μ, settled close nearby[16]. In surrounding space electric, $\mathbf{E}$, and magnetic, $\mathbf{B}$, fields arise. Although vectors $\mathbf{E}$ and $\mathbf{B}$ are constant the circulating flow of energy is created around such a system which is described by Poynting's vector

$$\mathbf{S} = \varepsilon_0 c^2 \mathbf{E} \times \mathbf{B}. \tag{26}$$

The volume density of angular momentum is given by

$$\mathbf{L} = \mathbf{r} \times \mathbf{P}, \tag{27}$$

where $\mathbf{P}$ is the volume density of linear momentum defined as

$$\begin{aligned} \mathbf{P} &= \mathbf{S}/c^2 = \mathbf{D} \times \mathbf{B} = \left(-\frac{q\mathbf{r}}{4\pi r^3}\right) \times \left\{ \frac{\left[3(\mu \cdot \mathbf{r})\mathbf{r}/\mathbf{r}^2 - \mu\right]}{4\pi\varepsilon_0 c^2 r^3} \right\} \\ &= \frac{q(\mathbf{r} \times \mu)}{(4\pi c)^2 \varepsilon_0 r^6} = \frac{r_0(\mathbf{r} \times \mu)}{(q/m)4\pi r^6}. \end{aligned} \tag{28}$$

Here $r_0 = q_e^2/4\pi\varepsilon_0 m_e c^2 -$ classical radius of electron. Substitution of (25) into (24) gives

$$\mathbf{L} = \mathbf{r} \times \mathbf{P} = (...)\mathbf{r} \times (\mathbf{r} \times \mu) = (...) \left[\mathbf{r}(\mathbf{r} \cdot \mu) - \mu r^2\right] \tag{29}$$

When we align μ with $Z-$ axis and use spherical polar coordinates, $z-$ component of $\mathbf{L}$ is

$$L_z = -(...)\mu r^2 \sin^2\theta. \tag{30}$$

The total angular momentum of the field extending from $r = a$ to infinity is given by the integral

$$S = \int_a^\infty L_z dV = -(2/3)\frac{r_0 \mu}{a(q_e/m_e)}. \tag{31}$$

For $a = (2/3)r_0$ we have

$$S = -\mu/(q_e/m_e), \tag{32}$$

which just happens to be the entire spin angular momentum of the electron. Moreover, Feynman showed that if we extend the classical coulomb field all the way down to 2/3 of the classical electron radius, r_0, the entire mass of the electron is contained in it's field.

By analogy we suppose that intersecting chromoelectric, $\mathbf{E}_{ch}$, and chromomagnetic, $\mathbf{B}_{ch}$, fields create around quark (antiquark) circulating flow of energy, the color analog of Poynting's vector

$$\mathbf{S}_{ch}=c^2\mathbf{E}_{ch} \times \mathbf{B}_{ch}. \tag{33}$$

Spin of quark (antiquark) then is defined by the integral

$$\mathbf{s} = \int_a^{\infty} \mathbf{r} \times (\mathbf{E}_{ch} \times \mathbf{B}_{ch}) d^3\mathbf{r}. \tag{34}$$

Obviously, lower limit of integration, a, is connected with the size of quark (antiquark). Now we must take into account correlated oscillations of quarks inside hadrons that lead to periodic changes in values of $\mathbf{E}_{ch}$ and $\mathbf{B}_{ch}$. At maximal displacement of quark from the origin of oscillation $\mathbf{E}_{ch}$ and $\mathbf{B}_{ch}$ around it are expanded to the maximum (size of constituent quark) and so is for circulating flow $\mathbf{S}_{ch}$. At minimum displacement $\mathbf{E}_{ch}$ and $\mathbf{B}_{ch}$, owing to destructive interference described in Section 2, are confined in small region around the quark and circulating flow of energy $\mathbf{S}_{ch}$ is concentrated in a narrow shell. Recalling that entire spin of quark is due total angular momentum of circulating flow we compare it with circular flow of ideal liquid and apply hydrodynamic considerations. Equations of hydrodynamics for ideal liquid flow read

$$\frac{\partial \Omega}{\partial t} + \nabla \times (\Omega \times \mathbf{v}) = 0, \tag{35}$$

$$\Omega = \nabla \times \mathbf{v}, \tag{36}$$

$$\nabla \cdot \mathbf{v} = 0, \tag{37}$$

where $\mathbf{v}$ is vector of liquid velocity and Ω is its vorticity defining circulation around unit area. Equation (35) implies angular momentum conservation law applied to liquid flow. In other words, if σ is the area of vortex field in ideal liquid then

$$\Omega \cdot \sigma = \oint_{\sigma} \mathbf{v} \cdot d\mathbf{r} = const, \tag{38}$$

that means the larger area of circulation the less the vorticity and vice versa. Following these considerations we assume that the spin of quark is conserved during oscillation and

$$\mathbf{s} = const = \Omega \cdot \sigma. \tag{39}$$

Rigorously, the case at hand is the conservation of quark spin module because quarks inside hadronic system can interact and what is conserved is the total angular momentum of the system. Whereas divergence of Ω, (36), is equal zero it follows that vortices tend to form closed loops (strings) in space. Thus, according to our approach quark spin is given by circulating flow of polarized vacuum (sea quarks and gluons) and recalling that entire mass of constituent quark (antiquark) is contained in it's color field one can consider quarks as to be vortices in vacuum. Now gauge field $A_\mu(x)$ in (12) contains along with the scalar part, φ, vector components, $\mathbf{A}$, as well. Whereas vortex is singularity (hole) in vacuum the quantization of spin follows from nonsimple-connectedness of vacuum structure. Spin quantization also goes from the fact that circulation of the velocity, (38), in superfluid liquid is quantized. Representation of quark spin as created by circulating flow is alternative solution to, so called, "spin crisis" observed in deep inelastic scattering with polarized beam and target[17].

Summing up we make a speculative statement: elementary particles are nothing but vortices in vacuum; their charges and spins originate from circulating flow of vacuum fluid. If a charge and magnetic bar settled nearby create circulating flow of vacuum fluid why not the reverse? H. Poincarĕ in his article "New Mechanics" wrote: "No matter exist, the only essences are the holes in ether". This idea was elaborated by several authors[18, 19].

This research was partly supported by the Russian Foundation of Basics Research, grants 99-07-90383 and 99-01-01103.

References

[1] G. Musulmanbekov, Proc. of XVII Int. Kazimierz Meeting on Particle Theory and Phenomenology, Iowa, 1995, World Scientific, 1996, p. 347–353.; Proc. of XXVI Int.Symp. on Multiparticle Dynamics, Sept. 1–5, Faro, 1996, World Scientific, 1997, p. 357–363; Nucl. Phys. Suppl. **B71**,1999, 117.

[2] G. Musulmanbekov, Proc. Int. Conf. on Elastic and Diffractive Scattering (VIIIth Blois Workshop), Protvino, 1999,World Scientific, 2000, p. 433–443.

[3] E. Schrodinger, "Der stetige Ubergang von der Mikro-zur Makromechanik", Naturwissenschaften **14**, 1926, 664.

[4] C.C. Yan, Am. J. Phys., **62** ,1994,) 147.

[5] J.J. Aubert et al., Nucl. Phys. **293B**, 1987, 740.

[6] A.C. Benvenuti et al., Phys. Lett., **237B**, 1990, 599.

[7] R. Rajaraman, Phys. Rep., **21C**,1975, 229.

[8] P. Vinciarelli, Acta Phys. Austriaca Suppl. **15**, 1976, 512.

[9] W. Troost, CERN Report, 1975 (unpublished).

[10] Y. Hsu, Phys. Rev. **D22**, 1980, 1394.

[11] T.A. Kontorova, Ya.I. Frenkel, Zh. Eksp. Teor. Fiz. **8**, 1938, 89; ibid **8**, 1938, 1340.

[12] R. Dashen, B.Hasslacher and a. Neveu, Phes. Rev. **D10**, 1974, 4114.

[13] F.J. Belinfante, Physica **6**, 1939, 887.

[14] H.C. Ohanian, Am. J. Phys., **54** ,1986, 500.

[15] J. Hiqbie, Am. J. Phys., **56**, 1988, 378.

[16] R.P. Feynman, R. B. Leighton and M. Sands, "The Feynman Lectures on Physics", Addison–Wesley Pub. Co. Inc., 1964, Vol.2, p.295.

[17] J. Ashman, et al., Nucl. Phys., **B328**, 1988, 1.

[18] W.M. Honig, Lett. Nouvo Cim., **19**, 1977, 137.

[19] P. Tewari, "Space Vortices of Energy and Matter", Adarsh Press, Aligarh, 1978.

ϕ^4-Field theory on a Lie group

M.V.Altaisky
[1]*Joint Institute for Nuclear Research, Dubna, 141980, Russia*
[2]*Space Research Institute, Profsoyuznaya 84/32, Moscow, 117810, Russia*
e-mail: altaisky@mx.iki.rssi.ru

1 Introduction

The ultra-violet (UV) divergences appearing in quantum field theory at small distances (high momentum $\Lambda \to \infty$) are well known to be intimately related to the properties of the theory with respect to the group of scale transformations. For a wide class of theories, known as ***multiplicatively renormalizable*** theories, the problem can be essentially simplified by the scale transformation of fields (ϕ) and coupling constants (g)

$$\phi_R = Z_\phi^{-1}\phi, \quad g = g_0 Z_g^{-1}.$$

The renormalized Green functions

$$G_n^R(x_1,\ldots,x_n;g,\Lambda) = Z_\phi^{-n} G_n(x_1,\ldots,x_n;g_0,\Lambda)$$

become finite in the limit $\Lambda \to \infty$, with all divergences hidden in infinite renormalization constants $Z_\phi(g,\Lambda), Z_g(g,\Lambda)$.

The independence of physical results on scale transformations

$$\Lambda' = e^l\Lambda, \quad x' = e^{-l}x \tag{1}$$

is known as renormalization group (RNG) equation.

The modern quantum field theory has become inconceivable without RNG methods. Most of the results obtained phase transitions, quantum electrodynamics, quantum chromodynamics etc. are direct consequences of RNG methods. Therefore, we may have a temptation to base the theory on some kind of covariance with respect to scale transformations sct from very beginning, not after facing the UV divergences problem.

The best way to study any physical system is to choose a functional basis with the symmetry properties as close to the symmetry of the system as possible. For this reason we choose the spherical functions to study the hydrogen atom, and for the same reason we use plane waves to describe a particle moving in homogeneous space. Of course it is also possible to apply plane waves to SO_3 symmetrical problem, but one can hardly expect any use of it.

It is important what is implied by "the symmetry of the problem". We assume that the system is described by a set of complex-valued functions ϕ^α defined on a manifold $\mathcal{M}$, $\phi^\alpha := \phi^\alpha(x), x \in \mathcal{M}$. A system is said to have a symmetry group G if the action of the group G on the independent variables (*coordinates*) and dependent variables (*fields*)

$$x \to x' = \hat{T}x, \quad \phi^{\alpha}(x) \to \phi'^{\alpha}(x') = \hat{M}^{\alpha}_{\beta}\phi^{\beta}(x),$$

where $\hat{T}$ and $\hat{M}$ are operators, does not change the action functional (or any other functional which is believed to determine the dynamics of the system).

If the transformation group does not affect the fields themselves ($\hat{M}^{\alpha}_{\beta} \equiv \hat{1}$), but only coordinates

$$\phi^{\alpha}(x) \to \phi'^{\alpha}(x') = \phi^{\alpha}(\hat{T}^{-1}x'),$$

the field ϕ^{α} is called a scalar with respect to the transformation group G.

The most important group of transformations used in physics is the Poincare group $x_{\mu}' = \Lambda^{\nu}_{\mu}x_{\nu} + b_{\mu}$. The wave functions of elementary particles - electrons, photons, quarks etc. - are not Poincare scalars. They have nontrivial transformation properties under Lorenz rotations Λ^{ν}_{μ} and are classified according to their spin. However, it is possible to consider certain simplistic models with scalar fields, which do have, or may have physical implications for real systems. One of the most known models is the scalar theory of critical behavior, where magnetization $\phi(x)$ is considered as a function of the coordinate in Euclidean space. The scalar field theory was application point of the Wilson renormalization group awarded by a Nobel prize in 1982. The scalar field theory in Euclidean space is an analytical continuation ($\tau = it$) of a field theory in Minkovski space, and is receiving a lot of attention.

In this paper we restrict ourselves to the theory of complex-valued scalar field. Usually, the scalar field theory is defined on the n-dimensional Euclidean space R^n, which is isomorphic to the group of translations

$$x' = x + b, \quad x, b \in R^n \tag{2}$$

The representation of the translation group tg on the space of square-integrable functions is given by $U(b)\phi(x) = \phi(x - b)$. The unitary representation of the translation group is defined on the space of periodic functions

$$U(b)e^{-imx} = e^{imb}e^{-imx}, \quad U(-b) = U^{*}(b).$$

Thus it is possible to decompose a function $\phi(x)$ with respect to the representations of translation group G

$$\phi(x) = \int_G e^{ixb}\hat{\phi}(b)db. \tag{3}$$

This is Fourier decomposition. Similarly, a function may be decomposed with respect to SO_3 rotations, Poincare group [1] and other groups.

Since the concept of the group is just more general than the concept of the Euclidean or Minkovski space, a question naturally arises: *For what groups it is physically meaningful to construct a decomposition like ft and use may we have of it in field theoretic calculations?*

From physical point of view, the coordinates (x) can not be measured with arbitrary high precession, and it seems more reasonable to speak about the values of the fields ϕ^{α} measured at a position x with the finite resolution Δx. *The claim of the present paper is that an adequate description of this situation, which inherits the RNG ideas, can be achieved if we use an analog of ft decomposition on the base of the affine group*

$$x' = ax + b, \tag{4}$$

where, as it will be seen later, a can be understood as resolution and b as a coordinate.

The goal of the present paper is to construct a ϕ^4 model where the scalar field $\phi(a,b)$ is defined on the affine group (in the sense that a, b a coordinates on the affine group ag) and to link the new model with renormalization properties of the known ϕ^4 model in R^n.

The paper is organized as follows. In *section 2* we review the basic formalism of ϕ^4 theory in R^n. In *section 3* we remind the technique of wavelet transform with respect to a locally compact Lie groups. In *section 4* the ϕ^4 theory on the affine group is presented.

2 ϕ^4 Field theory

The scalar field theory with the forth power interaction $\frac{\lambda}{4!}\phi^4(x)$ defined on Euclidean space $x \in R^n$ is one of the most instructive models any textbook in field theory starts with, see e.g. [2]. Often called a Ginsburg-Landau model for its ferromagnetic counterpart, the model describes a quantum field with the (Euclidean) action

$$S[\phi] = \int d^n x \frac{1}{2}(\partial_\mu \phi)^2 + \frac{m^2}{2}\phi^2 + \frac{\lambda}{4!}\phi^4(x). \tag{5}$$

in n-dimensional Euclidean space. Alternatively, the theory of quantum field in Euclidean space is equivalent to the theory of classical fluctuating field with the probability measure $\mathcal{D}P = e^{-S[\phi]}\mathcal{D}\phi$. In this case m^2 is the deviation from critical temperature $m^2 = |T - T_c|$, and λ is the fluctuation interaction strength. To some extent, the ϕ^4 model considered in this way describes a second type phase transition at zero external field in any system with one-component order parameter $\phi = \phi(x)$ and symmetry $\phi \to -\phi$.

The Green functions (correlation functions)

$$G_m(x_1, \dots x_m) \equiv \langle \phi(x_1) \dots \phi(x_m) \rangle = \frac{1}{W_E[J]} \left. \frac{\delta^n}{\delta J(x_1) \dots \delta J(x_m)} \right|_{J=0} W_E[J] \tag{6}$$

are evaluated as functional derivatives of the generating functional

$$W_E[J] = \mathcal{N} \int \mathcal{D}\phi \exp\left[-S[\phi(x)] + \int d^n x J(x)\phi(x) \right]. \tag{7}$$

(The formal constant $\mathcal{N}$ associated with functional measure $\mathcal{D}\phi$ dropped hereafter.)

The straightforward way to calculate G_m is to factorize the interaction part $V(\phi)$ of generation functional gf in the form

$$W_E[J] = \exp\left[-V\left(\frac{\delta}{\delta J} \right) \right] W_0[J], \tag{8}$$

where

$$W_0[J] = \int \mathcal{D}\phi \exp\left(J\phi - \frac{1}{2}\phi D\phi\right) = \exp\left(-\frac{1}{2}JD^{-1}J\right), \quad D = -\partial^2 + m^2 \quad (9)$$

is the free part of the generating functional. The perturbation expansion is then evaluated in k-space, where

$$\widehat{D}^{-1}(k) = \frac{1}{k^2 + m^2}.$$

The perturbative calculation of the correlation functions in $n > 2$ dimensions suffers ultra-violet divergence starting from one-loop approximation

$$I_1(n) = -\frac{\lambda}{2}\int \frac{d^n k}{(2\pi)^n} \frac{1}{k^2 + m^2}. \quad (10)$$

The ϕ^4 is renormalizable, i.e. the divergences can be eliminated by renormalization of the fields and parameters

$$\phi = Z_\phi \phi_R, \quad \lambda_0 = \lambda m^{2\epsilon} Z_\lambda, \quad m_0^2 = m^2 Z_m, \quad (11)$$

where all divergences are hidden in infinite renormalization constants Z_ϕ, Z_λ, Z_m.

Technically, the elimination of divergences is related to the evaluation of the loop integral 11i in the spherical domain in k-space limited from above $|k| < \Lambda$, with substitution of fixed coupling constant λ_0 to running coupling constant $\lambda = \lambda(\Lambda)$. In the case of ϕ^4 theory in the dimension $n = 4 - \epsilon$, the renormalization, as it was shown by K.Wilson [4], leads to the exact scaling of the coupling constant

$$\lambda(\Lambda) = \lambda_0 \Lambda^\epsilon \quad (12)$$

at the limit of infinite cutoff momentum $\Lambda \to \infty$. Similar type scaling takes place for other types of 4-th power interactions, say for Fermi interaction $G_0(\bar\psi\psi)^2$.

If we believe, that the power-law dependence of coupling constant on the cutoff momentum, really means the dependence of interaction strength on the scale $a = \Lambda^{-1}$, rather than a pure mathematical trick, we should find a way to incorporate this dependence at the level of the basic model, rather than at technical level. Doing so, after reviewing some necessary facts from group representation theory in next section, we will use the decomposition (often referred to as *wavelet transform*) with respect to the affine group for this purpose.

3 Partition of the unity

From the group theory point of view, the reformulation of the theory from the coordinate representation $\phi(x)$ to the momentum representation $\hat\phi(k)$ by means of Fourier transform ft, is only a particular case of decomposition of a function with respect to representation of a Lie group G. $G : x' = x + b$ for the case of Fourier transform, but other groups may be used as well, depending on the physics of a particular problem.

Let us remind briefly how the decomposition with respect to the given representation of a Lie group is performed [5, 6]. Let $\mathcal{H}$ be a Hilbert space. Let $U(g)$ be a square-integrable representation of a locally-compact Lie group G acting transitively on $\mathcal{H}$, $\forall \phi \in \mathcal{H}, g \in G : U(g)\phi \in \mathcal{H}$. Let there exist such a vector $|\psi\rangle \in \mathcal{H}$, that satisfies ***the admissibility condition***:

$$C_\psi = \|\psi\|^{-2} \int_{g\in G} |\langle\psi|U(g)\psi\rangle|^2 d\mu(g) < \infty, \tag{13}$$

where $d\mu(g)$ is the left-invariant Haar measure on G.

Then, any vector $|\phi\rangle$ of a Hilbert space $\mathcal{H}$ can be represented in the form:

$$|\phi\rangle = C_\psi^{-1} \int_G |U(g)\psi\rangle d\mu(g) \langle\psi U^*(g)|\phi\rangle, \tag{14}$$

The equation pu is also known as ***the partition of the unity*** with respect to a Lie group G and is often written in the form

$$\hat{1} = C_\psi^{-1} \int_G |U(g)\psi\rangle d\mu(g) \langle\psi U^*(g)|.$$

The basic vector ψ used to construct decomposition pu is often called ***fiducial vector***, or ***basic wavelet***.

3.1 Translation group: Fourier transform

The most familiar case of the unity partition is the decomposition with respect to the momentum eigenstates

$$|\phi\rangle = \int |k\rangle dk \langle k|\phi\rangle. \tag{15}$$

In the later case G is the group of translations and the Haar measure is simply dk. There is no need in explicit notation of any fiducial vector ψ there, because the group of translations is Abelian and the representation $U(k)$ is just a mapping between the vectors $k \in R^n$ and the eigenvectors of the momentum operator. However, for the case of other Lie groups we have to put ψ explicitly. Of course, the final physical results of the theory should be independent on fiducial vector ψ.

3.2 Affine group: Wavelet transform

For the case of affine group ag, $x, b \in R^n$, with the SO_n rotations dropped for simplicity, the left-invariant Haar measure is $d\mu(a,b) = a^{-n-1}dadb$, the representation induced by a basic wavelet ($\psi(x) \in L^2(R^n)$ for definiteness) is $U(g) = a^{-n/2}\psi((x-b)/a))$. So,

$$\begin{array}{rcl} \phi_a(b) & = & \int a^{-n/2}\bar{\psi}\left(\frac{x-b}{a}\right)\phi(x)d^n x, \\ \phi(x) & = & C_\psi^{-1}\int a^{-n/2}\psi\left(\frac{x-b}{a}\right)\phi_a(b)\frac{dadb}{a^{n+1}}, \end{array} \tag{16}$$

where

$$C_\psi = \int \frac{|\hat{\psi}(k)|^2}{|k|} d^n k. \tag{17}$$

See e.g. [7] for detailed explanation.

4 ϕ^4 model on the affine group

Let us turn to the fourth power interaction model with the (Euclidean) action functional

$$
\begin{aligned}
S[\phi] &= \tfrac{1}{2}\int \phi(x_1)D(x_1,x_2)\phi(x_2)dx_1dx_2 \\
&+ \tfrac{1}{4!}\int V(x_1,x_2,x_3,x_4)\phi(x_1)\phi(x_2)\phi(x_3)\phi(x_4)dx_1dx_2dx_3dx_4
\end{aligned} \quad (18)
$$

Using the notation

$$U(g)|\psi\rangle \equiv |g,\psi\rangle, \quad \langle\phi|g,\psi\rangle \equiv \phi(g), \quad \langle g_1,\psi|D|g_2,\psi\rangle \equiv D(g_1,g_2)$$

we can rewrite the generating functional gf for the field theory with action Sphi in the form

$$
\begin{aligned}
Z_G[J] &= \int \mathcal{D}\phi(g)\exp\Bigl(-\tfrac{1}{2}\int_G \phi(g_1)D(g_1,g_2)\phi(g_2)d\mu(g_1)d\mu(g_2) \\
&- \tfrac{\lambda_0}{4!}\int_G \tilde V(g_1,g_2,g_3,g_4)\phi(g_1)\phi(g_2)\phi(g_3)\phi(g_4)d\mu(g_1)d\mu(g_2)d\mu(g_3)d\mu(g_4) \\
&+ \int_G J(g)\phi(g)d\mu(g)\Bigr),
\end{aligned} \quad (19)
$$

where $\tilde V(g_1,g_2,g_3,g_4)$ is the result of the application of the transform

$$\tilde\phi(g) := \int \overline{U(g)\psi(x)}\phi(x)dx$$

to $V(x_1,x_2,x_3,x_4)$ in all arguments x_1,x_2,x_3,x_4.

Let us turn to the particular case of the affine group ag. The restriction imposed by the admissibility condition adc on the fiducial vector ψ (the basic wavelet) is rather loose: Only the finiteness of the integral C_ψ given by adcf is required. This practically implies only that $\int\psi(x)dx = 0$ and that $\psi(x)$ has compact support. For this reason the wavelet transform wtl2 can be considered as a microscopic slice of the function $\phi(x)$ taken at a position b and resolution a with "aperture" ψ. Each particular aperture $\psi(x)$ of course has its own view, but the physical observable should be independent on it. In practical applications of WT very often either of the derivatives of the Gaussian $\psi_n(x) = (-1)^n d^n/dx^n e^{-x^2/2}$ is used, but for the purpose of the present paper only the admissibility condition is important but not the shape of $\psi(x)$.

So, for the case of decomposition of scalar free field in R^n with respect to affine group, the inverse free field propagator matrix element is

$$
\begin{aligned}
\langle a_1,b_1;\psi|D|a_2,b_2;\psi\rangle &= \int d^nx(a_1a_2)^{-\frac{n}{2}}\bar\psi\left(\frac{x-b_1}{a_1}\right)D\psi\left(\frac{x-b_2}{a_2}\right) \\
&= \int \frac{d^nk}{(2\pi)^n}e^{ik(b_1-b_2)}(a_1a_2)^{\frac{n}{2}}\overline{\hat\psi(a_1k)}(k^2+m^2)\hat\psi(a_2k) \\
&\equiv \int \frac{d^nk}{(2\pi)^n}e^{ik(b_1-b_2)}D(a_1,a_2,k).
\end{aligned}
$$

Assuming the homogeneity of the free field in space coordinate, i.e. that matrix elements depend only on the differences $(b_1 - b_2)$ of the positions, but not the positions themselves, we can use (a, k) representation:

$$\begin{aligned} D(a_1,a_2,k) &= a_1^{n/2}\overline{\hat{\psi}(a_1k)}(k^2+m^2)a_2^{n/2}\hat{\psi}(a_2k) \\ D^{-1}(a_1,a_2,k) &= a_1^{n/2}\overline{\hat{\psi}(a_1k)}\left(\frac{1}{k^2+m^2}\right)a_2^{n/2}\hat{\psi}(a_2k) \\ d\mu(a,k) &= \frac{d^nk}{(2\pi)^n}\frac{da}{a^{n+1}}. \end{aligned} \tag{20}$$

So, we have the same diagram technique as usual, but with extra "wavelet" term $a^{n/2}\hat{\psi}(ak)$ term on each line and the integration over $d\mu(a, k)$ instead of dk.

Concerning the Lorenz covariance of the resulting theory (i.e. invariance under rotations, since Euclidean version of the theory is considered), the introduction of the new scale variable a, practically means that instead one scalar field $\phi(x)$, we have to deal with a collection of fields labeled by the resolution parameter $\{\phi_a(x)\}_a$. For each of them the invariance under rotations and translations holds of course. The things are so simple only if we assume quantization = functional integration in the space of numeric-valued functions only, without paying any special attention to possible commutation relations $[\phi_{a_1}(x), \phi_{a_2}(y)]$. What happens for the case of operator-valued functions is not clear enough [8].

Now, turning back to the coordinate representation wtl2, where a is the resolution ("window width") and recalling the power law dependence ws obtained by Wilson expansion, we can define the ϕ^4 model on affine group, with the coupling constant dependent on scale. The simplest case of fourth power interaction of this type is

$$V_{int} = \int \frac{\lambda(a)}{4!}\phi_a^4(b)d\mu(a,b), \quad \lambda(a) \sim a^\nu. \tag{21}$$

The one-loop order contribution to the Green function G_2 in the theory with interacion vint can be easily evaluated (for isotropic wavelet $\hat{\psi}(\mathbf{k}) = \hat{\psi}(\mathbf{k})$, otherwise the constants will be different) by integration over a scalar variable $z = ak$:

$$\int \frac{a^\nu a^n|\hat{\psi}(ak)|^2}{k^2+m^2}\frac{d^nk}{(2\pi)^n}\frac{da}{a^{n+1}} = C_\psi^{(\nu)}\int \frac{d^nk}{(2\pi)^n}\frac{k^{-\nu}}{k^2+m^2}, \tag{22}$$

where

$$C_\psi^{(\nu)} = \int |\hat{\psi}(\mathbf{k})|^2\mathbf{k}^{\nu-1}\mathbf{dk}.$$

Therefore, there are no ultra-violet divergences for $\nu > n-2$.

In the next orders of perturbation expansion each vertex will contribute $k^{-\nu}$ to the formal divergence degree of each diagram. This is quite natural from dimensional consideration, for a - is a "scale" (window width), and k is "inverse scale". So far, for the ϕ^N theory in n dimensions a diagram with E external lines and V vertexes has a formal divergence degree

$$D = n + E\left(1-\frac{n}{2}\right) + V\left(\frac{n}{2}(N-2) - N - \nu\right). \tag{23}$$

5 Conclusion

Of course, the power law behavior of the coupling constant $\lambda(a) = a^\nu$ is not quite realistic. It seems more natural if λ vanish outside a limited domain of scales. The physical meaning of considering a field theory on the affine group seems more important. Doing so we acquire two parameters: the *coordinate* b and the *resolution* a. The former is present in any field theory, but the later is not. Such model can be considered as a continuous counterpart of lattice theory, but now we consider the grid size, or scale, as a physical parameter of interaction and there is no need to get rid of it at the end of calculations.

***

The author is grateful to the Department of Mathematics, University of Alberta, where this research was partially done, for financial support and stimulating interest and also to Dr. T.Gannon and Prof. V.B.Priezzhev for useful discussions.

References

[1] J.R.Klauder and R.F.Streater, A wavelet transform for the Poincare group, J.Math.Phys., 32:1609–1611, 1991.

[2] P. Ramond, "Field Theory: A modern Primer", Benjamin/Cummings Publishing Company,Inc., Massachussets, 1981.

[3] N.N. Bogoliubov and D.V. Shirkov, "Introduction to the theory of quantized fields", John Wiley, New York, 1980.

[4] K.G. Wilson, Quantum field-theory models in less than 4 dimensions, Physical Review D, 7(10):2911–2927, 1973.

[5] A.L. Carey, Square-integrable representations of non-unimodular groups, Bull. Austr. Math. Soc., 15:1–12, 1976.

[6] M. Duflo and C.C. Moore, On regular representations of nonunimodular locally compact group, J. Func. Anal., 21:209–243, 1976.

[7] C.K.Chui, "An introduction to wavelets", Academic Press, Inc., San Diego, 1992.

[8] P. Federbush, A new formulation and regularization of Gauge theories using a non-linear wavelet expansion, Progr. Theor. Phys., 94:1135–1146,1995.

From Quantum Action to Quantum Chaos

H. Jirari[a], H. Kröger[a]*, G. Melkonyan[a], X.Q. Luo[b]
and K.J.M. Moriarty[c]
[a] Département de Physique, Université Laval,
Québec, Québec G1K 7P4, Canada
Email: hkroger@phy.ulaval.ca
[b] Department of Physics, Zhongshan University, Guangzhou 510275, China
Email: stslxq@zsu.edu.cn
[c] Department of Mathematics, Statistics and Computer Science,
Dalhousie University, Halifax, Nova Scotia B3H 3J5, Canada
Email: moriarty@cs.dal.ca

1 Introduction

This article is about the relation between classical mechanics and quantum mechanics. The question is asked: Can quantum mechanics be formulated in such a way that it looks like some sort of classical mechanics? Why do we ask such question in the first place? The answer is interesting from the point of view of interpretation of quantum mechanics. L. de Broglie [1] has pointed out that quantum mechanics has two faces: The particle interpretation and the wave interpretation. Maybe there is a third interpretation, where quantum mechanics has the face of classical mechanics. The answer is interesting also for the purpose of a proper definition, quantitative analysis and understanding of phenomena occuring in quantum physics, the definition of which comes from classical physics. Examples are quantum chaos and quantum instantons. The affirmative answer to the above question has been proposed recently in Refs.[2, 3], stating that quantum transition amplitudes can be expressed in terms of some action, called the quantum action, which has the form of the classical action but has modified parameters.

1.1 Bridges between classical mechanics and quantum mechanics

A general method to build bridges from classical to quantum physics is the path integral. By "bridges" we mean a relation, e.g., involving the quantum

*invited talk of H. Kröger

Frontiers of Fundamental Physics 4, Edited by Sidharth and Altaisky,
Kluwer Academic/Plenum Publishers, New York, 2001

transition amplitude and the classical action. In particular, starting from the path integral, the following "bridges" have been suggested:

(i) *Sum over classical paths.* Let us consider the Q.M. transition amplitude from x_{in}, t_{in} to x_{fi}, t_{fi} given by the path integral. In certain cases this path integral can be expressed as a sum over classical paths only,

$$G(x_{fi}, t_{fi}; x_{in}, t_{in}) = \sum_{\{x_{cl}\}} Z \exp\left[\frac{i}{\hbar}\, S[x_{cl}]|_{x_{in},t_{in}}^{x_{fi},t_{fi}}\right], \tag{1}$$

where $S[x_{cl}]$ is the classical action evaluated along the classical trajectory from x_{in}, t_{in} to x_{fi}, t_{fi}. This is true, e.g. for the harmonic oscillator. Unfortunately, such relation holds only in a few exceptional cases [4].

(ii) *Gutzwiller's trace formula.* Gutzwiller [5] has established a relation between the density of states of the quantum system and a sum over classical periodic orbits (periodic orbit quantisation). The trace formula reads (see Ref.[6])

$$\rho(E) = \rho_0(E) - \frac{1}{2\pi\hbar}\mathrm{Im}\sum_p T_p \sum_{n=1}^{\infty} \frac{\exp[in(\Phi_p(E)/\hbar - \mu_p\tau/2)]}{i\sin[n\lambda_p(E)/2]}, \tag{2}$$

where $\rho(E) = Tr[\delta(E-H)]$ denotes the density of states, and $\rho_0(E)$ is the average level density. The sum runs over all primitive periodic orbits p, the index n denotes repeated traversal of primitive periodic orbits and T_p is the traversal time of such an orbit. $\Phi_p(E)$ is the action of the periodic orbit p at energy E and $\lambda_p(E)$ denotes a Lyapunov exponent. Here μ_p is a constant characteristic for the orbit p. The trace formula has been applied successfully in the semi-classical regime (e.g. highly excited states of atom). Wintgen [7] applied it to the diamagnetic hydrogen system and was able to extract periodic orbit information from experimental level densities.

(iii) The *effective action* has been introduced in quantum field theory in such a way that it gives an expectation value $<\phi> = \phi_{class}$ which corresponds to the classical trajectory and which minimizes the potential energy (effective potential). Thus one can obtain the ground state energy of the quantum system from its effective potential. The effective action Γ [8, 9] is defined by

$$\begin{aligned} Z[J] &= e^{-iW[J]} \\ \frac{\partial}{\partial J(x)}W[J] &= -<0|\phi(x)|0>_J \\ \phi_{cl}(x) &= <0|\phi(x)|0>_J \\ \Gamma[\phi_{cl}] &= -W[J] - \int d^4y\ J(y)\phi_{cl}(y). \end{aligned} \tag{3}$$

An effective action has been also considered at finite temperature [10]. Because the effective action has a mathematical structure similar to the classical action, and the quantum effects are taken into account by parameters different from their classical counter parts, the effective action looks like the ideal way to bridge the gap from quantum to classical physics and eventually solve the quantum chaos and quantum instanton problem. However, there is a catch. The effective potential and the effective action in quantum mechanics has been computed using perturbation theory by Cametti et al. [11]. Consider the Lagrangian

$$\begin{aligned} L(q,\dot{q},t) &= \frac{m}{2}\dot{q}^2 - V(q) \\ V(q) &= \frac{m}{2}\omega^2 q^2 + U(q), \end{aligned} \tag{4}$$

and $U(q)$ is, say, a quartic potential $U(q) \sim q^4$. Then the effective action is obtained in doing a loop ($\hbar$) expansion

$$\begin{aligned} \Gamma[q] &= \int dt \Big(-V^{eff}(q(t)) + \frac{Z(q(t))}{2}\dot{q}^2(t) \\ &+ A(q(t))\dot{q}^4(t) + B(q(t))(d^2q/dt^2)^2(t) + \cdots \Big) \\ V^{eff} &= \frac{1}{2}m\omega^2 q^2 + U(q) + \hbar V_1^{eff}(q) + O(\hbar^2) \\ Z(q) &= m + \hbar Z_1(q) + O(\hbar^2) \\ A(q) &= \hbar A_1(q) + O(\hbar^2) \\ B(q) &= \hbar B_1(q) + O(\hbar^2). \end{aligned} \tag{5}$$

There are higher loop corrections to the effective potential V^{eff} as well as to the mass renormalisation Z. The most important property is the occurrence of higher time derivative terms. Actually, there is an infinite series of increasing order. Here comes the problem. When we want to interpret Γ as effective action, the higher time derivatives require more intial/boundary conditions than the classical action. This is a catastrophy. In the following we will present an alternative way to construct an action taking into acount quantum corrections.

2 Quantum Action

We want to construct a renormalized or quantum action from transition matrix elements, which involve the time evolution. In quantum physics the transition amplitude from x_{in}, t_{in} to x_{fi}, t_{fi} is given by

$$G(x_{fi}, t_{fi}; x_{in}, t_{in}) = \int [dx] \exp[\frac{i}{\hbar} S[x]] \Big|_{x_{in}, t_{in}}^{x_{fi}, t_{fi}}, \tag{6}$$

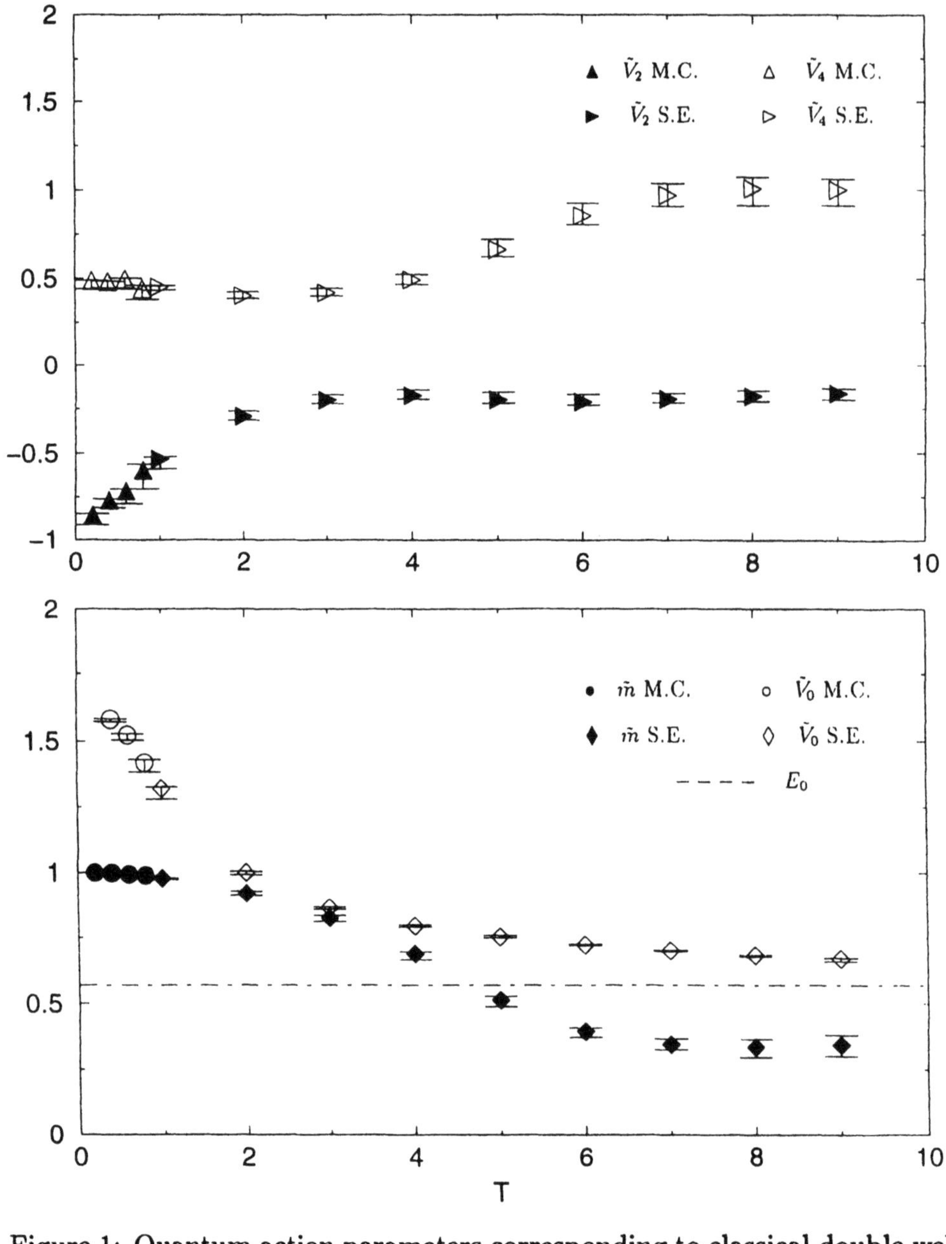

Figure 1: Quantum action parameters corresponding to classical double well potential $V(x) = \frac{1}{2}(x^2 - 1)^2$.

where S denotes the classical action. In Ref.[2, 3] we have proposed the existence of a quantum action which satisfies the following properties:
Conjecture: For a given classical action $S = \int dt \frac{m}{2}\dot{x}^2 - V(x)$ there is a quantum action $\tilde{S} = \int dt \frac{\tilde{m}}{2}\dot{x}^2 - \tilde{V}(x)$, which allows to express the transition amplitude by

$$G(x_{fi}, t_{fi}; x_{in}, t_{in}) = \tilde{Z} \exp[\frac{i}{\hbar} \tilde{S}[\tilde{x}_{cl}]\Big|_{x_{in},t_{in}}^{x_{fi},t_{fi}}]. \tag{7}$$

Here $\tilde{x}_{cl}$ denotes the classical path, such that the action $\tilde{S}(\tilde{x}_{cl})$ is minimal (we exclude the occurrence of conjugate points or caustics). $\tilde{Z}$ denotes the normalisation factor corresponding to $\tilde{S}$. Eq.(7) is valid with the *same* action $\tilde{S}$ for all sets of boundary positions x_{fi}, x_{in} for a given time interval $T = t_{fi} - t_{in}$. The parameters of the quantum action depend on the time T. The quantum action converges to a non-trivial limit when $T \to \infty$. Any dependence on x_{fi}, x_{in} enters via the trajectory $\tilde{x}_{cl}$. $\tilde{Z}$ depends on the action parameters and T, but not on x_{fi}, x_{in}.

One may ask: What is the difference between effective and quantum action? Conceptually, effective action and quantum action look quite similar. However, its technical definition is different and also its physical content. The effective action requires $< \phi >= \phi_{cl}$, while the quantum action does not. The effective action corresponds to infinite time and allows to obtain the ground state energy, but the quantum action is defined for arbitrary finite time T. In Euclidean formulation, the inverse time corresponds to temperature. Thus the quantum action allows to describe quantum physics at finite temperature including excited states (see below). However, the effective action can be defined also at finite temperature [10]. The effective action can be computed analytically by perturbation theory (loop expansion). However, this series is not convergent. Practically, it can be used only for some small number of loops and small values of the perturbation parameter. The quantum action can be computed non-perturbatively for all values of the coupling parameter. The effective action has the defect of generating higher order time derivatives. The quantum action is postulated to be free of higher time derivative terms. To construct the quantum action being sensitive to excited states, one needs transition matrix elements beyond the vacuum sector. We have chosen to use position states in Q.M. In Q.F.T. this corresponds to Bargman states.

2.1 Construction of quantum action

Suppose the classical action is given by

$$S = \int_0^T dt \frac{m}{2} \dot{x}^2 - v_4 x^4(t). \tag{8}$$

Then we make an ansatz for the quantum action

$$\tilde{S} = \int_0^T dt \frac{\tilde{m}}{2} \dot{x}^2 - \left\{ \tilde{v}_0 + \tilde{v}_1 x(t) + \cdots + \tilde{v}_N x^N(t) \right\}. \tag{9}$$

Then $\tilde{m}, \tilde{v}_0, \ldots, \tilde{v}_N$ are the renormalized parameters which take into account the quantum corrections. Their values are determined by making a global best fit to a number of transition amplitudes $G(x_j, T; x_i, 0)$ (which satisfies Eq.7), where x_i, x_j haven been taken from a set of points $\{x_1, \cdots, x_J\}$ and those points have been chosen to cover some interval $[-a, +a]$. More details are given in Refs.[2, 3]. As an example, the parameters of the quantum action corresponding to the double well potential (action $S = \int_0^T dt \frac{1}{2} m \dot{x}^2 - \{v_0 + v_2 x^2 + v_4 x^4\}$, $v_0 = \frac{1}{2}$, $v_2 = -1$, $v_4 = \frac{1}{2}$) as function of T is shown in Fig.[1]. One observes that the parameters of the quantum action vary with the transition time T. For small T (limit $T \to 0$) the parameters of the quantum action are consistent with those of the classical action. For sufficiently large time T, the parameters of the quantum action tend to converge asymptotically.

2.2 Quantum action at finite temperature

First we make a Wick rotation to imaginary time. The purpose is, first to make the path integral well defined (Wiener measure) allowing to apply Monte Carlo methods for its numerical computation. Secondly, the instanton is defined in imaginary time. One effect of this transformation is that it changes a relative sign between the kinetic term and the potential term of the action. Thus in the following we work with imaginary time (Euclidean) actions and Green's functions. Let us see how the quantum action is related to finite temperature physics. According to the laws of quantum mechanics and thermodymical equilibrium, the expectation value of some observable O, like e.g. average energy is given by

$$\begin{aligned} < O > &= \frac{Tr\,[O\ \exp[-\beta H]]}{Tr\,[\exp[-\beta H]]} \\ &= \frac{\int_{-\infty}^{+\infty} dx \int_{-\infty}^{+\infty} dy < x|O|y >< y| \exp[-\beta H] |x >}{\int_{-\infty}^{+\infty} dx < x| \exp[-\beta H] |x >}, \end{aligned} \tag{10}$$

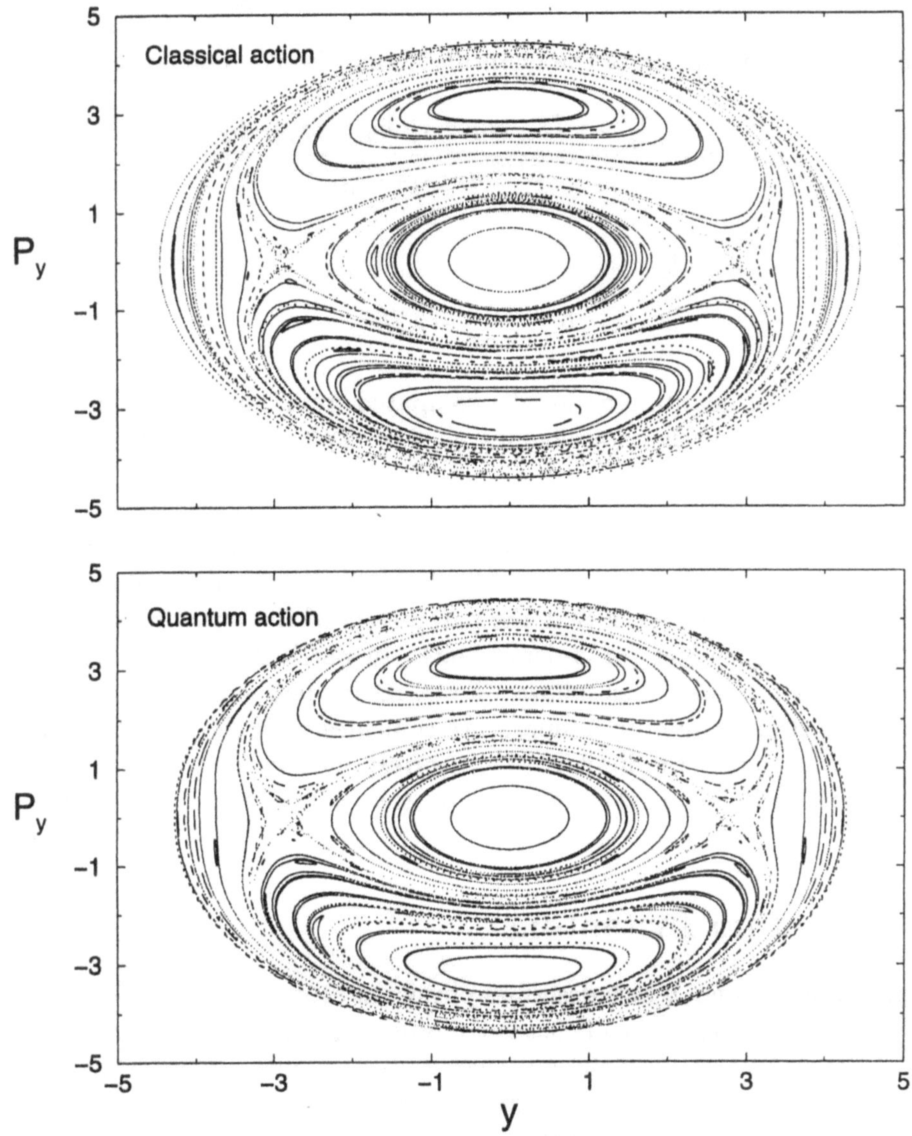

Figure 2: Classical and quantum Poincaré sections corresponding to classical potential $V(x,y) = \frac{1}{2}(x^2 + y^2) + 0.05x^2y^2$. Energy $E = 10$.

where β is related to the temperature τ by $\beta = 1/(k_B\tau)$. On the other hand the (Euclidean) transition amplitude is given by

$$G(x_{fi}, T; x_{in}, 0) = < x_{fi}| \exp[-HT/\hbar]|x_{in} > \tag{11}$$

Thus from the definition of the quantum action, Eq.(7), one obtains

$$< O >= \frac{\int_{-\infty}^{+\infty} dx \int_{-\infty}^{+\infty} dy < x|O|y > \exp[-\tilde{S}_\beta|_{x,0}^{y,\beta}]}{\int_{-\infty}^{+\infty} dx \exp[-\tilde{S}_\beta|_{x,0}^{x,\beta}]}, \tag{12}$$

if we identify

$$\beta = \frac{1}{k_B\tau} = T/\hbar. \tag{13}$$

As a result, the quantum action $\tilde{S}_\beta$ computed from transition time T, describes equilibrium thermodynamics at $\beta = T/\hbar$, i.e. temperature $\tau = 1/(k_B\beta)$.

In the case of the double well potential we have found that parameters of the quantum action vary as function of T (Fig.[1]). Translating this behavior into temperature, it means that the parameters of the quantum action are temperature dependent (or β-dependent). In particular, we can interpret the behavior for small T as follows. $T = 0$ means temperature $\tau = \infty$. The quantum action at infinite temperature coincides with the classical action. On the other hand, the limit $T \to \infty$ corresponds to temperature $\tau \to 0$. The quantum behavior is dominated by the ground state (Feynman-Kac formula).

3 Quantum Instantons

Quantum mechanics, which describes physics at atomic length scales can not be understood by the laws of classical physics valid at macroscopic length scales. Examples are: Heisenberg's uncertainty principle, quantum tunneling, Schrödinger's cat paradox, entangled states, Einstein-Rosen-Podolski paradox, quantum cryptology, quantum computing etc. On the other hand, in modern physics there are notions which have proven to be quite useful and which have their origin in classical physics. For example consider instantons. Instantons play a role in quantum chromodynamics (QCD), the standard model of strong interactions. They may be important for the mechanism of confinement of quarks. Presumably they play an important role in nuclear matter at high temperature and density, where a phase transition from the hadronic phase to the quark-gluon plasma has been predicted. Even a richer

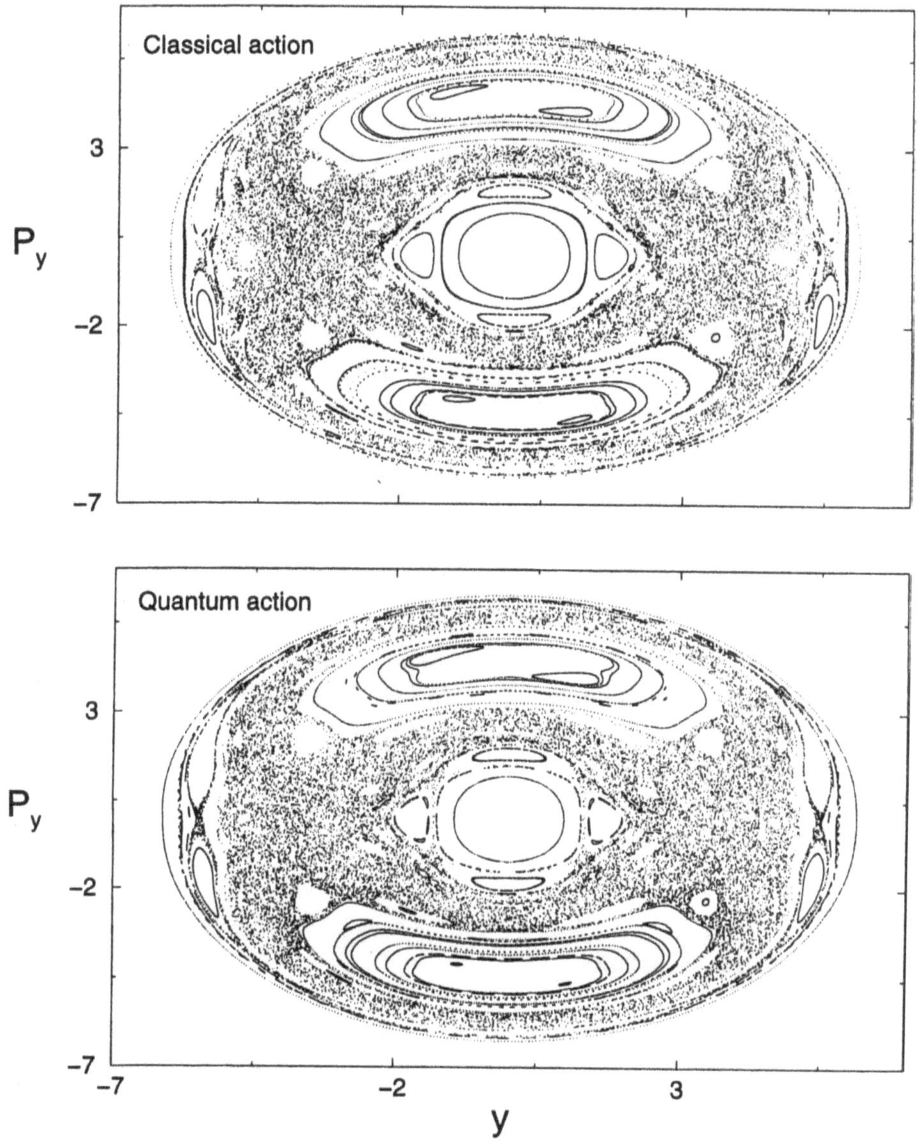

Figure 3: Same as Fig.[2]. Energy $E = 20$.

phase structure may exist [12]. Furthermore, in the inflationary scenario of the early universe, instantons are important. For a review see Ref.[13]. During inflation, quantum fluctuations of the primordial field expand exponentially and eventually end up as a classical field. The fluctuations are of the size of the horizon [14]. The classical fluctuations eventually lead to galaxy formation [15].

In quantum physics, an instanton solution is conventionally defined as the saddle point approximation of the (Euclidean) path integral. However, there is a problem with the proper definition of instantons in quantum physics: Let us consider a 1-D system in quantum mechanics with a particle of mass m moving in a potential $V(x) = A(x^2 - a^2)^2$. This potential has two minima at $x = \pm a$. The instanton $x_{inst}(t)$ is the solution of the classical equation of motion in imaginary time, with boundary conditions such that the particle starts at $x(t = -\infty) = -a$, $\dot{x}(t = -\infty) = 0$ and arrives at $x(t = +\infty) = +a$, $\dot{x}(t = +\infty) = 0$. The problem again is that quantum mechanics does not allow to specify both, position and momentum with zero uncertainty.

In Ref.[3] we have suggested to define a quantum instanton solution via the quantum action. This means to compute non-perturbatively the quantum action $\tilde{S}$ (in imaginary time) and analyze if the corresponding quantum potential $\tilde{V}$ has multiple degenerate minima (degenerate vacua). Then the quantum instanton is defined as the classical solution $\tilde{x}_{class}$ between those minima (there is no problem with boundary conditions). Such quantum instanton solutions have been computed in quantum mechanics for the 1-D double well potential in Ref.[3]. The quantum instantons were found to be "softer" than the classical instantons (potential minima were closer and the potential barrier was lower).

4 Quantum Chaos

Classical deterministic chaos has been observed in a huge number of phenomena in macroscopic i.e. classical physics. But chaotic phenomena were also found in systems ruled by quantum mechanics. For example, the hydrogen atom in a strong magnetic field shows strong irregularities in its spectrum [16]. Irregular patterns have been found in the wave functions of the quantum mechanical model of the stadium billard [17]. Billard like boundary conditions have been realized experimentally in mesoscopic quantum systems, like quantum dots and quantum corrals, formed by atoms in semi-conductors [18].

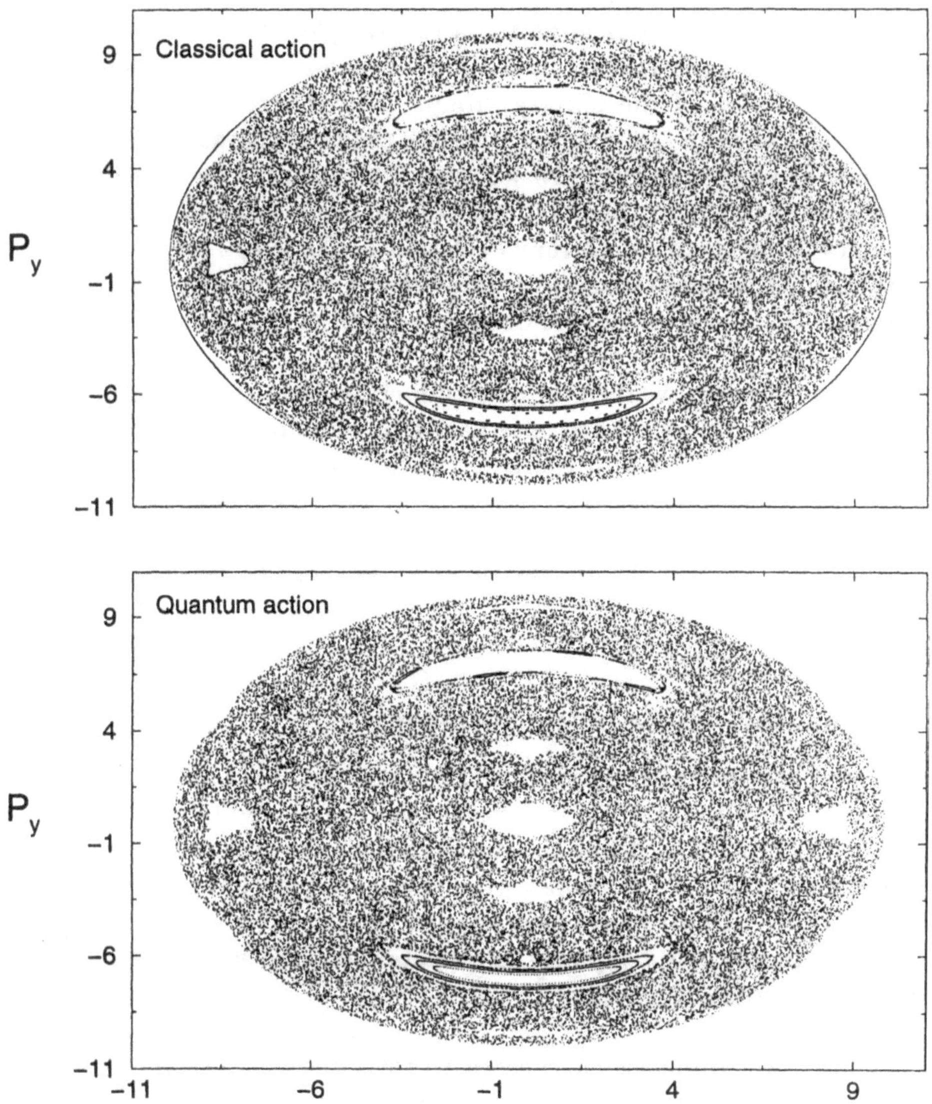

Figure 4: Same as Fig.[2]. Energy $E = 50$.

So what is the problem with chaos in quantum physics? It has to do with its proper definition. The underlying reason is due to the dynamical group of time evolution. In classical mechanics time evolution of a system can be viewed as an infinite sequence of infinitesimal canonical transformations. The corresponding dynamical group is the symplectic group. In quantum mechanics, a system governed by a time independent Hamiltonian, follows the time evolution of the unitary group. This difference has simple but drastic consequences: In classical physics, chaos is characterized, e.g. by Lyapunov exponents or Poincaré sections. This is based on identifying trajectories in phase space (position and conjugate momentum). In quantum mechanics, Heisenberg's uncertainty relation $\Delta x \Delta p \geq \hbar/2$ does not allow to specify a point in phase space with zero error! Consequently, the apparatus of classical chaos theory can not be simply taken over to quantum physics.

Due to this problem, workers in quantum chaos have tried to characterize such systems in different ways, alternative to those of classical chaos. One successful route has been to characterize the spectral density of quantum system with chaotic classical counterpart by Poisson versus Wigner distributions. There is a conjecture by Bohigas et al. [19], which says that the signature of a classical chaotic system is a the spectral density following a Wigner distribution.

4.1 Quantum chaos in 2 dimensions

As the problem with a proper definition of quantum chaos has the same root as the problem with quantum instantons, we suggest also to apply the same strategy of solution, i.e. define quantum chaos via the quantum action. Then the quantum action $\tilde{S}$ incorporates the effects of quantum physics, but has mathematically the structure of a classical action. The apparatus of classical chaos theory, like Lyapunov exponents, Poincaré sections etc. can be applied to the quantum action $\tilde{S}$.

As is well known 1-dimensional conservative systems with a time-independent Hamiltonian are integrable and do not produce classical chaos. An interesting candidate to consider is the K-system, corresponding to the potential $V = x^2y^2$. This decribes a 2-D Hamiltonian system, being almost globally chaotic, having small islands of stability [20]. However, from the numerical point of view more convient, but also showing classical chaos, is the following related system, investigated by Pullem and Edmonds [21], It is defined by the classical action

$$S = \int_0^T dt\ \frac{1}{2}m(\dot{x}^2 + \dot{y}^2) - (v_{2+2}(x^2 + y^2) + v_{22}x^2y^2). \tag{14}$$

As parameters of the classical action we use $m = 1$, $v_{2+2} = 0.5$, $v_{22} = 0.05$ and the convention $\hbar = k_B = 1$. The Poincaré sections corresponding to energies $E = 10, 20, 50$ are shown in the upper part of Figs.[2-4].

For the corresponding quantum action, we make the following ansatz, which is compatible with time-reversal symmetry, parity conservation and symmetry under exchange $x \leftrightarrow y$,

$$\begin{aligned} \tilde{S} &= \int_0^T dt\ \frac{1}{2}\tilde{m}_{2+2}(\dot{x}^2 + \dot{y}^2) + \frac{1}{2}\tilde{m}_{11}\dot{x}\dot{y} \\ &- \ \{\tilde{v}_0 + \tilde{v}_{11}xy + \tilde{v}_{2+2}(x^2 + y^2) + \tilde{v}_{22}x^2y^2 + \tilde{v}_{1+3}(xy^3 + x^3y) \\ &+ \ \tilde{v}_{4+4}(x^4 + y^4)\}. \end{aligned} \tag{15}$$

We have determined numerically the parameters of the quantum action for transition time $T = 0.5$, corresponding to temperature $\tau = 2$, and find

$$\begin{aligned} \tilde{m}_{2+2} &= 0.9998(2) \\ \tilde{m}_{11} &= 0.0000(3) \\ \tilde{v}_0 &= 1.1875(32) \\ \tilde{v}_{11} &= 0.0105(31) \\ \tilde{v}_{2+2} &= 0.5098(63) \\ \tilde{v}_{22} &= 0.0523(15) \\ \tilde{v}_{1+3} &= 0.0016(12) \\ \tilde{v}_{4+4} &= 0.0017(30). \end{aligned} \tag{16}$$

The data are consistent with vanishing parameters $\tilde{m}_{11}$, $\tilde{v}_{1+3}$ and $\tilde{v}_{4+4}$. The quantum action slightly modifies the parameters $\tilde{v}_{2+2}$ and the parameter $\tilde{v}_{22}$. We computed the Poincaré sections for the quantum action at temperature $\tau = 2$, corresponding to energies $E = 10, 20, 50$. They are shown in the lower part of Figs.[2-4]. One observes that the quantum system also displays chaos, and the Poincaré sections are slightly different from those of the classical action. One should note that the classical action at $T = 0$ is equivalent to a quantum action at temperature $\tau = \infty$.

5 Discussion

We have discussed the use of the quantum action, which can be considered as a renormalized classical action at finite temperature. We found that the quantum action solves the problem of proper definitions of quantum instantons and quantum chaos. As an example, we have considered harmonic oscillators with a weak anharmonic coupling ($V_{coupl} \sim x^2y^2$) and computed the quantum action at temperature $\tau = 2$. We compared Poincaré sections at temperature $\tau = \infty$ and $\tau = 2$ and found that both display chaos.

Acknowledgements
H.K. and K.M. are grateful for support by NSERC Canada. X.Q.L. has been supported by NSF for Distinguished Young Scientists of China, by Guangdong Provincial NSF and by the Ministry of Education of China. H.K. is grateful for discussions with L.S. Schulman.

References

[1] L. de Broglie, Nature 112, 1923, 540; Ann. de Physique (Paris) (10) 2, 1925.

[2] H. Jirari, H. Kröger, X.Q. Luo, K.J.M. Moriarty and S.G. Rubin, Phys. Rev. Lett. 86, 2001, 187.

[3] H. Jirari, H. Kröger, X.Q. Luo, K.J.M. Moriarty and S.G. Rubin, quant-ph/9910116, Phys. Lett. A, in press.

[4] L.S. Schulman, "Techniques and Applications of Path Integration", John Wiley&Sons, New York, 1981.

[5] M.C. Gutzwiller, "Chaos in Classical and Quantum Mechanics", Springer, Berlin, 1990, and references therein.

[6] A. Blümel and W.P. Reinhardt, Chaos in Atomic Physics, Cambridge Univ. Press, Cambridge, 1997.

[7] D. Wintgen, Phys. Rev. Lett. 61, 1988, 1803.

[8] G. Jona-Lasinio, Nuov. Cim. 34, 1964, 1790.

[9] S. Coleman and E. Weinberg, Phys. Rev. D7, 1973, 1888.

[10] L. Dolan and R. Jackiw, Phys. Rev. D9, 1974, 3320.

[11] F. Cametti, G. Jona-Lasinio, C. Presilla and F. Toninelli, quant-ph/9910065.

[12] E.V. Shuryak, Nucl. Phys. B(Proc. Suppl.)83-84, 2000, 103.

[13] A.D. Linde, Inflation and Quantum Cosmology, Acad. Press, San Diego, 1990.

[14] A. Starobinsky, Phys. Lett. B91, 1980, 99.

[15] M.Y. Khlopov, Cosmoparticle Physics, World Scientific, Singapore, 1998.

[16] H. Friedrich and D. Wintgen, Phys. Repts. 183, 1989, 37.

[17] S.W. McDonald and A.N. Kaufman, Phys. Rev. Lett. 42, 1979, 1189.

[18] H.J. Stöckmann., Quantum Chaos, Cambridge Univ. Press, Cambridge, 1999.

[19] O. Bohigas, M.J. Giannoni and C. Schmidt, Phys. Rev. Lett. 52, 1984, 1.

[20] P. Dahlquist and G. Russberg, Phys. Rev. Lett. 65, 1990, 2837.

[21] R.A. Pullen and A.R. Edmonds, J. Phys. A: Math. Gen. 14, 1981, L477.

Resolution of the Einstein-Podolsky-Rosen Nonlocality Puzzle

C. S. Unnikrishnan
Gravitation Group, Tata Institute of Fundamental Research,
Homi Bhabha Road, Mumbai (Bombay) - 400 005, India &
Centre for Non-accelerator Particle Physics,
Indian Institute of Astrophysics,
Koramangala, Bangalore - 560 034, India
Email: unni@tifr.res.in, unni@iiap.ernet.in

1 Introduction

1.1 The Einstein-Podolsky-Rosen problem

In 1935, Einstein, Podolsky and Rosen (EPR)[1] considered the question whether the quantum mechanical description of physical reality was complete, and argued that it wasn't. A theory is incomplete if its mathematical elements do not represent all possible physical realities that could exist in the physical system. The crucial and essential assumption in their argument was strict locality, in the spirit of special relativity. According to EPR, locality implies that a measurement on one system has no influence whatsoever on the result of another measurement on a system that is space-like separated from the first - the absence of spooky action-at-a-distance. Also, they had *defined* and included the concept of objective reality in the analysis. If the value of an observable was predictable with certainty without in any way disturbing the system, the observable had a physical reality according to EPR. They considered a quantum mechanical system consisting of two particles with well defined simultaneous values for their *total* momentum and their *relative* positions. The wavefunction

$$\Psi(q_1, q_2) = \int_{-\infty}^{\infty} dp \exp[\frac{i}{\hbar}(q_1 - q_2 + q_0)p] \tag{1}$$

where q_0 is a constant, is simultaneously the eigenfunction of the commuting operators $\hat{P} = \hat{p}_1 + \hat{p}_2$ and $\hat{Q} = \hat{q}_1 - \hat{q}_2$, with eigenvalue $p_0 = 0$, and q_0 respectively. The wavefunction is entangled, and the state of each particle is not characterized by a definite value for either position or momentum. Measuring any one of these variables on one particle enables a prediction with certainty of the value of the same variable on the other particle without a measurement. Then according to the EPR definition of reality, the second particle possesses a definite physical reality for this variable. Since the decision of which variable is to be measured on the first particle can be done after the two particles have propagated into space-like separated regions, and due to the assumption of strict locality by which the first measurement cannot influence the state of the second particle in any way superluminally, it is clear that the particles should possess objective reality for both variables. Therefore the quantum mechanical description of physical reality should be considered incomplete, since quantum mechanics does not allow the simultaneous objective reality of noncommuting observables.

1.2 Bohr's reply

Bohr had tried to counter the 'bolt from the blue' from EPR[2, 3]. A major point of his defence was that the analysis of every quantum measurement has to take into account of the whole experimental arrangement. According to him[2], "the procedure of measurements has an essential influence on the conditions on which the very definition of the physical quantities in question rests. Since these conditions must be considered as an inherent element of any phenomenon to which the term 'physical reality' can be unambiguously applied, the conclusions of EPR would not appear to be justified". In the more detailed paper[3] he said, "Of course...there is no mechanical disturbance of the system under investigation during the last critical stage of the measuring procedure. But even at this stage there is essentially the question of an influence on the very conditions which define the possible types of predictions regarding the future behaviour of the system". It seems that Bohr starts with attacking the notion of physical reality used by EPR, but ends up discarding the notion of locality as well. This, of course, is consistent with the Copenhagen interpretation that treats the whole multiparticle system as one quantum systems, described by a single monistic wavefunction.

1.3 Bell's analysis

Bell's analysis of the EPR problem in the early sixties established the Bell's inequalities obeyed in any local hidden variable theory for the correlations of entangled particles[4]. He used the remarkably insightful example discussed by Bohm employing the singlet pair – two particles entangled in spin, with total spin zero[5]. Quantum mechanical correlations calculated using the entangled wave-function and spin operators violate these inequalities. This means that any local realistic theory that assumes Einstein locality and the EPR kind of objective reality necessarily predicts results for correlation that does not agree with the quantum mechanical predictions. Subsequently, various experiments have established beyond doubt that there cannot be a viable local realistic description of quantum mechanics[6]. These results have been interpreted as evidence for nonlocal influences in quantum measurements involving entangled particles. In the realist view, measurement of an observable on one of the particles in an entangled pair seems to convey the result of this measurement instantaneously to the other particle resulting in the correct nonclassical correlations. In the quantum mechanical terminology, the measurement of an observable on one of the particles collapses the entire wave-function instantaneously and nonlocally. In both views, the second particle ***nonlocally acquires*** a definite value for the observable. The no-signalling theorems in this context prohibit any faster than light signalling using this feature, and therefore signal locality is not violated. But the stronger requirement of Einstein locality is violated. As Schrödinger stated[7], when he discussed the peculiarities of entangled quantum

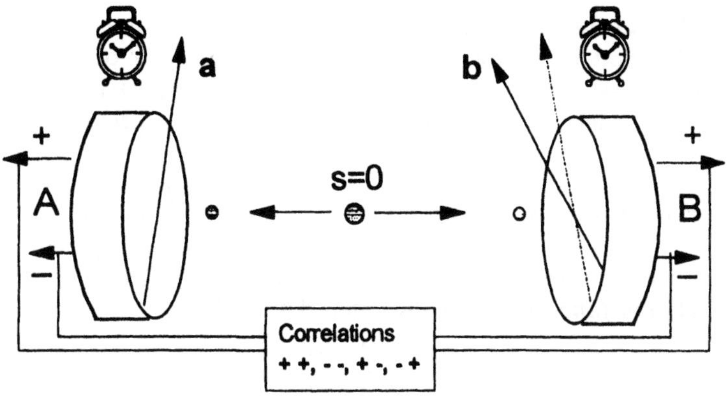

Figure 1: Spin correlation measurements on entangled particles by observers A and B. The outputs of the two analyzers are +1 or -1 and these are correlated in a coincidence unit for each pair of particles. **a** and **b** are the two analyzer directions. The dotted arrow on the right analyzer shows the direction **a** for reference.

systems in 1935, "It is rather discomforting that the theory should allow a system to be steered or piloted into one or the other type of state at the experimenter's mercy in spite of his having no access to it". This discomfort with nonlocality that goes against the spirit of special relativity and against the simplest notion of rationality in physics has been the central issue that required a physical and rational resolution.

In the standard Bohm version of the EPR problem[5], the two-particle state is described by the wave function

$$\Psi_S = \frac{1}{\sqrt{2}}\{|1,-1\rangle - |-1,1\rangle\} \tag{2}$$

where the state $|1,-1\rangle$ is short form for $|1\rangle_1 \, |-1\rangle_2$, and represents an eigenvalue of $+1$ for the first particle and -1 for the second particle *if measured* in *any* particular direction. Ψ_S is inherently nonlocal, describing both particles together, even when they are far apart in space-like separated regions. It is a superposition of two-particle product states and there is no specific state assigned to any of the particles individually. The standard quantum mechanical view is that a measurement on *one* particle changes the whole wave-function, collapsing the states of *both* particles to definite values.

Two observers A and B make measurements on these particles individually at space like separated regions with time stamps such that these results can be correlated later through a classical channel. The general scheme of such an experiment is outlined in Fig. 1.

Bell assumed locality in the sense of absence of instantaneous action at a distance. This is represented by restrictions on the outcomes A and B of measurements at the two locations.

$$A(\mathbf{a}, \mathbf{h}_1) = \pm 1, \quad B(\mathbf{b}, \mathbf{h}_2) = \pm 1 \tag{3}$$

A and B denote the outcomes $+1$ or -1 (written as $+$ or $-$) of measurements A and B, and $\mathbf{a}$ and $\mathbf{b}$ denote the settings of the analyzer or the measurement apparatus for the first particle and the second particle respectively. $\mathbf{h}_1$ and $\mathbf{h}_2$ are hidden variables associated with the outcomes. The Bell correlation function[4] is of the form

$$P(\mathbf{a},\mathbf{b}) = \int d\mathbf{h}\rho(\mathbf{h})A(\mathbf{a},\mathbf{h}_1)B(\mathbf{b},\mathbf{h}_2), \text{where} \int d\mathbf{h}\rho(\mathbf{h}) = 1 \tag{4}$$

This is an average over the product of eigenvalues. $P(\mathbf{a},\mathbf{b})$ is similar to the classical *correlation function of the outcomes* defined by

$$P(\mathbf{a},\mathbf{b}) = \frac{1}{N}\sum(A_i B_i) \tag{5}$$

The essence of Bell's theorem is that the function $P(\mathbf{a},\mathbf{b})$ has distinctly different dependences on the relative angle between the polarizers for a local hidden variable description and for quantum mechanics.

1.4 Where is the phase?

Unfortunately, the Bell correlation function misses out the most important physical aspect of individual systems in the microscopic world - the phase associated with their essential wave property. Wave nature exhibited by microscopic single particle systems is at the heart of all puzzling aspects of quantum phenomena. This means that every reasonable description of microscopic phenomena should have theoretical constructs at the single particle level that can represents the wave nature, and especially the phase. There is no quantity in $P(\mathbf{a},\mathbf{b})$ that reflects the phase information associated with the individual particle. So, the assumptions that went into the formulation of Bell's theorem has not encompassed the essential physical properties. The correct ingredients should have been locality and wave-particle unity[8]. I will show that it is possible to reproduce the correct correlations under the assumption of strict locality, by incorporating the correct physical ingredient - phase coherence at source[9].

In effect, Bell's theorem states that quantum correlations cannot be reproduced employing local quantities that describe purely particle like properties, or deterministic properties for all relevant observables. This is rather obvious since wave-particle unity is an essential element of microscopic behaviour.

The EPR definition of objective reality is strongly motivated by classical particle mechanics. This is reflected in the correlation function used by local realists and in Bell's analysis; It does not preserve the crucial phase information. For any physical system with wave properties, the wave amplitude and phase are the basic quantities. The EPR definition of objective reality was too restrictive. There exists objective reality in the microscopic world, even before measurements, but that is at the level of quantum phases and not at the level of eigenvalues.

There is another related issue, regarding the notion of locality. Many people have been using the separability of probability as a criterion for locality. But separability of probability is not a good criterion for locality. There is even a classical counter example from optics - a case where all physical aspects obey locality, and even the Bell's inequality, but the separability of probability is invalid - the Hanburry Brown-Twiss intensity correlations.

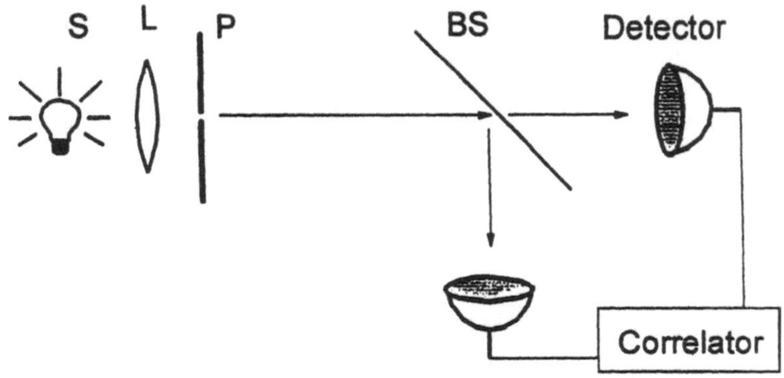

Figure 2: The basic scheme of the Hanburry Brown-Twiss experiment. S: thermal source, L: condensing lens, P: Pinhole, and BS: beam splitter. Alternatively, the two detectors could be 'looking' at a thermal source like a star.

1.5 Brown-Twiss intensity correlations, locality and separability

In the Brown-Twiss interferometer[10], two fast detectors measure the short-time averaged local intensities of light (or radio waves) incident on the detectors and a correlator multiplies these photocurrents and averages the product over a long duration. The measurement is done as a function of the separation of the two detectors, or equivalently as a function of different times of detection at the two detectors. Figure 2 shows the basic scheme of the experiment.

The source could be thermal, like a star, and there are intensity fluctuations that are recorded by both detectors. The long duration average of photocurrent at each detector is more or less a constant, and this is the local probability. But the joint probability is not a product of the local probabilities at the two detectors! The long time average of the intensities at space-time points (x_1, t_1) and (x_2, t_2) is given by

$$\begin{aligned} \langle I_1(x_1,t_1)I_2(x_2,t_2)\rangle &= \left(\frac{1}{2}\epsilon_0 c\right)^2 \left\langle |E(x_1,t_1)|^2\, |E(x_2,t_2)|^2 \right\rangle \\ &= \left(\frac{1}{2}\epsilon_0 c\right)^2 \langle E^*(x_1,t_1)E^*(x_2,t_2)E(x_2,t_2)E(x_1,t_1)\rangle \quad (6) \end{aligned}$$

This intensity correlation can be written as

$$\langle I_1(x_1,t_1)I_2(x_2,t_2)\rangle = \left(\frac{1}{2}\epsilon_0 c\right)^2 |\langle E^*(x_1,t_1)E(x_2,t_2)\rangle|^2 + \langle I_1\rangle \langle I_2\rangle \quad (7)$$

The first term is the interference term, written exactly as one would write for the Young's double slit experiment, and the second term is the product of local average intensities. There is an interference pattern with up to 50% visibility, in the product of the intensities that quantifies the coherence at source. The crucial physical aspect that gives rise to this intensity interference is the instantaneous phase factors in the two optical wave trains that are incident on the two detectors. Obviously there is no nonlocality in this case. Measurement of current with one detector does not affect the current in the other detector. Yet, there is correlation in the intensities, much like the correlations seen in quantum systems. The difference is that the visibility is not 100% as could be obtained in pure quantum systems (this is again another statement of Bell's inequality). The Brown-Twiss example can be formulated completely classically, and obeys the Bell's inequality, but violates the criterion of separability of probabilities.

1.6 Resolution of the EPR nonlocality puzzle

The widely debated EPR puzzle is resolved in a straightforward and simple way[9, 11, 12] when the wave nature of single particle systems is taken into account correctly. The point is that though the experimenter measures properties that are attributed to particles, like position, spin direction etc., the microscopic behaviour itself is governed by the inherent wave nature. The situation has some similarity to the phenomenon of interference. The measurement itself is at the level of intensities on the screen, but the physical process that forms the interference pattern is governed by wave-like properties or probability amplitudes. Therefore calculations of quantum correlations should use probability amplitudes, instead of average over product of eigenvalues. These amplitudes are purely local quantities describing the behaviour of each particle locally and separately. The squares of the amplitudes give the probabilities for local measurements. Correlations at source of several particles separated into space-like regions are encoded in the mutual coherence or relative phase of these probability amplitudes. The correlation is calculated by taking a suitable product of the probability amplitudes first and then squaring this amplitude correlation, instead of simply multiplying the eigenvalues and averaging as in the local realistic theories. The deviation from the standard quantum mechanical path is that I use local probability amplitudes instead of the multiparticle wave function that is inherently spread out and nonlocal.

In the following section I derive the correct quantum correlations assuming strict locality, without using negative probabilities. There is a class of theories, with strict locality, objective reality for an absolute phase variable (internal variable), and without negative probabilities, that do not violate the Bell's inequalities. Standard quantum mechanics of entangled particles can be recast in this form. This was (wrongly) considered impossible in the light of Bell's theorem.

2 Quantum correlations from local amplitudes

2.1 The basic formalism

Consider the breaking up of a singlet state as in the standard Bohm version of the EPR problem[5]. The two-particle state is described by the wave function

$$\Psi_S = \frac{1}{\sqrt{2}}\{|1,-1\rangle - |-1,1\rangle\} \tag{8}$$

where the state $|1,-1\rangle$ is short form for $|1\rangle_1 |-1\rangle_2$, and represents an eigenvalue of $+1$ for the first particle and -1 for the second particle *if measured* in any particular direction.

Two observers make measurements on these particles individually at space like separated regions. I assume that strict locality is valid at the level of probability amplitudes[9]. A measurement changes probability amplitudes only locally. ***Measurements performed in one region do not change the magnitude or phase of the complex amplitude for the companion particle in a space-like separated region.*** The local setting of the polarizers, Stern-Gerlach analyzers, detectors etc. (collectively denoted as analyzer) is represented by **a** and **b** for the two distant apparatus. Since we need to deal with correlated particles which may have a definite phase relationship at source (when the particles are produced together, for example) I introduce internal variables associated with each particle. I denote these variable as ϕ_1 and ϕ_2. Their values are unaltered once the particles are separated.

I now state the primary assumptions:

1. Locality: The local amplitude for the first particle C_1 that decide the passage of the particle through an analyzer depends only on the local variables **a** and ϕ_1. Similarly C_2 depends only on **b** and ϕ_2. If we denote the passage as + and the alternate outcome as −, then the statement of locality is for the relevant amplitudes is

$$C_{1\pm} = C_{1\pm}(\mathbf{a}, \phi_1), \quad C_{2\pm} = C_{2\pm}(\mathbf{b}, \phi_2) \tag{9}$$

   The square of the amplitudes give the corresponding probabilities ($C_{1+}C_{1+}^*$ gives the probability for transmission, for example).

   I will also state the locality at the level of the eigenvalues as

$$A(\mathbf{a}, \phi_1) = \pm 1, B(\mathbf{b}, \phi_2) = \pm 1 \tag{10}$$

   This is the same locality assumption as in the local realistic theories[4]. But, this has a meaning different from its meaning in standard local realistic theories. Here, this means that the outcomes, *when measured*, depend only on the local setting and the local internal variable.

2. Coherence at source: The correlations of the particles are encoded in the difference of the internal variables ϕ_1 and ϕ_2. If the particles have perfect correlations at source then all the pairs in the ensemble have the same value for the difference $|\phi_1 - \phi_2| = \phi_0$. (The value of ϕ can vary from particle to particle randomly, but the relative phase ϕ_0 between the two particles in all correlated pairs is constant).

3. The amplitude correlation function is a normalized inner product of the amplitudes. This is of the form

$$U(\mathbf{a}, \mathbf{b}) = \mathit{Real}(NC_iC_j^*) \tag{11}$$

   where N is the normalization constant related to the total joint probability. The square of the amplitude correlation function gives the joint probabilities for events of the form (++), (−−), (+−), and (−+). For a general case this can be calculated in a manner similar to the Brown-Twiss correlations. All probabilities are guaranteed to be positive definite in our formalism since the amplitude correlation function is real.

It is significant that our correlation function is not $\int d\phi\rho(\phi)A(\mathbf{a},\phi_1)B(\mathbf{b},\phi_2)$. The correlation of the eigenvalues $P(\mathbf{a},\mathbf{b}) = \frac{1}{N}\sum(A_iB_i)$ will be derived from the absolute square of $U(\mathbf{a},\mathbf{b})$ [9, 11, 12]. ***The crucial difference from local realistic theories is that the correlation is calculated from quantities which preserve the relative phases.***

We do not know the mapping between the local settings and the final outcomes, and therefore there is no determinism. The remarkable feature of my approach is that we do not need to know anything about this mapping. I do not assume any hidden variable that determines the results of the measurements. The variables ϕ_1 and ϕ_2 could be considered as hidden variables in a formal sense, but such initial undetermined phases are already part of the quantum formalism, though unused. Consider ϕ as a reference for the particles to determine the angle of a polarizer or analyzer encountered on their way, *locally*.

I denote the angle of the first analyzer θ_1 with reference to ϕ_1 as θ. Similarly, the second particle which has the internal phase angle $\phi_1 + \phi_0$, where ϕ_0 is a constant, encounters the second analyzer oriented at angle θ_2 at a space-like separated point. Let the orientation of this analyzer with respect to the internal phase angle of the second particle is θ'. We have $\theta - \theta' = \theta_1 - \theta_2 + \phi_0$.

An experiment in which each particle is analyzed by orienting the analyzers at various angles θ_1 and θ_2, as depicted in Fig. 1, is considered next. At each location the result is two-valued denoted by (+1) for transmission and (−1) for absorption of each particle, for any angle of orientation. The classical correlation function, which is also the experimenter's correlation function, $P(\mathbf{a},\mathbf{b}) = \frac{1}{N}\sum(A_iB_i)$ satisfies $-1 \leq P(\mathbf{a},\mathbf{b}) \leq 1$. We note that $P(\mathbf{a},\mathbf{b})$ denotes the average of the quantity (*number of detections in coincidence – number of detections in anticoincidence*), where 'coincidence' denotes both particles showing same value for the measurement and 'anticoincidence' denotes those with opposite values.

Now I state the expressions for the amplitudes and the amplitude correlation function for the widely discussed singlet state. These are similar to the Feynman amplitudes, for example the amplitude for the particle to go from a point x_1 to x_2.

The local amplitudes are $C_{1+} = \frac{1}{\sqrt{2}}\exp\{is(\theta_1 - \phi_1)\}$ for the first particle at the first polarizer and $C_{2+} = \frac{1}{\sqrt{2}}\exp\{is(\theta_2 - \phi_2)\}$ for the second particle at the second polarizer. The amplitudes C_{1-} and C_{2-} for the events denoted by '−' differ only in the phase for the maximally entangled state.

The explicit dependence of the amplitude on the spin of the particle is motivated by the fact that we are dealing with systems with phases and the phase associated with the spin rotations (a geometric phase) is a necessary input in this description[13]. The correlation at source is encoded in ϕ_0. The individual measurements at each end separately will now give the correct result for transmission for any angle of orientation. These probabilities are

$$C_1C_1^* = C_2C_2^* = \frac{1}{2} \tag{12}$$

Events of both types (++) and (−−) contribute to a "coincidence". The correlation function for an outcome of either (++) or (−−) of two maximally entangled particles is

$$U(\theta_1,\theta_2,\phi_o) = 2\,Re(C_1C_2^*) = \cos\{s(\theta_1 - \theta_2) + s\phi_o\}. \tag{13}$$

It is normalized such that its square will give the conditional joint probabilities of the type 'outcome + for the second particle, given that the outcome for the first particle is +, etc. All references to the individual values of the internal variable ϕ has dropped out.

I now derive the relation between this correlation function and the experimenter's correlation function $P(\mathbf{a},\mathbf{b}) = \frac{1}{N}\sum(A_i B_i)$. Since $U_{++}^2 = U_{--}^2$ for the maximally entangled state, $U^2(\theta_1, \theta_2, \phi_o)$ is the probability for a coincidence detection (++ or −−), and $(1 - U^2(\theta_1, \theta_2, \phi_o))$ is the probability for an anticoincidence (events of the type +− and −+). Since the average of the quantity (number of coincidences − number of anticoincidences) =

$$U^2(\theta_1, \theta_2, \phi_o) - (1 - U^2(\theta_1, \theta_2, \phi_o)) = 2U^2(\theta_1, \theta_2, \phi_o) - 1, \tag{14}$$

the correspondence between $P(\mathbf{a},\mathbf{b})$ and $U(\theta_1, \theta_2, \phi_o)$ is given by the expression,

$$\begin{aligned} P(\mathbf{a},\mathbf{b}) &= 2U^2(\theta_1, \theta_2, \phi_o) - 1 \\ &= 2\cos^2\{s(\theta_1 - \theta_2) + s\phi_o\} - 1 \end{aligned} \tag{15}$$

There is a remarkable difference between the correlation function used in Bell's analysis and the one given above. Our correlation function contains the wave properties inherent in quantum particles, whereas Bell's correlation function does not have this feature. Both analysis use the *same* assumptions on locality. Now I discuss some specific examples.

2.2 Spin-$\frac{1}{2}$ particles and Photons

Consider the singlet state breaking up into two spin-$\frac{1}{2}$ particles propagating in opposite directions to spatially separated regions. Since orthogonality of the two particles in any basis implies a relative angle of π for spinors, I set $\phi_o = \pi$. Then the correlation function and $P(\mathbf{a},\mathbf{b})$ calculated from this function are

$$\begin{aligned} U(\theta_1, \theta_2, \phi_o) &= \cos\{s(\theta_1 - \theta_2) + s\phi_o\} \\ &= \cos\{\frac{1}{2}(\theta_1 - \theta_2) + \pi/2\} \\ &= -\sin\frac{1}{2}(\theta_1 - \theta_2) \end{aligned} \tag{16}$$

$$\begin{aligned} P(\mathbf{a},\mathbf{b}) &= 2\sin^2(\frac{1}{2}(\theta_1 - \theta_2)) - 1 \\ &= -\cos(\theta_1 - \theta_2) = -\mathbf{a}\cdot\mathbf{b} \end{aligned} \tag{17}$$

This is identical to the quantum mechanical predictions obtained from the singlet entangled state and Pauli spin operators. We have reproduced the correct correlation function using local amplitudes.

For the case of photons entangled in orthogonal polarization states we get, by setting $s = 1$ and $\phi_o = \pi/2$ to represent orthogonal polarization,

$$\begin{aligned} U(\theta_1, \theta_2, \phi_o) &= \cos\{(\theta_1 - \theta_2) + \pi/2\} \\ &= -\sin(\theta_1 - \theta_2) \end{aligned} \tag{18}$$

$$P(\mathbf{a},\mathbf{b}) = 2\sin^2(\theta_1 - \theta_2) - 1 = -\cos(2((\theta_1 - \theta_2)) \quad (19)$$

which is the correct quantum mechanical correlation.

2.3 Three-particle GHZ correlations

One of the most remarkable results of the formalism presented here is that it reproduces the quantum correlations of multiparticle systems with more than two entangled particles in a physically transparent and insightful way[12, 14]. We consider the three particle GHZ state[15] defined as

$$|\Psi_{GHZ}\rangle = \frac{1}{\sqrt{2}}(|1,1,1\rangle - |-1,-1,-1\rangle) \quad (20)$$

where the eigenlabels in the kets are with respect to the z-axis basis.

The prediction from quantum mechanics for the measurement represented by the operator $\sigma_x^1 \otimes \sigma_x^2 \otimes \sigma_x^3$ is given by

$$\sigma_x^1 \otimes \sigma_x^2 \otimes \sigma_x^3 \, |\Psi_{GHZ}\rangle = -\,|\Psi_{GHZ}\rangle \quad (21)$$

Equivalently the joint probabilities for various outcomes in the x direction are

$$P(+,+,+) = P(-,-,+) = P(+,-,-) = P(-,+,-) = 0 \quad (22)$$
$$P(-,-,-) = P(+,+,-) = P(+,-,+) = P(-,+,+) = 1 \quad (23)$$

Local realistic theories predict that the product of the outcomes in the x direction for the three particles should be $+1$, i.e.,

$$P(+,+,+) = P(-,-,+) = P(+,-,-) = P(-,+,-) = 1 \quad (24)$$

The other four joint probabilities are zero. This contradicts Eqs. 21-23 and highlights the conflict between a local realistic theory and quantum mechanics.

I define the local amplitudes for the outcomes + and − at the analyzer (with respect to the x basis) for the first particle as $C_{1+} = \frac{1}{\sqrt{2}}\exp(i\theta_1)$, and $C_{1-} = \frac{1}{\sqrt{2}}\exp(i(\theta_1 + \pi/2))$. The amplitude C_{1-} contains the added angle $\pi/2$ because this amplitude is orthogonal to C_{1+}. Similarly, we have $C_{2+} = \frac{1}{\sqrt{2}}\exp(i\theta_2)$, and $C_{2-} = \frac{1}{\sqrt{2}}\exp(i(\theta_2 + \pi/2))$ for the second particle and $C_{3+} = \frac{1}{\sqrt{2}}\exp(i\theta_3)$, and $C_{3-} = \frac{1}{\sqrt{2}}\exp(i(\theta_3 + \pi/2))$ for the third particle[14].

Correlation function is obtained from $N\mathrm{Real}(C_1 C_2^* C_3^*)$, where N is a normalization constant, and its square is the relevant joint probability. (There is no unique definition of the amplitude correlation function. The final results are independent of the particular definition we use). Since we want $N\mathrm{Re}(C_{1-}C_{2-}^*C_{3-}^*) = \pm 1$, we choose $C_{1-}C_{2-}^*C_{3-}^*$ to be pure real. This gives

$$\frac{N}{2\sqrt{2}} Real(\exp i(\theta_1 - \theta_2 - \theta_3 - \pi/2)) = \pm 1 \quad (25)$$

$$\theta_1 - \theta_2 - \theta_3 - \pi/2 = 0 \text{ or } \pm\pi \quad (26)$$

We can choose the relevant relative phases to satisfy this condition. Then we get

$$P(-,-,-) = 1$$

Rest of the joint probabilities given in Eqs. 22 and 23 automatically follow, since flipping sign once rotates the complex number $C_{1-}C_{2-}^{*}C_{3-}^{*}$ through $\pi/2$. The square of $N\text{Real}(C_1C_2^*C_3^*)$ is then 1 for an odd number of $(-)$ outcomes and 0 for even number of $(-)$ outcomes.

Similar construction also applies to general multiparticle maximally entangled states.

Here it is worth making a comment regarding the information content in the N-particle quantum state[16]. I have constructed the 2^N joint probabilities in the N-particle system using $2N$ complex numbers. This is possible because the 2^N complex numbers describing the N-qbit state are not independent. The analogy with classical coin-tossing probabilities will make this very clear. Each coin can be specified by 2 real numbers, giving the probability for head (p) and probability for tail (q). For N coins, there are $2N$ such numbers (when probability is normalized for each coin, there are only N independent numbers since $p_i = 1 - q_i$). But there are 2^N joint probabilities for outcomes of the type $H_1H_2T_3H_4...T_N$. These 2^N numbers cannot be specified arbitrarily since the probabilities have to be correctly satisfied for all possible experiments including that on a smaller number of coins or on individual coins. One the other hand, all the 2^N joint probabilities can be generated from the $2N$ individual probabilities. The quantum situation is exactly same, except that instead of real number probabilities we have to use complex numbers as probability amplitudes.

2.4 Multiparticle interference

Similar considerations apply well for particles entangled in other sets of variables like momentum and coordinate, and energy and time. These cases of two-particle entanglement can be mapped on to the spin-$\frac{1}{2}$ singlet problem with two-valued outcomes[12]. Starting from the local amplitudes $C_1 = \frac{1}{\sqrt{2}}\exp(i\alpha k(x_1 - x_o)/2)$, and $C_2 = \frac{1}{\sqrt{2}}\exp(i\alpha k(x_2 - x_o)/2)$ we can derive the probability for coincidence detection as

$$P(x_1, x_2) = \cos^2(\alpha k(x_1, x_2)/2) = \frac{1}{2}(1 + \cos k\alpha(x_1 - x_2)) \tag{27}$$

This is the two photon correlation pattern with 100% visibility, *obtained without nonlocality*. x_1 and x_2 are the coordinates of the two detectors separated by a space-like interval. k is the wave vector and α is a scaling factor for the angle subtended by the two slits at the detectors, source etc. Clearly, none of the experiments using entangled down-converted photons in interferometers imply any nonlocality. In particular the statement, often made, that it is the *interference* between two photons that gives rise to the pattern given by eq. 27 is not correct. It is a correlation function. The example of multiparticle interference is only one quantum step away from the familiar Brown-Twiss intensity correlations. I have discussed this already.

3 Discussion

3.1 The virgin quantum state

There is a subtle but real difference between an unmeasured quantum particle that has never gone through a classical apparatus and another particle that has been prepared in a definite but random and unknown (to the experimenter) state. In the latter case there is reality to the eigenvalue, though unknown, whereas in the former case no objective reality can be ascribed at the level of eigenvalues. The state of each particle in an entangled multiparticle system is such a ***virgin state.*** This is described by the complex probability amplitudes employed in our analysis. The local probability amplitudes were complex numbers defined by $\langle\theta|\phi\rangle$ where $\langle\theta|$ symbolizes the final state defined by the detector orientation and $|\phi\rangle$ symbolizes the initial state of the individual particle. The unconventional aspect is that the state $|\phi\rangle$ does not represent any definite spin direction in space, since it is labelled by just an internal phase variable. (In fact, there is no definite state for a microscopic systems that has never gone through a classical apparatus.) The concept of a state that has not gone through any state preparation scheme and labelled by just the internal phase variable - a virgin state - is the new input in quantum mechanics and this brings the possibility of a complete, though indeterministic, description that preserves strict locality[8, 12]. A real measurement involves the interaction of a quantum state and a classical apparatus. The result can be used to predict the result of a similar measurement on the correlated particle, but our analysis shows that there is no nonlocal influence. Though the result, if measured, for the second particle can be predicted with certainty, its state is still a virgin state and not one with a definite eigenvalue as EPR thought. This means that quantum systems have their objective reality at a level deeper than that described by the EPR definition of reality. *So, predictability with certainty does not amount to a state reduction or measurement.*

3.2 Popper's experiment and proof against state reduction-at-a-distance

The uncertainty principle is the basic pillar of quantum theory, and it requires that state reduction is accompanied by inevitable dispersion in the conjugate variable. The preparation of a definite pure state in quantum mechanics is always associated with uncontrollable dispersions in all the conjugate noncommuting observables. Therefore, whether or not state reduction truly takes place can be tested only if the dispersion in the relevant variable is also tested, apart from testing for the definite value of the observable for which the state is prepared.

Clearly, if there is state-reduction-at-distance due to a quantum measurement on one of the particles of an entangled pair, the dispersion of the relevant observable for the companion particle should obey the uncertainty product. So, predictability with certainty of one observable is no indication of collapse to a definite state. There should be associated dispersion in the relevant noncommuting observables. This is ruled out from the Popper's experiment[17], first discussed by K. Popper in 1933, and tested recently experimentally[18]. Figure 3 schematically describes the experiment. Particles entangled in position and momentum originate at the

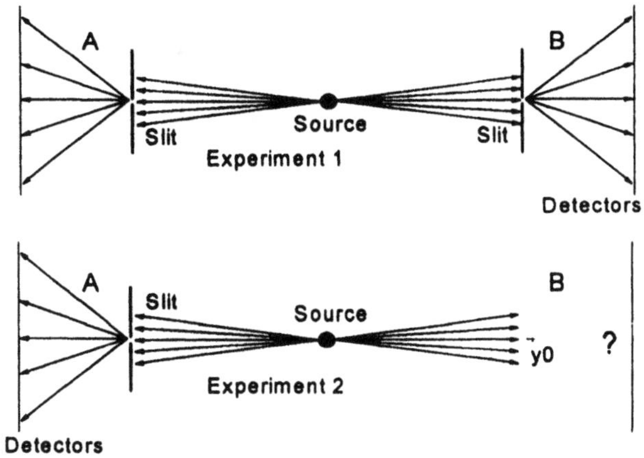

Figure 3: Popper's experiment as proof against state reduction at a distance. In experiment 1, there are real slits at A and B and both particles show momentum dispersion satisfying the uncertainty principle. In experiment 2, the dispersion happens only at the real slit and not at the virtual slit y0.

source and propagate in opposite directions. Putting a slit and a detector behind the slit is a measurement of the transverse position, and the position of the companion particle can be deduced with certainty without a disturbing measurement. In experiment 1, the slits are put in the paths of both particles and a standard diffraction pattern is expected on the screens. In experiment 2, only one of the particles encounters the slit, at A. Once this measurement is made, the position of the second particle near B can be deduced with certainty from momentum correlation, within about the same size as that of the real slit at A where a 'real' measurement is made. If 'one measurement and the consequent ability to predict with certainty' amounts to a measurement on the companion particle and its state reduction, then the transverse momentum of the second particle also should disperse as given by the uncertainty principle. It is easy to argue that this does not happen, as Popper himself had argued in a different context. The theoretical arguments based on signal locality and conservation laws[19], as well as the experimental results[18] show that the dispersion for the unmeasured companion particle does not change as a result of the measurement at A. This is not a violation of the uncertainty principle, but it is a solid proof that there is no real state-reduction-at-a-distance. There is no additional dispersion at B since there is no state reduction. Popper's experiment and its analysis forces us to radically change the currently held view on quantum nonlocality, and with it the interpretation of quantum teleportation, entanglement swapping etc., as I have mentioned elsewhere[8, 12]. *Bell state measurement does not collapse the distant particle into a definite state.* In fact, the interpretation of all experiments that uses the properties of Bell state measurements would drastically change once we abandon state reduction-at-a-distance. (By a remarkable coincidence, it is the Popper's experiment, ignored for a good part of a century, that supports the elimination of nonlocality from quantum correlations. Popper had argued passionately against the irrational notion of nonlocality. He declared that "the end of locality would mean the end of the rationality in physics"[20].)

Our analysis shows that quantum mechanical description is complete in our formalism that includes the virgin state, in the sense of EPR. There is a representation in the theory for every possible physical reality. Before a real measurement the maximum possible physical reality is represented by the probability amplitudes we used. After a state preparing measurement, the standard formalism of quantum mechanics applies and then our formalism is the same as the Feynman amplitude formalism. EPR's conclusion of incompleteness was based on their inadequate definition of reality that could not encompass the reality associated with wave-like characteristics.

3.3 Summary

I have constructed a correct theory of quantum correlations starting from local ingredients, without using negative probabilities. This clarifies many issues that have been debated for several decades and resolves the nonlocality puzzle in the EPR correlations. There is no nonlocal influence between correlated particles separated into space-like regions. There is no more any conceptual conflict with special relativity. The objective reality of unmeasured virgin quantum systems is at the level of an internal phase variable. Phase coherence at source accounts completely for the quantum correlations of subsystems propagated into space-like separated regions. The Bell correlations function as well the correlation function of traditional hidden variable theories did not incorporate the concept of phase, and therefore could not reproduce quantum correlations. Our solution has new physical and philosophical implications[21] regarding the nature of reality, measurement and state reduction in quantum systems. It also reveals new interpretations for entanglement swapping and quantum teleportation. The solution presented here has reliable empirical support from the results of the Popper's experiment.

Acknowledgments: I thank Sukant Saran, H. Stapp, V. Singh, L. K. Grover, F. Laloe, F. Perales, J. Robert, S. Braunstein, A. K. Pati, Andal Narayanan, R. Nityananda, R. Cowsik, and Martine Armand for numerous discussions that took this work forward. Partial financial support for this research comes from the Department of Science and Technology, Government of India.

References

[1] A. Einstein, B. Podolsky, and N. Rosen, *Phys. Rev.*, **47**, 777, 1935.

[2] N. Bohr, *Nature*, 136, 65, 1935.

[3] N. Bohr, *Phys. Rev.*, 48, 696, 1935.

[4] J. S. Bell, *Physics*, **1**, 195, (1965); "Speakable and unspeakable in quantum mechanics", Cambridge University Press, 1987.

[5] D. Bohm, "Quantum Theory", Prentice Hall, Englewood Cliffs, NJ, 1951.

[6] A recent resource for exhaustive search is the book, "The Einstein, Podolsky, and Rosen Paradox in Atomic, Nuclear and Particle Physics", A. Afriat and F. Selleri, Plenum Press, New York and London, 1999.

[7] E. Schrödinger, *Proc. Camb. Phil. Soc.* **31**, 555, 1935.

[8] C. S. Unnikrishnan, *Annales de la Fondation Louis de Broglie (Paris)*, **25**, 363, 2000.

[9] C. S. Unnikrishnan, *There is no spooky action-at-a-distance in quantum correlations: Resolution of the EPR nonlocality puzzle*, quant-ph/0001112, 2000.

[10] R. Hanburry Brown and R. O. Twiss, *Nature*, **178**, 1447, 1956.

[11] C. S. Unnikrishnan, quant-ph/0005103, to appear in the proceedings of the International Conference on Quantum Optics, Minsk, Belarus May, 2000.

[12] C. S. Unnikrishnan, *Current Science*, **79**, 195, 2000.

[13] C. S. Unnikrishnan, to be published; The geometric phase associated with spin rotations can be obtained directly from a generalization of the usual dynamical phase $\int p_i dx^i$, to include the spin as one of the momenta p_i and the angular coordinate as the corresponding dx^i.

[14] C. S. Unnikrishnan, "Three-particle GHZ correlations without nonlocality", quant-ph/0004089, 2000.

[15] D. M. Greenberger, M. Horne, and A. Zeilinger in "Bell's Theorem, Quantum Theory and Conceptions of the Universe", Ed. M. Kafatos, Kluwer, Dordrecht, 1989, pp 73-76 ; N. D. Mermin, *Phys. Today*, June 1990, p 9-11; D. M. Greenberger, M. Horne, A. Shimony, and A. Zeilinger, *Am. J. Phys.* **58**, 1131, 1990.

[16] This question was raised at various occasions during discussions with Luv Grover, Samuel Braunstein and Warren Smith.

[17] K. R. Popper, in "Open Questions in Quantum Physics", Eds. G. Tarozzi and A. van der Merwe, D. Reidel Publishing Co., 1985.

[18] Y-H. Kim and Y. Shih, *Found. Phys.*, **29**, 1849, 1999.

[19] C. S. Unnikrishnan, *Found. Phys. Lett.*, **13**, 197, 2000.

[20] Popper called for support for solutions that respect locality in his paper 'Bell's theorem: A note on locality', in "Microphysical Reality and Quantum Formalism", Eds.A. van der Merwe, F. Selleri, and G. Tarozzi, Kluwer Academic Publishers, 1988.

[21] Some of the philosophical implications of the results presented here will be examined in a detailed monograph in preparation.

Classical and Quantum Solutions of 2+1 Dimensional Gravity

M. Kenmoku, T. Matsuyama, R. Sato and S. Uchida

1 Introduction

Recently 2+1 dimensional gravity theory, especially AdS_3 has been studied extensively [1, 2]. It was shown to be equivalent to the 2+1 Chern-Simon theory [3] and has been investigated to understand the black hole thermodynamics, i.e. Hawking temperature [4] and others. The purpose of this report is to investigate the canonical formalism of the original 2+1 dimensional Einstein gravity theory instead of the Chern-Simon theory. For the spherically symmetric space-time, local conserved quantities (mass and angular momentum) are introduced and using them canonical quantum theory is defined. Constraints are imposed on state vectors and solved analytically. In order to extract the physical meaning to the wave function, we impose the de Broglie-Bohm interpretation [5, 6, 7] and derive the differential equation for the metrics, which include the quantum effect. After fixing the gauge choice, special solutions of the metrics are obtained. Especially the quantum effect of the closed de Sitter universe is obtained quantitatively. It is interesting to note that the birth of universe appeared as the real time tunneling in stead of the imaginary time tunneling in the WKB approximation by Vilenkin [8] and in the path integral method by Hartle and Hawking [9]. The strategy to obtain the solution is followed by our previous work [10].

2 Canonical Formalism

We start to consider the Einstein-Hilbert action with cosmological constant λ in 2+1 dimensional space-time,

$$I = \frac{1}{16\pi G_2}\int d^3x\sqrt{-^{(3)}g}(^{(3)}R - 2\lambda). \tag{1}$$

[0]1. Talk given at 4th Internatioal Symposium on Frontier of Fundamental Physics, Hyderabad, India, 9-13 December 2000.

2. Department of Physics, Nara Women's University, Nara 630-8506, Japan,
kenmoku@phys.nara-wu.ac.jp,
3. T. Matsuyama , Department of Physics, Nara University of Education, Takabatake-cho, Nara 630-8528, Japan,
matsuyat@nara-edu.ac.jp,
4. R. Sato and 5. S. Uchida,
Graduate School of Human Culture, Nara Women's University, Nara 630-8506, Japan,
reika@phys.nara-wu.ac.jp;satoko@phys.nara-wu.ac.jp

Frontiers of Fundamental Physics 4, Edited by Sidharth and Altaisky,
Kluwer Academic/Plenum Publishers, New York, 2001

The gravitational constant in 2+1 dimension is set to $G_2 = 1/4$ hereafter. The metrics in polar coordinate are expressed in ADM decomposition [11] as

$$ds^2 = -N^2dt^2 + \Lambda^2(dr + N^r dt)^2 + R^2(d\phi + N^\phi dt)^2 \\ +2C(dr + N^r dt)(d\phi + N^\phi dt), \tag{2}$$

where all metrics are assumed to be function of time t and radial coordinate r. In the following, dot and dash denotes the derivative with respect to t and r, respectively. The action in canonical formalism is in the form

$$\begin{aligned} I &= \int dt\,dr\,[P_\Lambda\dot{\Lambda} + P_R\dot{R} + P_C\dot{C} - (NH + N^r H_r + N^\phi H_\phi)] \\ &\quad - \int dt\,dr\left([(\Lambda P_\Lambda + CP_C)N^r]' + [(\frac{C}{\Lambda}P_\Lambda + R^2P_C)N^\phi]'\right), \end{aligned} \tag{3}$$

where canonical momenta are

$$P_\Lambda = \frac{\partial\mathcal{L}}{\partial\dot{\Lambda}} = \frac{2\Lambda R(N^rR' - \dot{R})}{N\sqrt{h}}, \tag{4}$$

$$P_R = \frac{\partial\mathcal{L}}{\partial\dot{R}} = \frac{2R[CN^{\phi\prime} + \Lambda\{(\Lambda N^r)' - \dot{\Lambda}\}]}{N\sqrt{h}}, \tag{5}$$

$$P_C = \frac{\partial\mathcal{L}}{\partial\dot{C}} = -\frac{(N^rC)' + R^2N^{\phi\prime} - \dot{C}}{N\sqrt{h}}, \tag{6}$$

and the Hamiltonian and the momentum constraints are defined as

$$H = -\frac{\sqrt{h}}{2}\left(\frac{P_\Lambda P_R}{\Lambda R} - P_C{}^2\right) - 2\left(-\frac{R'^2 + RR''}{\sqrt{h}} + \frac{RR'h'}{2h\sqrt{h}}\right) - 2\lambda\sqrt{h}, \tag{7}$$

$$H_r = P_RR' - CP_C{}' - \Lambda P_\Lambda{}', \tag{8}$$

$$H_\phi = -\left(\frac{C}{\Lambda}P_\Lambda + R^2P_C\right)'. \tag{9}$$

It is essential to introduce the local conservation quantities, the angular momentum J and the mass function M as follows.

$$J := -\int drH_\phi = \frac{C}{\Lambda}P_\Lambda + R^2P_C, \tag{10}$$

$$\begin{aligned} M &:= -\int dr\left(\frac{RR'}{\sqrt{h}}H + \frac{P_\Lambda}{\Lambda}H_r + P_CH_\phi\right) \\ &= \frac{1}{2}\left(P_\Lambda{}^2 + \frac{2CP_\Lambda P_C}{\Lambda} + R^2P_C{}^2 - \frac{(RR')^2}{h} - \lambda R^2\right) \end{aligned} \tag{11}$$

We make transformation from old variables Λ, R and C to new variables

$$\begin{pmatrix} \Lambda \\ R \\ C \end{pmatrix} \longrightarrow \begin{pmatrix} \bar{\Lambda} \\ \bar{R} \\ \bar{C} \end{pmatrix} = \begin{pmatrix} \sqrt{\Lambda^2 - C^2R^{-2}} \\ R \\ CR^{-2} \end{pmatrix}. \tag{12}$$

The corresponding momenta are transformed as

$$\begin{pmatrix} P_\Lambda \\ P_R \\ P_C \end{pmatrix} \longrightarrow \begin{pmatrix} P_{\bar{\Lambda}} \\ P_{\bar{R}} \\ P_{\bar{C}} \end{pmatrix} = \begin{pmatrix} \bar{\Lambda}\Lambda^{-1}P_\Lambda \\ C^2\Lambda^{-1}R^{-3}P_\Lambda + P_R + 2CR^{-1}P_C \\ C\Lambda^{-1}P_\Lambda + R^2P_C \end{pmatrix}. \tag{13}$$

3 Quantum Theory

Next we proceed the quantum theory in the Schrödinger picture and the quantized operators are denoted by the notation hat. Our strategy is to solve the eigenvalue equation for $\hat{J}$, $\hat{M}$ and the constraint equation for H_r step by step instead of solving the constraint equations $\hat{H}\Psi = 0, \hat{H}_r\Psi = 0$ and $\hat{H}_\phi\Psi = 0$.

Step 1: Angular momentum eigen equation
The eigenvalue equation of the local angular momentam (Eq. (10))

$$\hat{J}\Psi = \hat{P}_{\bar{C}}\Psi = j\Psi \tag{14}$$

For the eigenvalue j and the eigen function is obtained in the form

$$\Psi = e^{ij\Phi}u(\bar{\Lambda}, \bar{R}) , \tag{15}$$

with

$$\Phi = \int dr\bar{C}(r) . \tag{16}$$

Step 2: Momentum constraint equation
The radial momentum constraint equation

$$\hat{H}_r\Psi = (\bar{R}'\hat{P}_{\bar{R}} - \bar{\Lambda}(\hat{P}_{\bar{\Lambda}})')e^{ij\Phi}u(\bar{\Lambda}, \bar{R}) = 0 , \tag{17}$$

restricts the functional form of the wave function as

$$\Psi = e^{ij\Phi}u(Z) , \tag{18}$$

where we introduce variable Z

$$Z = \int dr\bar{\Lambda}f(\bar{R}, \chi) = \int dr \int^{\bar{\Lambda}(r)} d\bar{\Lambda}\bar{f}(\bar{R}, \chi), \tag{19}$$

with

$$\chi := R'^2\bar{\Lambda}^{-2} . \tag{20}$$

The arbitraly functions f and $\bar{f}$ are related to each other as:

$$f(\bar{R}, \chi) \quad = \quad -\int^{\chi} d\chi \frac{\bar{f}(\bar{R}, \chi)}{2\chi} . \tag{21}$$

Step 3: Mass eigen equation
The local mass operator $\hat{M}$ is defined as

$$\hat{M} - m = \frac{1}{2}A\hat{P}_{\bar{\Lambda}}A^{-1}\hat{P}_{\bar{\Lambda}} + \frac{1}{2}(-\chi + \hat{F}(\bar{R})) , \tag{22}$$

where

$$\hat{F}(\bar{R}) = 1 - 2m - \lambda\bar{R}^2 + \frac{1}{4}\hat{j}^2\bar{R}^{-2} , \tag{23}$$

and

$$A = A_Z(Z)\bar{A}(\bar{R},\chi)\ , \tag{24}$$

which is called ordering factor. We take $\bar{A}$ as

$$\bar{A} = \frac{\delta Z}{\delta \bar{\Lambda}} = \bar{f} = \sqrt{\chi - F_j(\bar{R})}\ , \tag{25}$$

where

$$F_j(\bar{R}) := \hat{F}\mid_{\hat{J}=j}(\bar{R})\ . \tag{26}$$

Then using the mass operator for each eigenvalue of angular momentum j

$$\hat{M}_j := \hat{M}\mid_{\hat{J}=j}\ , \tag{27}$$

the mass eigen equation

$$\hat{M}_j u_{j,m}(Z) = m u_{j,m}(Z)\ , \tag{28}$$

can be reduced to the equation with respect to Z

$$\frac{d^2 u_{j,m}(Z)}{dZ^2} - A_Z{}^{-1}\frac{\delta A_Z}{\delta Z}\frac{du_{j,m}(Z)}{dZ} + u_{j,m}(Z) = 0. \tag{29}$$

If we choose the remaining ordering factor as $A_Z = Z^{2\nu-1}$, the above equation becomes the Bessel equation

$$\frac{d^2 u_{j,m}(Z)}{dZ^2} - \frac{2\nu-1}{Z}\frac{du_{j,m}(Z)}{dZ} + u_{j,m}(Z) = 0, \tag{30}$$

and the solution is

$$u_{j,m}^{(\nu)}(Z) = Z^{\nu}[b_1 H_{\nu}^{(1)}(Z) + b_2 H_{\nu}^{(2)}(Z)], \tag{31}$$

where $H_\nu(Z)$ is the Hankel function. The argument Z is expressed using Eqs. (19) and (25) as

$$\begin{aligned} Z &= \int dr \int^{\bar{\Lambda}(r)} d\bar{\Lambda}\sqrt{\chi - F_j(\bar{R})} \\ &= \int dr \left(\bar{\Lambda}\sqrt{\chi - F_j(\bar{R})} - \bar{R}' \ln\left|\frac{\sqrt{\chi} + \sqrt{\chi - F_j(\bar{R})}}{\sqrt{\mid F_j(\bar{R})\mid}}\right|\right)\ , \end{aligned} \tag{32}$$

where χ and F_j are given in Eqs. (20) and (26).

4 de Broglie-Bohm Interpretation

We obtained the analytical wave function of the 2+1 dimensional gravity theory. The wave function is the functional of the metrics and its physical and geometrical meaning is not clear. We adopt the de Broglie-Bohm interpretation in order to extract geometrical information from the wave function. For this purpose, only 2nd kind Hankel function in Eq. (31) is chosen as the wave function for the Vilenkin boundary condition and is written in the polar coordinate

$$\Psi(Z) = Z^{\nu} H_{\nu}^{(2)}(Z) = \mid \Psi \mid e^{i\Theta}(Z)\ . \tag{33}$$

The dBB interpretation is defined as the momenta equal to the gradient of the phase of the wave function:

$$P_{\bar{\Lambda}} = \frac{\delta\Theta}{\delta\bar{\Lambda}} \ , \ P_{\bar{R}} = \frac{\delta\Theta}{\delta\bar{R}} \ , \ P_{\bar{C}} = \frac{\delta\Theta}{\delta\bar{C}} \ . \tag{34}$$

Inserting the original expressions of the momenta Eqs. (6) and (13) into Eq. (34) we obtain the differential equation for the metrics with respect to the time and space coordinates as

$$\frac{1}{N}(N^r\bar{R}' - \dot{\bar{R}}) \ = \ \bar{f}\frac{d\Theta}{dZ} \ , \tag{35}$$

$$\frac{1}{N}((\bar{\Lambda}N^r)' - \dot{\bar{\Lambda}}) \ = \ \frac{\bar{\Lambda}}{\bar{R}'}\bar{f}'\frac{d\Theta}{dZ} \tag{36}$$

$$-\frac{\bar{R}^3}{2N\bar{\Lambda}}((N^rC)' + (N^\phi)' - \dot{\bar{C}}) \ = \ j \ . \tag{37}$$

In the dBB interpretation, the quantum effects are expressed in the factor, which classical limit is 1, as

$$n_{\mathrm{dBB}} = -\frac{d\Theta}{dZ} \to 1 \quad \text{(classicallimit)} \ . \tag{38}$$

In order to integrate the equations (35-37), we fix the part of gauge freedom of the general coordinate trasformation as

$$N^r = N^\phi = 0 \ . \tag{39}$$

Under these impositions, metric $\bar{C}$ is determined from Eq. (37) as

$$\dot{\bar{C}} = \frac{N\bar{\Lambda}}{\bar{R}^3}j \tag{40}$$

We can derive the relation free from quantum effects taking the ratio of Eq. (36) over Eq. (36)

$$\frac{\dot{\bar{\Lambda}}}{\bar{\Lambda}} = \frac{\dot{\bar{R}}}{\bar{R}}\frac{\bar{f}'}{\bar{f}} \ . \tag{41}$$

We shall show some examples of the solution of the dBB interpretation in the next section.

5 Examples

Example 1: BTZ black hole

We further impose the coordinate condition as

$$\bar{R}' = 0 \ \text{and} \ \dot{\bar{R}} = 1, \tag{42}$$

which give the relation $\bar{R} = t$. Then we find the from Eqs. (35-36)

$$ds^2 = \frac{1}{F_j} n_{\text{dbb}} dt^2 - F_j dr^2 + t^2(\bar{C}^2 dr + d\phi)^2 \ , \tag{43}$$

where $F_j = 1 - 2m + j^2/(4t^2) - \lambda t^2$. The classical limit of the metrics under the space-time transformation $t \leftrightarrow r$, $\bar{C} \leftrightarrow N^\phi$, represents the BTZ black [2] .

Example 2: de Sitter universe

Next we show the example of de Sitter universe and set $N = 1$ and $m = j = 0$. We make the ansatz for metrics as

$$\bar{R} = a(t)r \ \text{ and } \ \bar{\Lambda} = b(t)/\sqrt{1 - Kr^2} \ , \tag{44}$$

where $K = 0$ and 1 represent the open and closed universe respectively. From Eq. (41) we obtain $a(t) = b(t)$ and from eq. (35) we obtain the equation for the scale factor of the universe as

$$\dot{a} = \sqrt{\lambda a^2 - K} \ n_{\text{dbb}} \ . \tag{45}$$

The corresponding classical solutions are open and closed de Sitter universe

$$a = \begin{cases} \exp(\sqrt{\lambda} t) & \text{for} \quad K = 0 \\ \lambda^{-1/2} \cosh(\sqrt{\lambda} t) & \text{for} \quad K = 1 \ . \end{cases} \tag{46}$$

Special interest is closed de Sitter universe because the finite quantum effect can be estimated. The classical closed de Sitter universe creates finite size at $t = 0$ and expands exponentially as $t \gg 0$. The interesting quantum scale factor can be estimated for the case of index of the Hankel function $\nu = 1/3$ and equation for the scale factor becomes

$$\dot{a} = \sqrt{\lambda a^2 - 1} \ \frac{2}{\pi Z \mid H^{(2)}_{1/3} \mid^2} \ . \tag{47}$$

We analyze this equation and obtain the result that the universe creates from $a = 0$, expands through the classical forbidden region ($0 < a < \sqrt{\lambda}$) by real time tunneling and approaches to the classical scale factor gradually for $a \gg 0$.

6 Summary

We have studied 2+1 dimensional spherically symmetric gravity theory and obtained the results.

(1) Quantum theory is defined through local conserved quantities $\hat{J}(r)$ and $\hat{M}(r)$.

(2) The de Broglie-Bhom interpretation is applied for the analytic wave function of universe.

(3) The differential equations of the dBB interpretation are solved and black hole metrics as well as expanding universe metrics are obtained. Especially the real time tunneling through the classically forbidden region occurs as quantum effect. It is interesting to compare with other approaches that the universe creates form nothing through imaginary time tunneling using WKB approximation by Vilenkin [8] and the path integral method by Hartle and Hawking [9].

References

[1] S. Deser, R. Jackiw and G.t' Hooft, Ann. Phys. **152**, 220, 1984.

[2] M. Banados, C. Teitelboim and J. Zanelli , Phys. Lett. **69**, 1849, 1992.

[3] E. Witten, Nucl. Phys. **B311**, 46, 1988.

[4] S.W. Hawking, Commun. Math. Phys. **43**, 199, 1975.

[5] D. Bohm, Phys. Rev. **85**, 166(1952); **85**, 180, 1952.

[6] J.S. Bell, "Speakable and Unspeakable in Quantum Mechanics" Cambridge University Press, Cambridge, 1987.

[7] P.R. Holland, "Quantum Theory of Motion" Cambridge University Press, Cambridge, 1993, Cambridge University Press, Cambridge, 1987.

[8] A. Vilenkin, Phys. Lett. **117B**, 25(1982); Phys. Rev. **D27**, 2848, 1983; Phys. Rev. **D37**, 888, 1988.

[9] J.B. Hartle and S. Hawking, Phys. Rev. **D28**, 2960, 1983.

[10] M. Kenmoku, H. Kubotani, E. Takasugi and Y. Yamazaki , Phys. Rev. **D59**, 124004, 1999.

[11] R. Armowitt, S. Deser and C.W. Misner, in "Gravitation: An Introduction to Current Research", edited by L.Witten Wiley, New York, 1962.

Gravitational Collapse

A. Beesham
Department of Mathematical Sciences, University of Zululand
Private Bag X1001, Kwa-Dlangezwa 3886 (South Africa)
Email: abeesham@pan.uzulu.ac.za

and

S.G. Ghosh
Department of Mathematics
Science College, Congress Nagar, Nagpur - 440 012 (India)
Email: sgghosh@yahoo.com; sgghosh@hotmail.com

1 Introduction

What is the fate of a star after it has exhausted its nuclear fuel? It can no longer remain in equilibrium, and must undergo gravitational collapse. If the star is not too massive it may settle down to a white dwarf or a neutron star. However, if the star is more massive than a few solar masses, then we have to invoke the general theory of relativity to study its final state. This theory is valid up to the Planck length (10^{-33} cm), below which quantum effects becomes important. In the early seventies, it was shown that singularities develop as a result of gravitational collapse where quantities such as the energy density of the collapsing matter and the curvature of the spacetime diverge [1]. However, the singularity theorems only predict the existence of singularities, but do not give any information about their nature or causal structure, e.g., can such a singularity communicate with a distant observer in the universe.

The conjecture that such a singularity from a regular initial surface must always be hidden behind an event horizon is called the cosmic censorship hypothesis (CCH) (see [2] for reviews on the status of the conjecture) and was proposed by Penrose [3]. Such a singularity is called a black hole and is hidden from distant observers. However, the singularity theorems do not state that collapse must result in a black hole. Another possibility is that the singularity could be visible to a distant observer, in which case it is called a naked singularity. In the case of a black hole, the so-called event horizon starts forming before the eventual collapse of the star to the singularity, whereas in the case of a naked singularity the formation of the horizon gets delayed sufficiently so that information can in principle be sent out to distant observers. The CCH forbids the existence of naked singularities. This conjecture remains as the central open issue in classical general relativity.

A violation of the CCH will have major implications for general relativity. The study of black holes is based on the validity of the CCH, e.g., the area theorem. Further, if naked singularities occur, this will result in a breakdown of predictability in general relativity, and as put by Wald [4], a mad scientist could conceivably destroy the universe. In the formulation of the CCH, one assumes that one is dealing with suitable matter, which satisfies certain conditions, such as one of the energy conditions. Secondly the CCH states that generically, no naked singularities will occur. In order to define precisely what is meant by generic, one requires a much deeper insight into Einstein's equations than is currently available. Perhaps if naked singularities do occur, they are confined to a set of measure zero, but again it is not clear what topology or measure should be imposed.

In the absence of a concrete formulation of the CCH, examples showing the existence of naked singularities (which appear to violate the conjecture) remain important. There are now many studies which have been carried out studying collapse in different situations. The pattern that seems to be emerging is that, depending upon the initial conditions, one can end up with either a black hole or a naked singularity. It is instructive to study an example of a spacetime that admits naked singularities.

2 Charged Vaidya-deSitter Spacetime

In recent years significant attention has been given in studying gravitational collapse with a non zero cosmological constant [6]-[11]. Penrose [12] conjectured that the cosmological constant may be related to a violation of the CCH. Markovy and Shapiro [6] constructed Oppenheimer-Snyder-deSitter models which were extended by Lake [7] to include inhomogeneous and degenerate cases. The Vaidya solution [13] is most commonly used as a testing ground for various forms of the CCH [5]. The Vaidya metric generalized to a spherically symmetric spacetime with a non zero cosmological constant $\Lambda > 0$ is called the Vaidya-deSitter spacetime [14] and is known to admit naked singularities [9, 10, 11]. The ingoing charged Vaidya solution [15] represents a radial infall of a stream of charged photons. Lake and Zannias [16], under the assumption of homothecity, found that naked singularities can be formed in the self-similar charged Vaidya spacetime. The usefulness of the model is that rich structure is exhibited. However, self similarity is a strong geometric condition on the spacetime and thus gives rise to a possibility that the naked singularity could be an artifact of a geometric condition rather than the gravitational dynamics of matter therein. So, it is useful to construct examples which are not self similar and which develop naked singularities. In this context, it is worthwhile to examine the gravitational collapse of charged radiation shells in an expanding deSitter background. The metric for this purpose is already known [17]. We find that strong curvature singularities do arise in charged Vaidya-deSitter spacetime providing an explicit counter- example to the CCH. An interesting point to note is that by introduction of charge, naked singularities cannot be avoided.

The charged Vaidya-deSitter metric in (v, r, θ, ϕ) coordinates is [17]

$$ds^2 = -\left(1 - \frac{2m(v)}{r} + \frac{e^2(v)}{r^2} - \Lambda\frac{r^2}{3}\right) dv^2 + 2dvdr + r^2 d\Omega^2 \qquad (1)$$

where $d\Omega^2 = d\theta^2 + sin^2\theta d\phi^2$, v represents advanced Eddington time, in which r is decreasing towards the future along a ray $v = const.$, the two arbitrary functions $m(v)$ and $e(v)$ (which are restricted by the energy conditions), represent, respectively, the mass and electric charge at advanced time v, and Λ is the cosmological constant. This metric (1) represents a solution to the Einstein equations for a collapsing charged null fluid in an expanding deSitter background.

The energy momentum tensor can be written in the form

$$T_{ab} = T_{ab}^{(n)} + T_{ab}^{(m)} \tag{2}$$

where

$$T_{ab}^{(n)} = \frac{1}{4\pi r^3}\left[r\dot{m}(v) - e(v)\dot{e}(v)\right] k_a k_b \tag{3}$$

with the null vector k_a satisfying $k_a = -\delta_a^v$ and $k_a k^a = 0$. The part of the energy momentum tensor $T_{ab}^{(m)}$ is related to the electromagnetic tensor F_{ab}:

$$T_{ab}^{(m)} = \frac{1}{4\pi}\left(F_{ac}F_b^c - \frac{1}{4}g_{ab}F_{cd}F^{cd}\right) \tag{4}$$

which satisfies Maxwell's field equations

$$F_{[ab;c]} = 0 \, and \, F_{ab;c}g^{bc} = 4\pi J_a \tag{5}$$

where J_a is the four-current vector.

Clearly, for the weak energy condition to be satisfied we require the bracketed quantity in eq. (3) to be non negative. We note that the stress tensor in general may not obey the weak energy condition. In particular, if $dm/de > 0$ then there always exists a critical radius $r_c = e\dot{e}/\dot{m}$ such that when $r < r_c$ the weak energy condition is always violated. However, in realistic situations, the particle cannot get into the region $r < r_c$ because of the Lorentz force and so the energy condition is still preserved [17, 18].

The Kretschmann scalar ($K = R_{abcd}R^{abcd}$, R_{abcd} is the Riemann tensor) for the metric (1) reduces to

$$K = \frac{48}{r^6}\left[m^2(v) - \frac{2}{r}e^2(v)m(v) + \frac{7}{6}\frac{e^4(v)}{r^2}\right] + \frac{8}{3}\Lambda^2 \tag{6}$$

So the Kretschmann scalar diverges along $r = 0$, establishing that metric (1) is scalar polynomial singular. The Weyl scalar ($C = C_{abcd}C^{abcd}$, C_{abcd} is the Weyl tensor) reads

$$C = \frac{48}{r^6}\left[m^2(v) - \frac{2m(v)e^2(v)}{r} + \frac{e^4(v)}{r^2}\right] \tag{7}$$

which also diverges along $r = 0$.

3 Existence and Nature of Naked Singularities

The physical situation is that of a radial influx of charged null fluid in an initially empty region of the deSitter universe. The first shell arrives at $r = 0$ at time $v = 0$ and the final at $v = T$. A central singularity of growing mass developed at $r = 0$. For $v < 0$ we have $m(v) = e(v) = 0$, i.e., an empty deSitter metric, and for

$v > T$, $\dot{m}(v) = \dot{e}(v) = 0$, $m(v)$ and $e^2(v)$ are positive definite. The metric for $v = 0$ to $v = T$ is charged Vaidya-deSitter, and for $v > T$ we have the Reissner Nordström deSitter solution.

Let $K^a = dx^a/dk$ be the tangent vector to the null geodesic, where k is an affine parameter. The geodesic equations, on using the null condition $K^a K_a = 0$, take the simple form

$$\frac{dK^v}{dk} + \frac{1}{r}\left[\frac{m(v)}{r} + \frac{e^2(v)}{r^2} - \frac{\lambda}{3}r^2\right](K^v)^2 = 0 \tag{8}$$

$$\frac{dK^r}{dk} + \left[\frac{\dot{m}(v)}{r} - \frac{e(v)e'(v)}{r^2}\right](K^v)^2 = 0 \tag{9}$$

Radial (θ and $\phi = const.$) null geodesics of the metric (1) must satisfy the null condition

$$\frac{dr}{dv} = \frac{1}{2}\left[1 - \frac{2m(v)}{r} + \frac{e^2(v)}{r^2} - \Lambda\frac{r^2}{3}\right] \tag{10}$$

Clearly, the above differential equation has a singularity at $r = 0$, $v = 0$. The nature (a naked singularity or a black hole) of the collapsing solutions can be characterized by the existence of radial null geodesics coming out from the singularity.

In order to get an analytical solution, we choose $m(v) \propto v$ and $e^2(v) \propto v^2$ [19]. To be specific we take

$$m(v) = \begin{cases} 0, & v < 0, \\ \lambda v (\lambda > 0) & 0 \leq v \leq T, \\ m_0 (> 0) & v > T. \end{cases} \tag{11}$$

and

$$e^2(v) = \begin{cases} 0, & v < 0, \\ \mu^2 v^2 & 0 \leq v \leq T, \\ e_0^2 & v > T. \end{cases} \tag{12}$$

When $\Lambda = 0$, the spacetime is self similar, admitting a homothetic Killing vector and the singularities have been analyzed in [16]. However if $\Lambda \neq 0$, the basic requirement of self similarity [20] is not met. So, the line element (1) does not admit any proper conformal Killing symmetries. Eq. (10), upon using eqs. (11) and (12) turns out to be

$$\frac{dr}{dv} = \frac{1}{2}\left[1 - 2\lambda X + \mu^2 X^2 - \Lambda\frac{r^2}{3}\right] \tag{13}$$

where $X \equiv v/r$ is the tangent to a possible outgoing geodesic. In order to determine the nature of the limiting value of X at $r = 0$, $v = 0$ on a singular geodesic, we let

$$X_0 = \lim_{r\to 0\; v\to 0} X = \lim_{r\to 0\; v\to 0} \frac{v}{r} \tag{14}$$

Using (13) and L'Hôpital's rule we get

$$X_0 = \lim_{r\to 0\; v\to 0} X = \lim_{r\to 0\; v\to 0} \frac{v}{r} = \lim_{r\to 0\; v\to 0} \frac{dv}{dr} = \frac{2}{1 - 2\lambda X_0 + \mu^2 X_0^2} \tag{15}$$

which implies,

$$\mu^2 X_0^3 - 2\lambda X_0^2 + X_0 - 2 = 0 \tag{16}$$

This algebraic equation governs the behavior of the tangent vector near the singular point. Thus by studying the solution of this algebraic equation, the nature of the singularity can be determined. The central shell focusing singularity is at least locally naked (for brevity we have addressed it as naked throughout this paper), if eq. (16) admits one or more positive real roots. When there are no positive real roots to eq. (16), the central singularity is not naked because in that case there are no out going future directed null geodesics from the singularity. Hence in the absence of positive real roots, the collapse will always lead to a black hole. Thus, the occurrence of positive real roots implies that the CCH is violated. We now examine the condition for the occurrence of a naked singularity.

An interesting point note from eq. (16) is that it admits at least one positive root for $\lambda > 0$ and $\mu^2 > 0$ and no negative roots [21], e.g., eq. (16) has the positive root $X_0 = 3.50976$ for $\lambda = \mu = 1/2$ and $X_0 = 1$ for $\lambda = 0$, $\mu = 1$. It is easy to see that eq. (16) admits all three positive roots if $\lambda^2 + 18\lambda\mu^2 \geq 16\lambda^3 + \mu^2 + 27\mu^4$. This happens when $\lambda \leq 0.082$ and $\mu \leq 0.094$ (correct to three decimal places). It is then easy to check that positive roots of Eq. (16) $X_0 = 4.839$, 6.318 and 7.403 correspond to $\lambda = 0.082$ and $\mu = 0.094$, whereas for $\lambda = 0.05$, $\mu = 0.05$, the roots are $X_0 = 2.661$, 11.748 and 25.591. For all such values of X_0, the singularity will be naked. It follows that the gravitational collapse of a charged null fluid in the de-Sitter background must lead to a naked singularity irrespective of the values of the parameter (λ, μ).

The charged Vaidya metric can be obtained by taking $\Lambda = 0$ in (1). However eq. (16) remains unchanged. The results of collapsing shells in an expanding deSitter background are similar to that of collapsing shells of radiation in a Minkowskian background, as should have been expected, since when $r \to 0$ the cosmological term $\lambda r^2/3$ is negligible. The global nakedness of the singularity can then be seen by making a junction onto Reissner Nordström deSitter spacetime.

When $\mu = 0$, the metric (1) is Vaidya-deSitter, and eq. (16) admits positive roots when $0 < \lambda \leq 1/16$ and hence singularities are naked for $0 < \lambda \leq 1/16$ [9, 10], which have been shown to be gravitationally strong in [11]. When $\mu = 0$ and $\Lambda = 0$, we get the Vaidya metric, and the singularities are again naked for $0 < \lambda \leq 1/16$ (see [2], for a review). Hence our unified treatment reduces under appropriate conditions to the Vaidya, charged Vaidya and Vaidya-deSitter cases which have been discussed previously.

Strength of Naked Singularities: The strength of a singularity is an important issue because there have been attempts to relate it to stability [22]. A singularity is termed gravitationally strong or simply strong, if it destroys by crushing or stretching any object which falls into it. If a singularity is strong, then it is believed that the spacetime cannot be extended through it, in contrast to a weak singularity. Recently, Nolan [23] gave an alternative approach to check the nature of singularities without having to integrate the geodesic equations. It was shown in [23] that a radial null geodesic which runs into $r = 0$ terminates in a gravitationally weak singularity if and only if $\dot{r}$ is finite in the limit as the singularity is approached (this occurs at $k = 0$), the over-dot here indicates differentiation along the geodesics. So assuming a weak singularity, we have

$$\dot{r} \sim d_0 \qquad r \sim d_0 k \tag{17}$$

Using the asymptotic relationship above and Eqs. (11), (12) , the geodesic equations yield

$$\ddot{v} \sim -(\lambda X_0 d_0^{-1} k^{-1} - \mu^2 X_0^2 d_0^{-1} k^{-1} - \frac{\lambda}{3} d_0 k) d_0^2 X_0^2 \tag{18}$$

But this gives

$$\ddot{v} \sim c k^{-1}, \tag{19}$$

where $c = (\lambda - \mu^2 X_0) X_0 d_0^{-1}$, which is inconsistent with $\dot{v} \sim d_0 X_0$, which is finite. Thus if the coefficient c of k^{-1} is non-zero, the singularity is gravitationally strong. This may be false in the case $c = 0$, which is equivalent to $X_0 = \lambda/\mu^2$. But inserting this into the root eq. (16) gives

$$\mu^2 = \frac{\lambda}{4}\left(1 \pm (1 - 8\lambda)^{1/2}\right) \tag{20}$$

Thus $c = 0$ corresponds to a set (in fact a closed curve) of measure zero in (λ, μ) parameter space and so is not of physical significance. Therefore, one may say that generically, the naked singularities is gravitationally strong in the sense of Tipler [24]. (We are grateful to Brien Nolan for the above argument.)

Having seen that the naked singularity in our model is a strong curvature singularity, we check to see if it is a scalar polynomial one. The Kretschmann scalar with the help of eqs. (11) and (12), takes the form

$$K = \frac{48}{r^4}\left(\lambda^2 X^2 - 2\lambda\mu^2 X^3 + \frac{7}{6}\mu^4 X^4\right) + \frac{8}{3}\Lambda^2 \tag{21}$$

which diverges at the naked singularity and hence the singularity is a scalar polynomial singularity. Next, we now turn our attention to the Weyl tensor. It can be noted that the Weyl scalar is zero in the deSitter region ($v < 0$). Eq. (7) can be written as

$$C = \frac{48}{r^4}\left[\lambda^2 X^2 - 2\lambda\mu^2 X^3 + \mu^4 X^4\right] \tag{22}$$

Thus the Weyl scalar also diverges at the naked singularity. The effect of the energy momentum tensor on the geometry can be found by evaluating the Ricci scalar ($R = R_{ab}^{ab}$, R_{ab} the Ricci tensor) and comparing it with the Weyl scalar [25]. The Ricci scalar for the metric (1) is

$$R = 4\left(\Lambda^2 + \frac{e^4(v)}{r^8}\right) = 4\left(\Lambda^2 + \frac{\mu^4 X^4}{r^4}\right) \tag{23}$$

In contrast to the uncharged Vaidya space time [25], in this case, the Ricci scalar diverges at the same rate as the Weyl scalar. So the Weyl scalar does not dominate in our case. The behavior of the Ricci scalar is greatly affected by the presence of charge. Therefore, one can say that the role played by initial conditions is equally important to that of the energy momentum tensor whenever a naked singularity occurs. Further, the Weyl tensor is that part of the curvature of spacetime that is not locally determined by matter, and hence the divergence of the Weyl scalar at the naked singularity implies that the singularity is not associated with the local matter distribution and hence should be taken seriously (see [25], for more details).

4 Unresolved issues

We have looked at an example showing the existence of naked singularities. However, this is not a physically realistic example. There have been many analytical and numerical studies carried out in spherically symmetric collapse involving different types of fluid, such as dust, perfect fluids, imperfect fluids and massless scalar fields (see, e.g., [2]). In all cases, the general conclusion seems to be the same, i.e., either a black hole or a naked singularity forms depending upon the initial conditions. These conditions are the density and velocity profiles and the rate of infall of the collapsing matter. The matter content of these models respects the energy conditions, and the collapse starts from regular initial data. We now discuss some unresolved issues related to the CCH.

4.1 Non-spherical collapse

Most of the studies in gravitational collapse have been carried out for the spherically symmetric case. There has been very little work in the collapse of non-spherical matter. It seems that naked singularities also occur in this case [4, 26]. However, more work remains to be done on non-spherical collapse.

4.2 Stability and genericity

Starting from physically realistic matter, do naked singularities form generically, i.e., given the set of all possible initial data, is there a set of nonzero measure which will lead to naked singularities. This is a difficult problem to address, as there is no indication of what measure or topology to impose on the initial data. This question is related to the notion of stability. Again, there is no generally accepted notion of stability in general relativity theory. The question usually addressed is if a naked singularity occurs, is it stable in some sense against perturbations of the initial data. Investigations have been made in this regard [27, 28, 29, 30, 31], but the general conclusion is still not clear.

4.3 Properties of naked singularities

Studies thus far show the existence of naked singularities. Further, there exists a set of non-spacelike trajectories emerging from the singularity, and this set is of nonzero measure, in some sense. In most cases, the naked singularity is a strong curvature singularity, so that it is not possible to extend the spacetime beyond the singularity. Some authors, such as [4], believe that fluids should not be used to study collapse as they admit shell crossing singularites and shocks.

4.4 Quantum effects

As collapse proceeds and one reaches the Planck length, quantum effects come into play, and one frequently hears of the statement that the singularities will be smeared out. However, by the time that one reaches the Plank length, one is in a very strong gravitational field, and for all practical purposes, one may regard this as a classical singularity. Hod and Piran [32] have carried out a very interesting investigation in which charge is slowly lowered into a black hole. They showed that classically, it is possible to destroy the horizon of the black hole, thereby leading to a naked singularity, and a violation the CCH. However, when quantum effects were taken into account, the CCH was enforced. This seems to suggest that the CCH may only be enforced by quantum gravity. On the other hand, if naked singularities do occur, then one could possibly test the laws of quantum gravity by studying collapse.

4.5 Thermodynamics

Even if naked singularities are admitted by general relativity, there may be some physical principle which forbids their occurrence. An argument along these lines was put forward by Barve and Singh [25]. It is known that the initial gravitational entropy during the early universe was low, and Penrose suggested that a suitable measure of this entropy is the Weyl curvature tensor. He proposed the Weyl curvature hypothesis, which states that the Weyl curvature should be zero, or at least negligible compared to the Ricci tensor, at the initial cosmological singularity.

Barve and Singh then suggested that a naked singularity should be compared to the initial singularity, and that the Weyl scalar should be negligible in the approach to a naked singularity. However, by looking at some examples, they found that the Weyl scalar diverges. Hence, they propose that naked singularites are ruled out by the second law of thermodynamics.

4.6 Observational consequences

Naked singularities could emit gravity waves [29]. Other obsevational consequences have been discussed recently by Krolak [33] in terms of the spin angular momentum per unit mass. These are observations of pulsars around a compact object and measurements of X-ray binary spectra from accretion discs. Naked singularities may also provide an origin for gamma ray bursts [34].

5 Conclusion

We have given a brief review of the CCH, studied an example, and indicated some unresolved issues. The CCH remains one of the most important unsolved problems in classical general relativity theory. Whilst the answer is still unknown, there has been some progress. Some of the earlier attempts to formulate the conjecture have been shown not to work. For example, naked singularities still occur when the energy conditions are satisfied, when allowance is made for pressure and when a reasonable equation of state is employed. Further, it was earlier thought that if naked singularities do occur, they will be gravitationally weak and removable, or that they will form a set of zero measure. These have been shown to be false. A proper formulation of the CCH, and its proof, perhaps using quantum gravity, remains one of the most challenging problems today.

References

[1] S. W. Hawking and G. F. R. Ellis, "The Large Scale Structure of Spacetime", Cambridge University Press, Cambridge, 1973.

[2] P. S. Joshi, "Global Aspects in Gravitation and Cosmology", Cambridge University Press, Cambridge, 1993; C. J. S. Clarke, Class. Quantum Grav., **10**, 1375, 1993; T. P. Singh, in "Classical and Quantum Aspects of Gravitation and Cosmology", eds. G. Date and B. R. Iyer, Institute of Mathematical Sciences Report 117, 1996; T. P. Singh, gr-qc/9805066; P. S. Joshi, gr-qc/0006101; S Jhingan and G. Magli, gr-qc/9903103.

[3] R. Penrose, Riv. del Nuovo Cim., **1**, 252, 1969.

[4] R. M. Wald, gr-qc/9710068.

[5] see, for example, Joshi (Ref. 2).

[6] D. Markovic, S. L. Shapiro, Phys. Rev., **D 61**, 084029, 2000.

[7] Lake, K. gr-qc/0002044 (unpublished)

[8] R. Cai, L. Qiao and Y. Zhang, Mod. Phys. Lett. **A 12**, 155, 1997.

[9] J. P. S. Lemos, Phys. Rev. D **59**, 044020, 1999.

[10] S. M. Wagh, S. D. Maharaj, Gen. Rel. Grav., **31**, 975, 1999.

[11] S. G. Ghosh and A. Beesham, Phys. Rev. D, **61**, 067502, 2000.

[12] S. W. Hawking and R. Penrose, "The Nature of Space and Time", Princeton University Press, Princeton, 1996.

[13] P. C. Vaidya, Proc. Indian Acad. Sci., **A33**, 264, 1951; Reprinted, Gen. Rel. Grav., **31**, 119, 1999.

[14] P. C. Vaidya, and K. B. Shah, Proc. Nat. Inst. Sci., **23**, 534, 1957.

[15] W. B. Bonnor and P. C. Vaidya, Gen. Rel. Grav., **1**, 159, 1970.

[16] K. Lake and T. Zannias, Phys. Rev. **D 43**, 1798, 1991.

[17] A. Wang, and Y. Wu, Gen. Rel. Grav., **31**, 107, 1999; A. Patino and H. Rago, Phys. Lett., **A 121**, 329, 1987.

[18] A. Ori, Class. Quantum Grav., **8**, 1559, 1991.

[19] $m \propto v$ was introduced by A. Papapetrou (see Ref 5 above) and subsequently used by many authors, $e^2 \propto v^2$ has been examined by Lake and Zannias (see Ref 7 above).

[20] A spherical symmetric space-time is self similar if $g_{tt}(ct, cr) = g_{tt}(t, r)$ and $g_{rr}(ct, cr) = g_{rr}(t, r)$ for every $c > 0$. A self similar spacetime is characterized by the existence of a homothetic Killing vector.

[21] First we note that eq. (16) being a cubic equation, it must have one real root. From theory of equation it is easy to see, for $\lambda > 0$ and $\mu^2 > 0$, any real root of the eq. (16) must be positive as negative values of X_0 do not solve this equation.

[22] S. S. Deshingkar, P. S. Joshi and I. H. Dwivedi, Phys. Rev., **D 59**, 044018, 1999.

[23] B. C. Nolan, Phys. Rev. D, **60**, 024014, 1999.

[24] F. J. Tipler, C. J. S. Clarke, and G. F. R. Ellis, in "General Relativity and Gravitation", ed. A. Held, Plenum, New York, 1980.

[25] S. Barve and T. P. Singh, Mod. Phys. Lett., **A12**, 2415, 1997.

[26] S. L. Shapiro and S. A. Teukolsky, Phys. Rev. Lett., **66**, 994, 1991; ibid. Phys. Rev., **D45** 2006, 1992; P. S. Joshi and A. Krolak, Class. Quantum Grav., **13** 3069, 1996; T. Nakamura, M. Shibata and K. Nakao, Prog. Theor. Phys., **89** 821, 1993.

[27] D. Christodoulou, Ann. Math., **140**, 607, 1994; **149**, 183, 1999.

[28] M. W. Choptuik, Phys. Rev. Lett., **70**, 9, 1993; C. Gundlack, gr-qc/0001046.

[29] H. Iguchi, K. Nakao and T. Harada, Phys. Rev., **D57**, 7262, 1998; Prog. Thoer. Phys., **101**, 1235, 1999.

[30] R. V. Saraykar and S. H. Ghate, Class. Quantum Grav., **16**, 281, 1999.

[31] F. Mena, R. K. Tavakol and P. S. Joshi, gr-qc/0002062.

[32] S. Hod and T. Piran, gr-qc/0011003.

[33] A. Krilak, Prog. Theor. Phys. Suppl., **136**, 45, 1999.

[34] P. S. Joshi, N. Dadhich and R. Maartens, Mod. Phys. Lett. A, **15**, 991, 2000; S. K. Chakravarti and P. S. Joshi, Int. J. Mod. Phys. D., **3**, 647, 1994.

Gravitational Wave and Spiral Galaxy (Gravito-Radiative Force)

Masataka Mizushima
Department of Physics, University of Colorado
Boulder, Colorado 80309 U.S.A.
e-mail:mizushima@colorado.edu

1 Gravito-Radiative Force

Einstein[1] proposed a general relativistic wave equation, which reduces to

$$\left(\nabla^2 - \left(\frac{\partial}{c\partial t}\right)^2\right)\phi_i^j = \frac{16\pi G}{c^4}T_i^j, \tag{1}$$

when we introduce gravitational potentials through

$$g_{00} = 1 - \frac{1}{2}(\phi_{00} + \phi_{11} + \phi_{22} + \phi_{33}), \tag{2}$$

$$g_{11} = 1 - \frac{1}{2}(\phi_{00} + \phi_{11} - \phi_{22} - \phi_{33}), \tag{3}$$

$$g_{22} = -1 - \frac{1}{2}(\phi_{00} - \phi_{11} + \phi_{22} - \phi_{33}), \tag{4}$$

$$g_{33} = -1 - \frac{1}{2}(\phi_{00} - \phi_{11} - \phi_{22} + \phi_{33}), \tag{5}$$

and

$$g_{01} = \phi_{01}, g_{12} = \phi_{12}, etc., \tag{6}$$

and assume that

$$\frac{\partial \phi_j^i}{\partial x^i} = 0 \tag{7}$$

and that all quantities nonlinear in ϕ_j^i are negligible.
In an empty space where all components of T_i^j are zero, (1) predicts that the gravitational potentials, ϕ_i^j, propagate through with speed c. This is the gravitational wave.
In eq.(1), T_i^j is the stress-energy tensor of a distributed mass. Assuming a point mass M, moving with velocity v, located at a distance r from the observation point p for the source, eq. (1) can be solved to obtain

$$\phi_{00}(p,t) = \frac{4GM}{c^2 r'}|t', \tag{8}$$

Frontiers of Fundamental Physics 4, Edited by Sidharth and Altaisky,
Kluwer Academic/Plenum Publishers, New York, 2001

$$\phi_{0\alpha}(p,t) = -\frac{4GMv_\alpha}{c^3 r'}|t', \tag{9}$$

$$\phi_{\alpha\beta}(p,t) = \frac{4GMv_\alpha v_\beta}{c^4 r'}|t', \tag{10}$$

where $\alpha, \beta = 1, 2,$ or 3, and

$$r' = r + (\mathbf{v}\cdot\mathbf{r}/c) \quad t' = t - (r/c). \tag{11}$$

Equation (11) shows a retardation effect, but we will neglect that in the following. Assuming a continuity between the mass source and the gravitational waves produced, Landau and Lifshitz[2] derived an expression of the rate of the energy loss, $-(d\mathcal{E}/dt)$, of the mass source due to the emission of the gravitational waves as

$$-\frac{d\mathcal{E}}{dt} = \frac{2G}{45c^3}(\sum M(3v_\alpha v_\beta - \delta_{\alpha\beta} v\dot{v})^2). \tag{12}$$

Taylor and Weisberg[3] found that the decay of the orbital motion of binary star PSR $1913+16$ follows this theoretical pprediction. Thus the gravitational wave is established.

Einstein[1] also proposed an equation of motion for a test particle with coordinates x^i as

$$\frac{d^2x^i}{ds^2} = -\Gamma^i_{jk}\frac{dx^j}{ds}\frac{dx^k}{ds} \equiv F^i/(mc^2), \tag{13}$$

where Γ^j_{jk} is a component of Christoffel symbols. The last expression of this equation defines the Einstein force, F^i, to be discussed below, and m is the mass of the test particle.

If we take the gravitational potentials instead of the metric, as eqs.(2) through (6) give, in the linear theory, we obtain[4] $s = ct$ and

$$\frac{d^2\mathbf{r}}{dt^2} = \mathbf{F}_E/m, \tag{14}$$

where

$$\mathbf{F}_E/m = -\frac{c^2}{4}\nabla\phi_{00} - c\frac{\partial\phi}{\partial t} + c\mathbf{V}\times(\nabla\times\phi)$$
$$+\frac{\partial[\mathbf{v}(\mathbf{V}\cdot\phi)]}{c\partial t} + \left[(\mathbf{V}.\nabla)\phi - \frac{1}{2}\nabla(\phi\cdot\mathbf{V})\right]\frac{\mathbf{V}\cdot\mathbf{v}}{c} \tag{15}$$

is the space component of the Einstein force introduced in eq. (13). The first term of the Einstein force, $\mathbf{F}_E$, is the Newtonian gravitational force. The first term in the second line of eq. (15) is called the gravito-radiative force (acceleration). In this term, $\mathbf{V}$ is the velocity of the test particle, and we do not differentiate $\mathbf{V}$ by time, t. This and the last term of eq. (15) are due to the tensor potentials introduced in eq. (10).

Taking the expressions of the gravito-vector potentials given by eq. (9), the gravito-radiative force (acceleration) term can be written as

$$\mathbf{F}_{rad}/m = -\frac{4G}{c^4 r}\sum M_a(\mathbf{v}_a(\mathbf{V}\cdot\dot{\mathbf{v}}_a) + \dot{\mathbf{v}}_a(\mathbf{v}_a\cdot\mathbf{V})), \tag{16}$$

where the summation is over all source masses M_a.

2 Formation of Spiral Galaxies

More than half of the galaxies, including our Milky Way galaxy, are spiral galaxies. If gravitational radiation (a gravito-radiation force) is due to a head-on collision between two black holes in a galactic center, the acceleration $\dot{\mathbf{v}}_a$ and the velocity $\mathbf{v}_a$ of M_a would be parallel, or antiparallel, to each other. Those for the partner, namely $M_b, \dot{\mathbf{v}}_b$ and $\mathbf{v}_b$, would also be parallel, or antiparallel, to each other, and $M_a\dot{\mathbf{v}}_a = -M_b\dot{\mathbf{v}}_b$ because of Newton's third law. Taking the coordinate system comoving with the galactic nucleus, this also means that $M_a\mathbf{v}_a = -M_b\mathbf{v}_b$. In the galactic nucleus it is likely that $v/c \approx 1$. Putting these assumptions together, we see that eq. (16) can be simplified to

$$\mathbf{F}_{rad}/m = \pm\frac{8G\Delta t}{c^4 r}M\dot{\mathbf{v}}(\dot{\mathbf{v}}\cdot\mathbf{V}), \tag{17}$$

where we assumed that $\mathbf{v} = \pm\dot{\mathbf{v}}\Delta t$. The double sign corresponds to the possibility of v being either parallel or antiparallel to $\dot{\mathbf{v}}$. It is likely that the galactic explosion we are discussing takes place in about

$$\Delta t = r_0/c \approx 10s, \tag{18}$$

where r_0 is a dimension of the skin of the black hole of mass M.
We assume that the galaxy was a circular flat disk with stars orbiting around the galactic nucleus at distance r with speed

$$V(r, \text{before}) = \sqrt{\frac{GM}{r}} \tag{19}$$

If the acceleration $\dot{\mathbf{v}}$ was in this plane of the circular orbit, the stars moving in the part of the circular orbit of radius r where $\mathbf{V}$ is parallel to $\dot{\mathbf{v}}$ will be accelerated by $(\mathbf{F}_{rad}/m)$ for the duration time Δt given by eq. (18). At the end of this impulse, the speed of a star orbiting at this side of the galactic center increases by

$$V(r,\text{after}) \pm V(r,\text{before}) = (\mathbf{F}_{rad}\Delta t/m)$$
$$= \frac{8GMv\dot{v}(\Delta t)}{c^4 r}V(r,\text{before}) = \frac{8(GM)^{3/2}}{c^2 r^{3/2}}, \tag{20}$$

where the double sign corresponds to the possibility of $\mathbf{F}_{rad}$ being either parallel or antiparallel to the local $\mathbf{V}$, that is, $\mathbf{V}$ and $\dot{\mathbf{v}}$ are parallel or antiparallel to each other. In eq. (20) we assumed that

$$v\dot{v}\Delta t = (\dot{v}\Delta t)^2 = c^2. \tag{21}$$

After the star receives this excess kinetic energy from the gravitational wave, that star goes out of the orbit until it comes to a distance r_{final} where its speed $V(r_{final},$ satisfies

$$V(r_{final}) = \sqrt{\frac{GM}{r_{final}}} \tag{22}$$

just like eq. (19). This means that the final orbit is circular. The total energy per unit mass is assumed to be conserved as

$$c^2 + \frac{1}{2}(V(r,\text{after}))^2 - \frac{GM}{r} = c^2 - \frac{GM}{2r_{final}}, \tag{23}$$

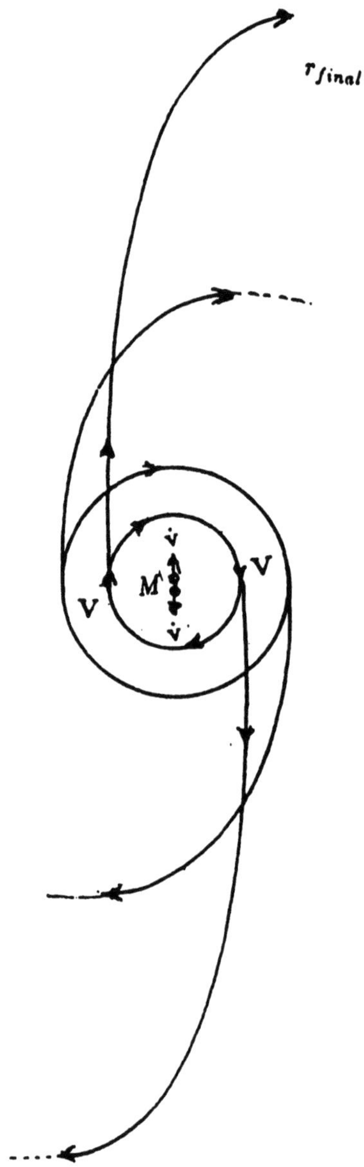

Figure 1: Typical orbits for stars before and after the galactic nuclear explosion, taking $\mathbf{F}_{rad}/m = (8G\Delta t/c^4 r)M\dot{\mathbf{v}}(\dot{\mathbf{v}} \cdot \mathbf{V})$.

using eq. (22). If, for simplicity, we assume that $V(r_{final})$ is much smaller than $V(r, \text{after})$, then from eqs. (20) and (23), we see

$$\frac{1}{r_{final}} = \frac{2}{r} - \frac{32(GMv\dot{v}\Delta t)^2}{c^8 r^3} = \frac{2}{r} - \frac{32(GM)^2}{c^4 r^3} \tag{24}$$

In this rough estimation ($v\dot{v}\Delta t = (\dot{v}\Delta t)^2 = c^2$), we see that if $M = 10^{39}kg, r_{final}$ calculated using eq. (23) becomes infinity if the initial orbital radius, $r_{critical}$, is

$$r_{critical} = \frac{4GM}{c^2} = 3 \times 10^{12}m, \text{if} \quad Mv\dot{v}\Delta t = 10^{56}J, \tag{25}$$

Those stars that were orbiting with a radius less than $3 \times 10^{12}m$ are lost into outer space, whereas those stars that were orbiting with a radius larger than $3 \times 10^{12}m$ now stay in new larger orbits of the galaxy after the explosion.

The spiral arms formed in this way would be almost radial at the beginning. The tangential orbiting speed of a star at the final radius is given by eq.(24) after the explosion. The angular speed, ω, of each star orbiting at r_{final} is given by

$$\omega = \frac{V(r_{final})}{r_{final}} = \frac{(GM)^{1/2}}{r_{final}^{3/2}} \tag{26}$$

A typical spiral galaxy is $NGC7742$. We do not know its central mass M, nor its diameter, but let us take $M = 10^{39}kg$, and take $r = 10^{19}m$ as the bottom of its spiral arm, making about a 2π angle with respect to the end where ω looks to be about zero. Then we estimate that the explosion took place about 10^8 years ago in this galaxy.

A much more important prediction of the present theory, which is fulfilled in the NGC7742, NGC2997, and many other spiral galaxies, is that each of them has two spiral arms symmetrically extended from the center.

In the original circular orbit around the galactic center, those stars on one side of the circle have velocities $\mathbf{V}$ parallel to $\dot{\mathbf{v}}$ of the black hole at the explosion, resulting in the spiral arm extending from the galactic center, as we discussed above. For the other stars on the opposite side of the original circular orbit, the velocities $\mathbf{V}$ are opposite in direction and the same in magnitude as the above-discussed stars. Therefore, the second group of stars is emitted from the circular orbit in the direction opposite to that of the first group of stars, as shown in Fig. 1. The second group of stars thus results in a spiral arm extending from the galactic center in the direction opposite to that of the first group of stars. If the original orbit is circular, the resulting two spiral arms are symmetric under the 180° rotation. Other groups of stars moving in the parts of the original circular orbit between the two parts just discussed have $\mathbf{V}$ perpendicular to $\dot{\mathbf{v}}$. According to eq.(17), those stars are not influenced by the gravitation wave and stay in the original circular orbit.

Some spiral galaxies, such as NGC1232, have more than one pair of spiral arms but still maintain the 180° rotational symmetry. The Milky Way galaxy has four pairs of arms, and the 180° rotational symmetry we are discussing is recognizable. The age of each arm seems to be of the order of 10^8 years. These complex galaxies have experienced more than one galactic nuclear explosion.

If $\dot{\mathbf{v}}$ is perpendicular to the galactic plane, then such an explosion would not produce any spiral arm in the galaxy.

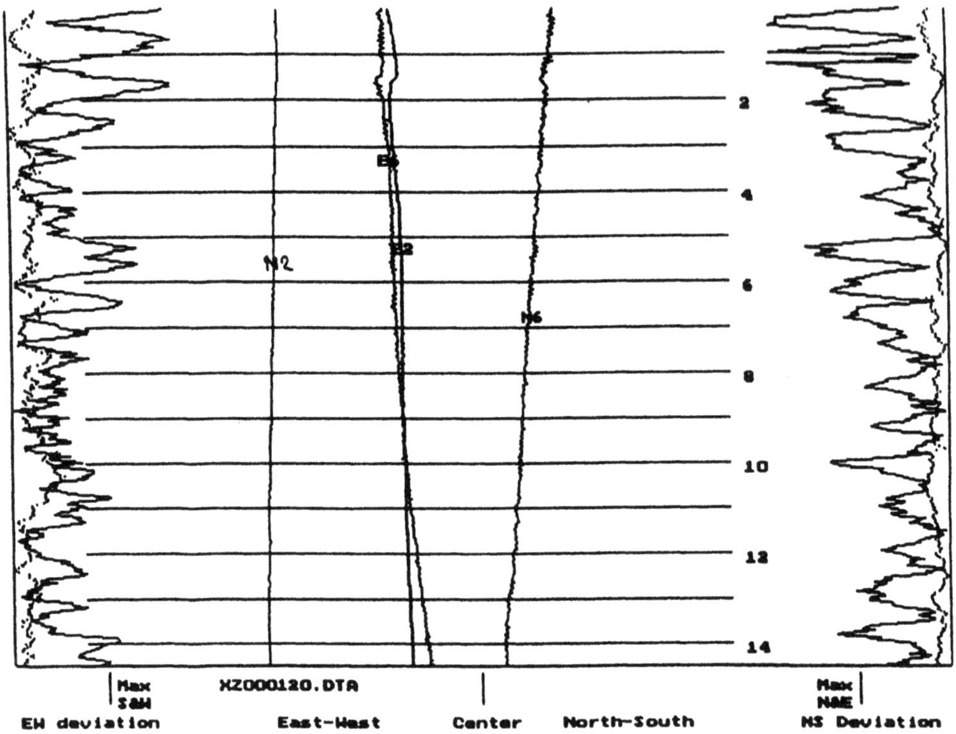

Figure 2: Coincidence of the average displacements of the depdulum bobs of the first two verticity meters (2-meter and 6-meter), observed at 1:45 universal time of January 20, 2000. Right-nad side is East and North, respectively, for the average displacement lines.

3 Detection of Gravitational Waves

Violent collisions between black holes, which create spiral arms, may have taken place only four times in the Milky Way galaxy, but many other collisions between black holes may be taking place, emitting gravito-radiative forces. In fact, the creation of spiral arms can take place only when **v** and $\dot{\mathbf{v}}$ are in the galactic plane, but many black-hole collisions may not satisfy such a condition.

As a gravitational wave (gravito-radiative force) comes to Earth, it is coupled with the rotational velocity **V** of Earth's crust making the tidal components of the gravito-radiative force[5]. Earth's crust at latitude Θ is moving in the East direction with

$$V = 464 cos\Theta m/s \tag{27}$$

with respect to the center of Earth.

Take the z axis along the North direction and x and y axes in the Earth-West plane, which contains an Earth-bound observation point. Let the spherical angles that $\mathbf{v}_a$ and $\dot{\mathbf{v}}$ make in the xyz system be Θ, ψ, and Θ', ψ', respectively. If the longitude of the observation point is Φ, then we obtain the East component of the gravito-radiative force due to the source as

$$F_{rad}(E)/m = -\frac{4GV}{c^4 r_g} \sum M_a v_a \dot{v}_a sin\Theta sin\Theta' sin(\Phi - \psi) sin(\Phi - \psi'), \qquad (28)$$

where r_g is the distance from the galactic center. The North component, which depends on the latitude of the observation point, Θ, also exists in general[6].
From eq. (28), we see that the tidal component of the gravito-radiative force depends on the longitude, Φ, of the observation point, or it tries to induce a mass quadrupole moment on Earth's crust. Earth's crust, however, is rigid because of the electromagnetic force with the quantum effect (chemical bond), and resists the gravitational effects. It is estimated that the gravito-radiative force can change a distance D between two points on the surface of earth by only $10^{-36}D$. A pendulum-bob, rotating around the center of Earth with Earth's crust, is free to move horizontally, following the gravito-radiative force, relative to the rigid Earth's crust[6].
A verticity meter is a motionless spherical pendulum with a device to measure and record the horizontal displacement of its bob. In Boulder, Colorado, we installed two verticity meters, 2 meter and 6 meters long, in 1997. Since then we detected at least 14 gravitational wave signals as coincidental impulses between them[6, 7, 8]. At the end of 1999, two more verticity meters were built, and coincidences among them were observed on December 18, 20, and 29, 1999; and January 13 and 20, 2000[8]. As an example, Fig. 2 shows a computer record of the displacements of the bobs of the two older verticity meters on Jan. 20, 2000. The center two lines, designated as $E2$ and $E6$, are the two-minute averages of the East-West displacements of the 2-meter and 6-meter verticity meters, respectively (left-hand side is West). The total width of the diagram corresponds to about $80\mu m$ towards the West at 1 : 45 universal time. The lines in the original record are colored to more easily distinguish the $E2$ and $E6$ lines. The other two vertical lines are $N2$ and $N6$ lines, showing North (right-hand side) and South displacements of the bobs of the 2-meter and 6-meter verticity meters, respectively, and no remarkable coincidental displacements are seen there. Fig. 3 shows more distinctive signals, coincidental to the signals shown in Fig. 2, in the third 2-meter verticity meter (#3), which was about four times more sensitive than the data shown in Fig. 2. The data shown in Fig. 3 are much noisier than those in Fig. 2, because #3 was more sensitive and more recently installed with insufficient air-shield. Another #4 verticity meter was installed by that time, but was not ready that day. The lines (dots) shown on the side of the diagrams are the mean square deviations of the two-minute interval. Again, the lines in the original diagrams are colored so that one can tell which mean square deviation is for which average line. Earthquakes and other noises are seen in the mean square deviations[9].
The kinks in the average lines $E2$ and $E6$ of Fig. 2 at 1 : 30, Jan. 20, 2000 are interpreted as due to the gravitation wave (impulse). Its duration time is much less than 2 minutes agreeing with our expectation, eq. (18). The following displacements for about $1\mu m$ toward West show a constant momentum of $10^{-8}m/8$. The potential for our pendulum-bob is assumed to be almost a flat for a small region of $1\mu m$ or less[9].

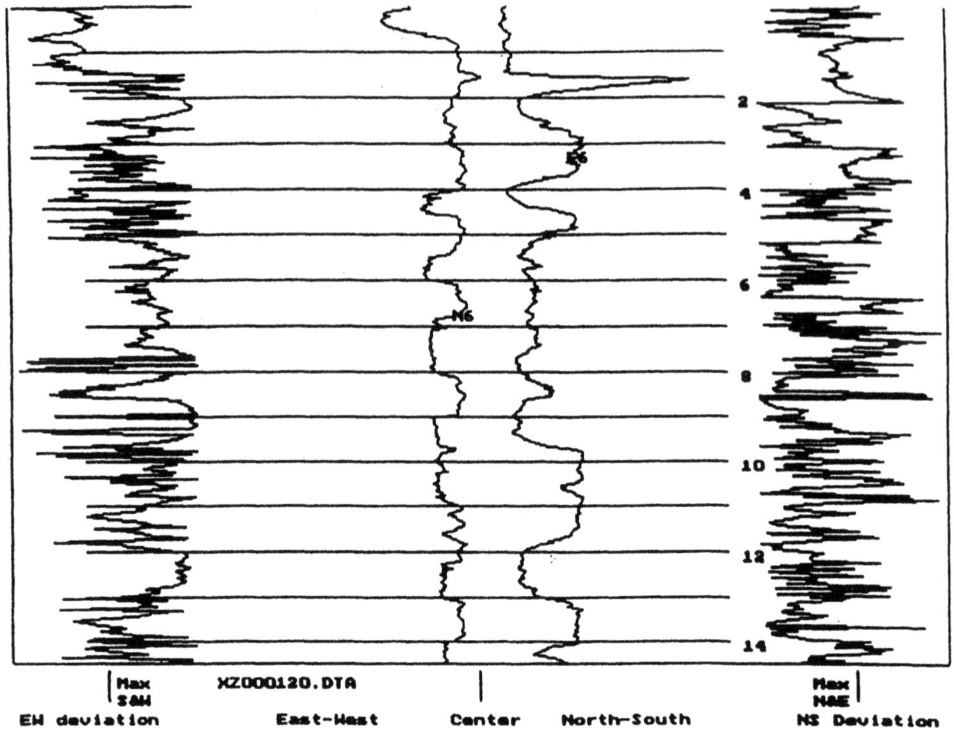

Figure 3: Coincidence of the average displacements of the pendulum bob of #3 verticity meter, to the signals shown in Fig. 2, at 1:45 universal time of January 20, 2000. The sensitivitly of #3 verticity meter is about four times that of the first verticity meters of Fig. 2. In this case, the right-hand side is West and South for the E6 and N6 lines, respectively, opposite to those of Fig. 2.

Taking eq.(28) as $F_{rad}(E)\Delta t/m = 10^{-8}m/8$, we obtain

$$Mv\dot{v}\Delta t = 10^{53}J, \tag{29}$$

assuming $r_g = 10^{20}m$ and $\Theta = \Theta' \approx \pi/2$, because the corresponding North-South displacements are very small. This numerical result is typical of all other experimental results we observed so far.

The sun must have received an impulse of 10^{-5} at the same time. This theory is also applicable to the orbital motions of a planet around the sun, the moon around Earth, and artificial satellites around Earth.

4 Is the Milky Way Galaxy Going to Explode?

In section 2, we assumed that $Mv\dot{v}\Delta t = Mc^2 = 10^{56}J$. In section 3, we reported that we observed gravitational waves in our laboratory with strengths corresponding to $Mv\dot{v}\Delta t = 10^{53}J$. This number is close to the above number. But the Milky Way galaxy did not explode on January 20, 2000, when we received a gravitational wave. When the galactic nucleus has mass M, the stars in a circular orbit with radius r must have a velocity given by eq.(19) in the nonrelativistic approximation. But when the resulting V is equal to c, the effective mass of such a star would be infinity, and stars cannot orbit inside this radius. This radius, called r_{rel}, is

$$r_{rel} = \frac{GM}{c^2} = 0.7 \times 10^{12} m; \text{if} \quad M = 10^{39} kg. \tag{30}$$

We saw in section 2 that the orbital radius of the stars that are kicked out into infinity by the gravito-radiative force must be less than $3 \times 10^{12} m$, as stated in the discussion below eq.(25). This radius is larger than the radius of the forbidden zone given by eq.(30).

Only those stars that have nonzero radial components of velocities can exist in this forbidden zone, but these stars have to collide with the galactic nucleus soon and cannot survive. It is noticed that although most of the spiral galaxies such as Andromeda and the Sombrero have bright center bulges, the Milky Way galaxy has no center bulge. It may mean that the forbidden zone of the Milky Way galaxy is almost empty.

Thus the same theory we presented in section 2, but with $Mv\dot{v}\Delta t = 10^{53} J$ as we obtained in the section 3, gives

$$r_{critical} = 3 \times 10^9 m$$

Assuming $M = 10^{39} kg$, this is inside the forbidden zone. If the forbidden zone were empty, there would be no star to be kicked out.

We still cannot answer whether an explosion will happen in the Milky Way galaxy. But we understand that one did not happen on January 20, 2000, nor at those times we have observed gravitational waves until now.

References

[1] A. Einstein, Ann.Phys., **49**, 769, 1916.

[2] L. Landau and E. Lifshitz, "The Classical Theory of Fields", Addison-Wesley, Cambridge, MA, 1951.

[3] H. Taylor and J.W. Weisberg, Astrophys. J. **345**, 134, 1989.

[4] M. Mizushima, Hadronic J, **17**, 97, 1994.

[5] M. Mizushima, "Verticity" of New Frontiers in Relativities, ed. L. Gill, Hadronic Press, Palm Harbor, FL, 1996, pp.171-271.

[6] M. Mizushima, "The Gravito-Radiative Force", in the proceedings of the Conference of Geometry and its Applications in Technology, Aristotle University of Thessaloniki, Greece, June 1999.

[7] M. Mizushima and R.W. Zimmerer, Hadronic J, **20**, 163, 1997.

[8] M. Mizushima, "Detection of Gravitational Waves", in the proceedings of the Third World Congress of Nonlinear Analysts, Catania, Italy, July 2000.

[9] M. Mizushima and R.W. Zimmerer, Canad.J.Phys. submitted.

Cosmology as a Format of Perceiving Reality

Masafumi Seriu
Department of Physics, Fukui University
Fukui 910-8507 (Japan)
Email:mseriu@edu00.f-edu.fukui-u.ac.jp

1 Introduction

We here study one of the fundamental issues in cosmology: To what extent a certain cosmological model traces the real history of the Universe. This significant problem has been raised and studied for a long time and it is often called the "model-fitting problem in cosmology" or the "averaging problem in cosmology" [1]. We have not yet reached a satisfactory understanding of this issue.

This problem is rooted in the very nature of cosmology: The total quantity of the geometrical information contained in our Universe is too enormous to be perceived by human intelligence, and we are forced to truncate the information to suit our brain capacity. In a sense this is what cosmology is all about; it is our attempt to grasp the reality, which is full of complexity, through a format of cognition, viz. a model. In cosmology, thus, there is in principle an inevitable discrepancy between the real Universe and its model.

Now the true problem arises when we attempt to guess the past or future of the Universe: There is no *a priori* guarantee that the model remains a valid one when it is evolved in time backward or forward according to the Einstein equation, a set of highly non-linear differential equations. We need a guarantee that the inevitable discrepancy between the reality and its model does not develop so much within the time scale of concern. Unless this uneasiness is cleared we cannot safely state anything meaningful for the past or future of our Universe. One might symbolically state the problem depicted here as *"Whether cosmology is possible?"*

There are several points to be clarified when tackling this problem. Among them, we start our analysis by focusing on

(a) How to quantify "discrepancy" or "closeness".

(b) How to take into account the influence of the "coarseness" of truncating information.

(c) How to take into account the types of observations performed for collecting information.

In a series of work, we have constructed a formalism for analyzing geometrical structures which is especially suitable for studying the type of problems such as the issue of reality and its model. The formalism naturally takes care of the above (*a*)-(*c*) also. In Section 2, we will summarize the basic setting of this formalism, or the

Frontiers of Fundamental Physics 4, Edited by Sidharth and Altaisky,
Kluwer Academic/Plenum Publishers, New York, 2001

spectral scheme for simplicity. In Section 3, the spectral evolution equations, one of the fundamental ingredients of the spectral scheme, are introduced. In Section 4, the spectral scheme is applied to analyze the issue of reality and model. We derive the discrepancy equations for two nearby geometries and formulate the criteria for a model to be a suitable one. Section 5 is devoted for discussions.

2 Spectral scheme

We here summarize the spectral representation scheme for discussing the issue of models in cosmology in later sections.

The basic idea of the spectral scheme is easily understand: To characterize the geometry of a space, we use the vibration modes of a field on the space. One may associate this idea with the well-known phrase "Can one hear the shape of the drum?" [2] Imitating this phrase, our idea may be symbolically stated as "Let us hear the shape of the Universe!" [3]. However, it requires a series of non-trivial constructions to realize this primitive idea in the form of a scheme. The spectral scheme mainly consists of the following three elements:

(i) Spectral distance d_N [3, 4]

(ii) Spectral space (*"Space of Spaces"*) $\mathcal{S}_N$ [5]

(iii) Spectral evolution equations [6]

Now they are explained one by one.

To be specific, let us consider an eigenvalue problem of the Laplacian Δ on a compact Riemannian space (Σ, h), $\Delta f = -\lambda f$.[1] We then get the ***spectra***, namely the set of eigenvalues $\{\lambda_n\}_{n=0,1,2,\cdots} := \{0 = \lambda_0 < \lambda_1 \le \lambda_2 \le \cdots \le \lambda_n \le \cdots\}$. We also obtain the set of real-valued eigenfunctions corresponding to the spectra, $\{f_n\}_{n=0,1,2,\cdots}$, normalized in a usual manner.

It can be said that *Issue (a)* in Sec.1 had been the main obstacle to make progress in the analysis of the relation of the reality and its model; we first need a suitable mathematical *language* to talk about the discrepancy or closeness of two shapes. Therefore the first thing to do is to introduce such a measure of closeness between two geometries.

Let $\mathcal{G}$ and $\mathcal{G}'$ be two spatial geometries. Let $\{\lambda_m\}_{m=0}^{\infty}$ and $\{\lambda'_n\}_{n=0}^{\infty}$ be the spectra for $\mathcal{G}$ and for $\mathcal{G}'$, respectively. Then the measure of closeness between $\mathcal{G}$ and $\mathcal{G}'$ can be introduced by comparing $\{\lambda_m\}_{m=0}^{\infty}$ and $\{\lambda'_n\}_{n=0}^{\infty}$ [5, 7]:

$$d_N(\mathcal{G}, \mathcal{G}') := \sum_{n=1}^{N} \mathcal{F}\left(\frac{\lambda'_n}{\lambda_n}\right) \ , \tag{1}$$

where $\mathcal{F}(x)$ $(x > 0)$ is any continuous function which satisfies

$$\begin{aligned} \mathcal{F}(1) &= 0 \ , \\ \mathcal{F}(y) &> \mathcal{F}(x) \ \text{for} \ y > x \ge 1 \ , \\ \mathcal{F}(1/x) &= \mathcal{F}(x) \ . \end{aligned} \tag{2}$$

[1] Note that (Σ, h) is a space, and not a spacetime.

For practical applications, it is preferable to impose further assumptions on $\mathcal{F}$ that it is smooth at $x = 1$ and $\mathcal{F}''(1) > 0$ (See Eq.(11) below). On dimensional grounds, the lower (higher) spectrum reflects the larger (smaller) scale properties of the geometry. Comparing only the low-lying spectra up to the N-th spectrum λ_N, thus, corresponds to comparing $\mathcal{G}$ and $\mathcal{G}'$ in a coarse-grained manner, down to the scale $O(\lambda_N^{-1/2})$, neglecting the difference in the smaller scale behavior. Thus the measure of closeness $d_N(\mathcal{G}, \mathcal{G}')$ introduced in this way naturally takes into account the scale-dependent property of observed geometry (*Issue (b)* in Sec.1).

We also note that the elliptic operator (e.g. Δ) is in principle interpreted as the observational apparatus for measuring geometry. Thus the choice of the operator decides the aspects of geometry to be compared in defining the closeness between the spaces. In this manner, our measure of closeness automatically takes care of the apparatus-dependent nature of observed geometry, too (*Issue (c)* in Sec.1).

It is convenient to set $\mathcal{F}$ to be $\mathcal{F}_1(x) = \frac{1}{2}\ln\frac{1}{2}(\sqrt{x} + 1/\sqrt{x})$. Then Eq.(1) becomes [3, 4]

$$d_N(\mathcal{G}, \mathcal{G}') = \frac{1}{2}\sum_{n=1}^{N} \ln\frac{1}{2}\left(\sqrt{\frac{\lambda'_n}{\lambda_n}} + \sqrt{\frac{\lambda_n}{\lambda'_n}}\right) \quad . \tag{3}$$

We mainly focus on this form hereafter.

It turns out that $d_N(\mathcal{G}, \mathcal{G}')$ in Eq.(3) does not satisfy the triangle inequality [3] so that one further step is required to justify regarding it "distance". It is achieved as follows.

Let us consider the collection of all D-dimensional compact Riemannian manifolds, *Riem*, equipped with d_N. In the present setting, it is natural to identify the so-called isospectral manifolds [8]. From the viewpoint of spectral scheme, there is no surprise in the isospectral manifolds. It is obvious that geometry cannot be perfectly specified by just a single kind of observational apparatus (Δ)[5, 4]. Now we introduce a space

$$\mathcal{S}_N = (Riem, d_N)/_\sim \quad , \tag{4}$$

where $/_\sim$ indicates the identification of the isospectral manifolds[2]. Now one can prove that the space $\mathcal{S}_N$ forms a metrizable space, so that d_N is essentially regarded as a distance [5]. Therefore we are entitled to call d_N the ***spectral distance*** of order N; $\mathcal{S}_N$ may be called the ***spectral space*** of order N, or the space of spaces.

We note that $\mathcal{S}_N$ provides us with the concept of "closeness" between spaces as a 'function' of the observational scale (N) and the observational apparatus (Δ). In this manner, the idea of the "scale-dependent topology" [9, 10], which is a fascinating, but difficult-to-formulate concept, is automatically materialized in the spectral scheme.

With its several desirable properties, the space of spaces $\mathcal{S}_N$ may serve as a basic arena for spacetime physics; it is particularly suitable for analyzing situations when more than two spaces are involved.

With the help of the space of spaces $\mathcal{S}_N$, it is now possible to picture the relation between the real Universe and its model as follows: Each point in $\mathcal{S}_N$ represents a certain geometry viewed with the apparatus Δ by the scale-limit λ_N. A set of models is represented by the set of points distributed throughout $\mathcal{S}_N$. The optimal model

[2]More rigorously, $\mathcal{S}_N$ is defined as the completion of the right-hand side of Eq.(4) by means of the latter's metrizable-space structure. [5]

$\mathcal{G}'$ for reality $\mathcal{G}$ is viewed as the point among them closest to the point representing $\mathcal{G}$ [7]. Thus, the real geometry $\mathcal{G}$ and its model geometry $\mathcal{G}'$ are represented by two points that are very close to each other in $\mathcal{S}_N$. What we should do now is, then, to study the time evolution of the spectral distance $d_N(\mathcal{G}, \mathcal{G}')$ between them. To do so, we need the evolution equations for the spectra of the universe since $d_N(\mathcal{G}, \mathcal{G}')$ is defined solely in terms of the spectra.

In this way, the spectral scheme is completed by supplying the spectral evolution equations, which will be explained in the next section.

3 Spectral distance and its time evolution

Being the spectral scheme at hand, we are now free to discuss the discrepancy between two arbitrary universes.

Let us investigate the time evolution of the spectral distance between two universes $\mathcal{G}$ and $\mathcal{G}'$. We need to choose the time-slicing for each of two spacetimes. First it is notable that the metrizable structure of $\mathcal{S}_N$ induces a metrizable structure on $Slice(\mathcal{M}, g)$, the space of all spatial slices of a given spacetime $(\mathcal{M}, g)$ [7]. In other words, we can talk about the distance between two different slicings of the same spacetime, by means of the metrizable structure of $\mathcal{S}_N$. Now suppose we want to compare two spacetimes $(\mathcal{M}, g)$ and $(\mathcal{M}', g')$. By a similar procedure as above, it is possible to select an optimal time-slicing for each spacetime, such that $(\mathcal{M}, g)$ and $(\mathcal{M}', g')$ look as similar as possible (as a time sequence of spatial geometries). We assume that this procedure has been already completed so that a suitable time t for each spacetime has been selected (and identified with each other).

Now, by taking the time derivative of Eq.(3), we get

$$\dot{d}_N(\mathcal{G}, \mathcal{G}') = \frac{1}{4} \sum_{n=1}^{N} \frac{\frac{\lambda'_n}{\lambda_n} - 1}{\frac{\lambda'_n}{\lambda_n} + 1} \left(\ln \frac{\lambda'_n}{\lambda_n} \right)^{\cdot} \quad . \tag{5}$$

In order to proceed further, thus, the evolution equations for the spectra are needed. They can be derived by the following consideration: The spatial geometry (Σ, h) is evolved in time by evolving the initial value (h_{ab}, K_{ab}) according to the Einstein equations in the Hamiltonian form. (Here h_{ab} and K_{ab} are the spatial metric and the extrinsic curvature, respectively.) Then the spectra of (Σ, h) also evolve in time accordingly. What we need to do is essentially to study the response of the spectra to the change of the spatial metric h_{ab}. In this manner the evolution equations for the spectra can be derived, which has been achieved in Ref.[6]. The spectral evolution equations so obtained are regarded as the spectral representation of the Einstein theory.

Just like the original Einstein equations, the spectral evolution equations are complicated system of equations. They are in the form of hierarchy equations as is expected from the fact that the spectra are not local quantities but they reflect the global information of spatial geometry. This new kind of representation of the Einstein theory should have several advantages for analyzing a certain type of problems; further investigations on the properties of the spectral evolution equations are awaited.

Before going further, we here introduce convenient notations that are very similar to those in quantum theory: Let $\mathcal{A}$ and $\mathcal{A}_{ab}$ be any function and any symmetric tensor field, respectively, on a spatial geometry (Σ, h). Then, we define $\langle\mathcal{A}\rangle_{mn} := \int_\Sigma f_m \mathcal{A} f_n \sqrt{h}$ and $\langle\mathcal{A}_{ab}\rangle_{mn} := \frac{1}{\sqrt{\lambda_m\lambda_n}} \int_\Sigma \partial^a f_m \mathcal{A}_{ab} \partial^b f_n \sqrt{h}$. Furthermore, $\langle\mathcal{A}\rangle_n := \langle\mathcal{A}\rangle_{nn}$ and $\langle\mathcal{A}_{ab}\rangle_n := \langle\mathcal{A}_{ab}\rangle_{nn}$.

Here we only show a typical equation among the system of spectral evolution equations (see Ref.[6] for more details):

$$(\ln\lambda_n)^{\cdot} = -2H_n \quad . \tag{6}$$

Here

$$H_n := H + \iota_n + \alpha_n \quad , \tag{7}$$

with the quantities

$$\begin{aligned} H : &= \frac{1}{D-1}\frac{\dot{V}}{V} \quad , \\ \iota_n : &= \frac{1}{D-1}\langle K - K_{av}\rangle_n - \frac{D-3}{4(D-1)}\frac{1}{\lambda_n}\langle\Delta K\rangle_n \quad , \\ \alpha_n : &= \langle\epsilon_{ab}\rangle_n \quad . \end{aligned} \tag{8}$$

(D= the spacetime dimension; $K = K_a{}^a$; $\sum'$ indicates that the zero-mode is not included in the summation; Q_{av} denotes the spatial average of a scalar quantity Q.) Here H corresponds to the Hubble parameter; ι_n and α_n can be interpreted as the inhomogeneity and the anisotropy of the geometry at the scale $\lambda_n^{-1/2}$. Thus, H_n is interpreted as the effective Hubble parameter at the scale $\lambda_n^{-1/2}$.

4 Discrepancy equations for two nearby geometries

Spectral scheme in principle allows the analysis of the discrepancy of two arbitrary geometries. Here, however, we focus on the simplest case when two universes are very close to each other in $\mathcal{S}_N$. We only consider the case when the discrepancy between two geometries $\mathcal{G}$ and $\mathcal{G}'$ is captured by a small difference in their spatial metrics,

$$\gamma_{ab} := h'_{ab} - h_{ab} \quad , \tag{9}$$

where h_{ab} and h'_{ab} are the spatial metrics for $\mathcal{G}$ and $\mathcal{G}'$, respectively. This case simulates the relation of the real Universe with its model.

The discrepancy γ_{ab} in their metrics induces the discrepancy of their n-th spectra [6],

$$\delta\ln\lambda_n := \frac{\lambda'_n - \lambda_n}{\lambda_n} = -\langle\overline{\gamma}_{ab}\rangle_n - \frac{1}{2}\langle\gamma\rangle_n \quad , \tag{10}$$

where $\gamma := h^{ab}\gamma_{ab}$ and $\overline{\gamma}_{ab} := \gamma_{ab} - \frac{1}{2}\gamma h_{ab}$. Then the leading term in Eq.(1) becomes [6]

$$d_N(\mathcal{G}, \mathcal{G}') = \frac{1}{2}\mathcal{F}''(1)\sum_{n=1}^{N}\left(\langle\overline{\gamma}_{ab}\rangle_n + \frac{1}{2}\langle\gamma\rangle_n\right)^2 \quad . \tag{11}$$

We note that the leading term of $d_N(\mathcal{G}, \mathcal{G}')$ is not affected by the choice of the function $\mathcal{F}(x)$ except for the unimportant numerical factor. Whenever two geometries are very close to each other in $\mathcal{S}_N$, the leading term of the spectral distance

$d_N(\mathcal{G},\mathcal{G}')$ is universally given by Eq.(11), *irrespective either of the detailed form of the spectral distance or of the gravity theory.*

It is convenient to introduce a vector-like notation $\vec{\gamma}$ with its n-th component being

$$\gamma_n := \langle \bar{\gamma}_{ab} \rangle_n + \frac{1}{2} \langle \gamma \rangle_n \quad .$$

Then Eq. (11) is expressed in a neat form as

$$d_N(\mathcal{G},\mathcal{G}') = \frac{1}{16} \vec{\gamma} \cdot \vec{\gamma} \quad , \tag{12}$$

where a standard Euclidean inner-product is implied and we have set $\mathcal{F}''(1) = 1/8$ corresponding to Eq.(3).

Taking the time-derivative of Eq.(12), we get

$$\dot{d}_N(\mathcal{G},\mathcal{G}') = \frac{1}{4} \vec{\gamma} \cdot \delta \vec{H} \quad , \tag{13}$$

where $\delta \vec{H}$ is a vector-like quantity whose n-th component is $\delta H_n := H_n' - H_n$, namely the difference of the effective Hubble parameter between $\mathcal{G}'$ and $\mathcal{G}$ at the scale $\lambda_n^{-1/2}$. [3]

Equation (13) shows that *the relative directions of $\vec{\gamma}$ and $\delta \vec{H}$ (in $\mathbf{R}^N$) determine the sign of $\dot{d}_N(\mathcal{G},\mathcal{G}')$.*

Taking the time-derivative once more, Eq.(13) yields

$$\ddot{d}_N(\mathcal{G},\mathcal{G}') = \frac{1}{2} \delta \vec{H} \cdot \delta \vec{H} + \frac{1}{4} \vec{\gamma} \cdot \delta \dot{\vec{H}} \quad . \tag{14}$$

We find out that $\delta \vec{H}$, *the discrepancy of the effective Hubble parameters up to order N, always induces a repulsive effect and that the sign and the magnitude of the quantity $\vec{\gamma} \cdot \delta \dot{\vec{H}}$ are crucial for the sign of $\ddot{d}_N(\mathcal{G},\mathcal{G}')$.* Equation (14) can be modified further. After some manipulations, Eq.(14) reduces to [6]

$$\begin{aligned} \ddot{d}_N(\mathcal{G},\mathcal{G}') \simeq \quad & \frac{1}{2} \delta \vec{H} \cdot \delta \vec{H} - \frac{1}{4} (\sum_{n=1}^{N} \gamma_n) \delta\{(1+q)H^2\} \\ & - \frac{1}{4} \vec{\gamma} \cdot \delta \left\{ 2H\vec{\iota} + (D-1)H\vec{\alpha} + \vec{c} + \frac{(D-3)}{2(D-2)} \frac{1}{\alpha} \vec{\mathcal{M}} \right\} \quad . \end{aligned} \tag{15}$$

Here, $q := -(1 + \frac{\dot{H}}{H^2})$ is the deceleration parameter; $\vec{\iota}$ and $\vec{\alpha}$ are the vectors whose n-th components are ι_n and α_n, respectively; $\vec{c}$ is the vector whose n-th component is $\langle r_{ab} \rangle_n$ with $r_{ab} := R_{ab} - \frac{1}{D-1} R h_{ab}$; $\alpha := \frac{c^3}{16\pi G}$; $\vec{\mathcal{M}}$ is the vector whose n-th component is $\mathcal{M}_n$ defined by

$$\mathcal{M}_n := \frac{1}{D-1} \{ \langle \rho - \rho_{av} \rangle_n + \frac{D-1}{D-3} \langle p - p_{av} \rangle_n \} - \frac{D-3}{4(D-1)} \frac{1}{\lambda_n} \{ \langle \Delta \rho \rangle_n + \frac{D-1}{D-3} \langle \Delta p \rangle_n \} \quad ,$$

[3] For definiteness, we define δQ as Q for $\mathcal{G}'$ (the second entry of $d_N(\cdot,\cdot)$) minus Q for $\mathcal{G}$ (the first entry of $d_N(\cdot,\cdot)$). Whole of the theory is so constructed as invariant under the exchange of the roles of $\mathcal{G}$ and $\mathcal{G}'$, as is inferred from Eqs.(1) and (2). When appropriate, we write $\delta\{\cdot\}$ (rather than $\delta(\cdot)$) to avoid a confusion with the δ-function.

which measures the inhomogeneity of the matter density (ρ) and its pressure (p) at the scale of $\lambda_n^{-1/2}$.

For two nearby geometries $\mathcal{G}$ and $\mathcal{G}'$ given in the form of initial data, we can formulate the criteria for them to form a pair of "similar" histories: Let τ be the typical time-scale we are interested in. Then, the criteria become

(C1) $d_N(\mathcal{G}, \mathcal{G}')$ is small.

(C2) $\tau \dot{d}_N(\mathcal{G}, \mathcal{G}')$ is negative, or at least, non-negative and small.

(C3) $\tau^2 \ddot{d}_N(\mathcal{G}, \mathcal{G}')$ is negative, or at least, non-negative and small.

It is suggestive to imagine that $\mathcal{G}$ and $\mathcal{G}'$ represent the real Universe and its model. It is regarded as a great progress achieved by the spectral scheme that we can now talk about the relation of two different cosmological histories like in the above criteria, though of course these criteria are not written in the form directly connected to our daily cosmological observations.

We can do more explicit analysis by setting definite models. Let $\mathcal{G}$ be a regular T^3 and $\mathcal{G}'$ be a T^3-space obtained by perturbing $\mathcal{G}$. We consider only the terms linear in γ_{ab} (See Eq.(9)). In this case we can make use of the standard linear perturbation theory for structure formation and directly estimate the time evolution of $d_N(\mathcal{G}, \mathcal{G}')$, without resorting to the discrepancy equations Eqs.(13) and (14).

We get the estimation [11],

$$d_N(\mathcal{G}, \mathcal{G}') \sim \quad A\xi(t)\, t^{-\frac{2(1-\nu)}{1+\nu}} + B{k_N}^2\, \eta(t)\, t^{-\frac{2(1-3\nu)}{3(1+\nu)}} \quad . \tag{16}$$

Here, $\nu = p/\rho$; A and B are appropriate positive numerical factors that make the expression dimension-free; $\xi(t) := 1 - Ct^{\frac{1+3\nu}{1+\nu}}$ and $\eta(t) := 1 - C't^{-\frac{2(1+3\nu)}{3(1+\nu)}}$ with C and C' being appropriate positive numerical factors that make $\xi(t)$ and $\eta(t)$ are positive for $t > \exists t_0$. We note that $\xi(t) \to 0$ and $\eta(t) \to 1$ as $t \to \infty$.

This is just the consideration of the linear regime, but it is notable that the spectral scheme enables us to reach explicit estimations such as Eq.(16).

5 Discussion

In this paper we analyzed the issue of the relation of the real Universe and its model in the framework of the spectral scheme. Though the analysis we have just sketched is still on a primitive stage, it is already clear that the spectral scheme provides us with a very powerful tool for dealing with problems that require quantitative description of geometrical information, or problems in which more than two geometries are involved. Indeed we have derived several important properties regarding the relation of the reality and its model which would not have been available without the spectral scheme. Here we realize the importance of preparing a suitable language first before talking about an entangled situation.

The problem we have just analyzed turns to a deeper issue than the original one when we put it in a wider context, as the issue of the relation among the triad (*Reality*, *Model*, *Dynamics*), which arises quite universally in various fields of science. Putting emphasis on the procedure of truncating information, it also can be regarded as the issue of classifying the objects into types, which may be our typical strategy of perceiving the complexity of reality. It may be concisely stated as the ***issue of types***. The analysis presented here can be the first step for discussing the issue of types in natural sciences.

* * *

The author thanks the Ministry of Education, the Government of Japan and Inamori Foundation, Japan for financial support.

References

[1] For instance, G.F.R. Ellis, in "Proceedings of the Tenth International Conference on General Relativity and Gravitation", edited by B. Bertotti, F. De Felice and A. Pascolini, Reidel, Dordrecht, 1984. See also, A. Krasiński, "Inhomogeneous Cosmological Models", Cambridge University Press, Cambridge, 1997, Chapter 8, and the references therein.

[2] M. Kac, Am. Math. Mon. **73**(4), 1966, 1.

[3] M. Seriu, Physical Review **D53**, 1996, 6902.

[4] M. Seriu, in "Proceedings of the 8th Workshop on General Relativity and Gravitation", K,Oohara et. al. (eds.), Niigata University, 1999.

[5] M. Seriu, Communications in Mathematical Physics **209**, 2000, 393.

[6] M. Seriu, Physical Review **D62**, 2000, #023516.

[7] M. Seriu, General Relativity and Gravitation **32**, 2000, 1473.

[8] See e.g., I. Chavel, "Eigenvalues in Riemannian Geometry", Academic Press, Orland, 1984.

[9] M. Visser, Physical Review **D41**, 1990, 1116.

[10] M. Seriu, Physics Letters **B319**, 1993, 74.

[11] M. Seriu, "Evolution of the discrepancy between the real Universe and its model", submitted for publication.

Schwarzschild metrics, quasi-universes and wormholes

A. G. Agnese and M. La Camera
Dipartimento di Fisica dell'Università di Genova
Istituto Nazionale di Fisica Nucleare, Sezione di Genova
Via Dodecaneso 33, 16146 Genova, Italy
E-mail: agnese@ge.infn.it ; lacamera@ge.infn.it

1 Introduction

It is well known [1] that the three-dimensional space

$$d^{(3)}s^2 = \frac{dr^2}{1 - \dfrac{r^2}{R^2}} + r^2(d\vartheta^2 + \sin^2\vartheta d\varphi^2) \tag{1}$$

of some models of closed homogeneous and isotropic universes has an especially simple geometry which can be seen best introducing a angular coordinate $0 \leq \chi \leq \pi$ via $r = R \sin\chi$ and transforming the line element (1) into the form

$$d^{(3)}s^2 = R^2 \left(d\chi^2 + \sin^2\chi \, d\Omega^2\right) \tag{2}$$

where

$$d\Omega^2 = d\vartheta^2 + \sin^2\vartheta d\varphi^2 \tag{3}$$

The metric (2) is that of a three-dimensional hypersurface of radius R which can be represented in a flat, four-dimensional Euclidean embedding space.
Our purpose is to employ a similar angular variable to describe the geometry of the exterior Schwarzschild solution and to investigate such a description of the interior solution also when $\chi > \pi/2$, a possibility which appears to have been ignored in the literature. In this way we can introduce the concept of "quasi-universe" and show that the the Einstein-Rose bridge is nothing else that an "extreme wormhole connecting two quasi-universes". Moreover it can be seen, in the framework of Brans-Dicke theory, that the Einstein-Rosen bridge becomes a traversable wormhole.

Frontiers of Fundamental Physics 4, Edited by Sidharth and Altaisky,
Kluwer Academic/Plenum Publishers, New York, 2001

2 The exterior Schwarzschild solution

The exterior spherically symmetric vacuum solution, which by Birkhoff's theorem is also static, will be written in standard coordinates as

$$ds^2 = \frac{dr^2}{1 - \frac{2m}{r}} + r^2 d\Omega^2 - N^2(t)\left(1 - \frac{2m}{r}\right) dt^2 \tag{4}$$

The term $N^2(t)$ allows the matching between exterior and interior values of g_{tt} when the interior solution is not static and the observer is below the radius r_1 of the body; of course in the static cases $N^2(t)$ reduces to a constant. Such a constant shall be written as $(1 - 2m/r_0)^{-1}$ if the observer is placed at r_0 above the radius r_1; so the light will appear to him red-shifted if received from $r < r_0$ and blue-shifted if received from $r > r_0$.

Coming back to the line element (4), we want to replace the radial coordinate r with an angular coordinate ψ; because of the covariance of Einstein's equations there are infinite ways to accomplish the replacement. We choose to define an angular coordinate ψ given by

$$r = \frac{2m}{\cos^2 \psi} \qquad -\frac{\pi}{2} \leq \psi \leq \frac{\pi}{2} \tag{5}$$

when $r > 2m$, and analytically continued to

$$r = \frac{2m}{\cosh^2 \psi} \qquad -\infty < \psi < \infty \tag{6}$$

when $r < 2m$. The line element (4) becomes, in the region $r > 2m$

$$ds^2 = \frac{16m^2}{\cos^6 \psi} d\psi^2 + \frac{4m^2}{\cos^4 \psi} d\Omega^2 - \frac{\sin^2 \psi}{\sin^2 \psi_0} dt^2 \tag{7}$$

Here the event horizon is placed at $\psi = 0$, while infinity is reached at $\psi = \pm\pi/2$. The metrical relations in the surface t = constant, $\vartheta = \pi/2$ are illustrated by means of the surface of revolution $f(r) = \sqrt{8m(r - 2m)}$ (remember the representation of the Flamm's paraboloid with the Einstein-Rosen bridge). In the extended region $r < 2m$ one has instead the line element

$$ds^2 = -\frac{16m^2}{\cosh^6 \psi} d\psi^2 + \frac{4m^2}{\cosh^4 \psi} d\Omega^2 + \frac{\sinh^2 \psi}{\sinh^2 \psi_0} dt^2 \tag{8}$$

which describes the interior of a black hole joined to the exterior by the event horizon placed at $\psi = 0$. It is worth noticing that the introduction of the ψ coordinate provides a division of the maximally extended Schwarzschild spacetime in four regions with two singularities corresponding to an equal gravitational mass, just as described by Kruskal-Szekeres coordinates. These singularities are placed at $\psi = \pm\infty$, being now ψ a time coordinate.

3 The interior Schwarzschild solution

The gravitational field inside a celestial body, say a star, modelled on an ideal fluid medium with energy-momentum tensor

$$T_{\mu\nu} = (\rho + p)u_\mu u_\nu + p g_{\mu\nu} \tag{9}$$

is given, for static distribution of matter and pressure and moreover under the hypotheses of spherical symmetry and constant mass density, by

$$ds^2 = \frac{dr^2}{1 - \frac{r^2}{R^2}} + r^2 d\Omega^2 - \left[A - B\sqrt{1 - \frac{r^2}{R^2}}\right]^2 dt^2 \tag{10}$$

Here A and B are integration constants to be determined by the matching conditions. We use this simple and rather unrealistic solution as a toy model uniquely to illustrate the employ, which is new if $\chi > \pi/2$, of the angular coordinate χ. If we now define this angular coordinate through

$$\frac{r}{R} \equiv \sin\chi \qquad\qquad 0 \leq \chi \leq \pi \tag{11}$$

the line element (10) becomes

$$ds^2 = R^2\left(d\chi^2 + \sin^2\chi d\Omega^2\right) - [A - B\cos\chi]^2 dt^2 \tag{12}$$

From Einstein's equations the pressure p and the mass density ρ are

$$p = \frac{1}{8\pi R^2}\left[\frac{3B\cos\chi - A}{A - B\cos\chi}\right], \qquad\qquad \rho = \frac{3}{8\pi R^2} \tag{13}$$

In formulating the matching conditions to connect the exterior and interior Schwarzschild solutions, continuity of the metric and its derivatives are to be taken into account. However in our simple example we rest on physical plausibility considerations, so we require that the metric is continuous for $\sin\chi_1 = \frac{r_1}{R}$, where r_1 is the radius of the body, that the pressure p vanishes on its surface and that the observer is in the interior at an angle χ_0.
As a result one obtains

$$\sin\chi_1 = \left(\frac{2m}{R}\right)^{1/3}, \quad A = \frac{3\cos\chi_1}{3\cos\chi_1 - \cos\chi_0}, \quad B = \frac{1}{3\cos\chi_1 - \cos\chi_0} \tag{14}$$

where m is the gravitational mass. The line element (10) can now be written

$$ds^2 = R^2\left(d\chi^2 + \sin^2\chi d\Omega^2\right) - \left[\frac{3\cos\chi_1 - \cos\chi}{3\cos\chi_1 - \cos\chi_0}\right]^2 dt^2 \tag{15}$$

So the observer receives the frequency of light red-shifted when coming from inside and blue-shifted when coming from outside. The matching to the exterior solution requires that

$$N^2 = \left[\frac{2\cos\chi_1}{3\cos\chi_1 - \cos\chi_0}\right]^2 \left(1 - \frac{2m}{R\sin\chi_1}\right)^{-1} \tag{16}$$

If the observer is at the exterior the previous values of A and B change accordingly. The pressure becomes

$$p = \frac{3}{8\pi R^2}\left[\frac{\cos\chi - \cos\chi_1}{3\cos\chi_1 - \cos\chi}\right] \tag{17}$$

and is obviously observer independent. Because of definition (11) two cases are now to be considered, depending whether for a given value of r_1 one chooses $\chi_1 < \pi/2$ or $\chi_1 > \pi/2$. In the former case, while the mass density ρ is constant, the pressure p, which is zero at the surface, increases inwards; the solution is non singular as long as p is finite. At $r = 0$ where p takes its maximum value, this is only possible for $\chi_1^{(1)} < \arccos(1/3) \approx 0.39\pi$, that is, as known [2], for $r_1/(2m) > 9/8$. In the latter case, the pressure p takes negative values in the interior, and the solution is non singular at $r = 0$ for $\chi_1^{(2)} > \pi/2$. In both cases, the weak energy condition

$$\rho \geq 0\,, \qquad \rho + p \geq 0 \tag{18}$$

is always satisfied. We would also point out that while the surface area $S = 4\pi R^2 \sin^2\chi_1$ is the same in the two cases, independently of the choice made for χ_1, things are different in calculating volumes, given by the formula

$$V = 4\pi R^3 \int_0^{\chi_1} \sin^2\chi\, d\chi = \pi R^3(2\chi_1 - \sin 2\chi_1) \tag{19}$$

To make an example let us consider two bodies having the same gravitational mass and the same density ρ but different values of χ_1 given respectively by $\chi_1^{(1)}$ and $\chi_1^{(2)} = \pi - \chi_1^{(1)}$ (and so the same value of $\sin\chi_1$). The ratio $V^{(2)}/V^{(1)}$ of their volumes is

$$\frac{V^{(2)}}{V^{(1)}} = \frac{2(\pi - \chi_1^{(1)}) + \sin 2\chi_1^{(1)}}{2\chi_1^{(1)} - \sin 2\chi_1^{(1)}} \tag{20}$$

Therefore while the volume $V^{(1)}$ encloses a star whose matter is endowed by the usual properties ($\rho > 0$, $p > 0$), the volume $V^{(2)}$ may be so large to be considered as a "quasi-universe", so named because it is an universe deprived of a spherical void, containing matter with properties ($\rho > 0$, $p < 0$, but $\rho + p > 0$); we do not call such a matter exotic, because it satisfies the weak energy condition and so also the null energy condition [3]. The connection between a body and a quasi-universe through a suitable part of the Flamm paraboloid is schematically represented in Figure 1. A different possibility is shown in Figure 2 where now two quasi-universes are joined through an Einstein-Rosen bridge (with throat at $\psi = 0$) which can be renamed "extreme wormhole"; here the matching conditions to be fulfilled for the second junction are the same already seen for the first, analogous quantities being now renamed with the same letter primed. Because the throat is in the vacuum, the null energy condition is not violated; so, according to the Morris-Thorne analysis [5] it is not seen as traversable by an observer placed in a fixed forwarding station. The Einstein-Rosen bridge (or extreme wormhole) can also be considered as a limiting case, when the post-Newtonian parameter $\gamma \to 1^+$, of the corresponding Brans-Dicke solution [6] (see in the following). Finally, because of the necessary equality of the gravitational masses in the three joined solutions, one obtains the following relation between the densities of the two quasi-universes ρ and ρ_1':

$$\frac{\rho}{\rho_1'} = \left(\frac{\sin\chi_1}{\sin\chi_1'}\right)^2 \tag{21}$$

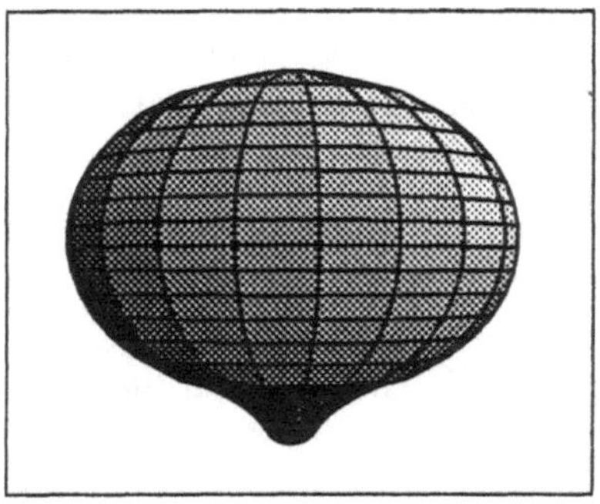

Figure 1: The connection between a body and a quasi-universe.

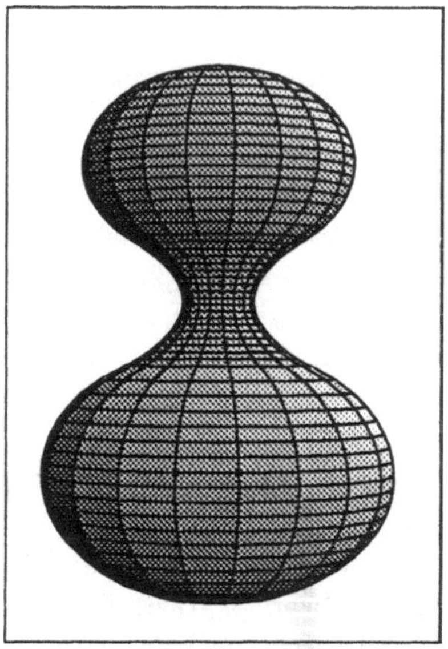

Figure 2: The connection between two quasi-universes.

To summarize the above results, the metrics corresponding to the exterior and interior Schwarzschild solutions have been rewritten replacing the usual radial coordinate with an angular one. With respect to the exterior solution, it covers four different regions of the space-time. With respect to the interior solution, it has been extended from the case $\chi < \pi/2$ (first-type solution) to the case $\chi > \pi/2$ of a quasi-universe (second-type solution). A second-type solution can be joined either to a first-type or to a second-type solution respectively through a suitable part of the Flamm's paraboloid or through a particular Einstein-Rosen bridge (extreme wormhole) provided the gravitational masses are equal.
Let us now consider Equations (7) and (8) in the limiting case when the exterior Schwarzschild solution goes over all the remaining space (asymptotic flatness). It is our opinion that the following unions ($\bigcup$) of two of the four regions - named hereafter I, II, III, IV according to the customary nomenclature [4] - of the Kruskal-Szekeres diagram give rise to distinct solutions:

1) $I \bigcup II$: there is a singularity corresponding to a gravitational mass m at $\psi = -\infty$ and a quasi-universe of gravitational mass m and density $\rho = 0$ at the boundary $\psi = \pi/2$. The two regions are separated at $\psi = 0$ by an event horizon.

2) $III \bigcup IV$: there is a singularity corresponding to a gravitational mass m at $\psi = +\infty$ and a quasi-universe of gravitational mass m and density $\rho = 0$ at the boundary $\psi = -\pi/2$. The two regions are separated at $\psi = 0$ by an event horizon.

3) $I \bigcup III$: there are two quasi-universes with gravitational mass m and density $\rho = 0$ at the boundaries $\psi = \pi/2$ and $\psi = -\pi/2$, connected by an extreme wormhole.

4) $II \bigcup IV$: the universe consists of two equal masses placed respectively at $\psi = -\infty$ and at $\psi = +\infty$ with a cosmological horizon at $\psi = 0$.

The Penrose diagram for the maximally extended Schwarzschild spacetime is a representation of the set of the four solutions.

More in general, one could consider expanding quasi-universes, which are universes with cavities [7],[8],[9],[10]. Inside one of such voids there is a body whose inertial mass is, by the equivalence principle, equal to its own gravitational mass and consequently, broadening the above considerations, also to the gravitational mass of the corresponding quasi-universe.

4 Brans-Dicke wormholes

We start by considering the static spherically symmetric vacuum solution of the Brans-Dicke theory of gravitation [11].

The related calculations were performed by us in Ref. [6], working in the Jordan frame, where the action is given (in units $G_0 = c = 1$) by

$$S = \frac{1}{16\pi} \int d^4x\sqrt{-g}\left[\Phi R - \frac{\omega}{\Phi}\nabla^\alpha\Phi\nabla_\alpha\Phi\right] \tag{22}$$

and with a suitable choice of gauge. Here we quote only the results relevant for the following.

The line element can be written as

$$ds^2 = e^{\mu(r)}dr^2 + R^2(r)d\Omega^2 - e^{\nu(r)}dt^2 \tag{23}$$

where $d\Omega^2 = d\vartheta^2 + \sin^2\vartheta d\varphi^2$ and, in the selected gauge:

$$R^2(r) = r^2\left[1 - \frac{2\eta}{r}\right]^{1-\gamma\sqrt{2/(1+\gamma)}} \tag{24a}$$

$$e^{\mu(r)} = \left[1 - \frac{2\eta}{r}\right]^{-\gamma\sqrt{2/(1+\gamma)}} \tag{24b}$$

$$e^{\nu(r)} = \left[1 - \frac{2\eta}{r}\right]^{\sqrt{2/(1+\gamma)}} \tag{24c}$$

Here γ is the post-Newtonian parameter

$$\gamma = \frac{1+\omega}{2+\omega} \tag{25}$$

and

$$\eta = M\sqrt{\frac{1+\gamma}{2}} \tag{26}$$

Finally the scalar field is given by

$$\Phi(r) = \Phi_0 \left[1 - \frac{2\eta}{r}\right]^{(\gamma-1)/\sqrt{2(1+\gamma)}} \tag{27}$$

while the effective gravitational coupling $G(r)$ equals

$$G(r) = \frac{1}{\Phi(r)} \frac{2}{(1+\gamma)} \tag{28}$$

the factor $2/(1+\gamma)$ being absorbed, as in Ref. [11], in the definition of G. Departures from Einstein's theory of General Relativity appear only if $\gamma \neq 1$, a possibility consistent with experimental observations which estimate it in the range $1 - 0.0003 < \gamma < 1 + 0.0003$ corresponding to the dimensionless Dicke coupling constant $|\omega| > 3000$.

When $\gamma < 1$ and $r \to 2\eta$, then $R(r)$, $e^{\nu(r)}$ and $G(r)$ go all to zero. Therefore we have a singularity with infinite red-shift and gravitational interaction decreasing while approaching the singularity.

When $\gamma = 1$ exactly, one has Schwarzschild solution of General Relativity

$$ds^2 = \frac{dr^2}{1 - \frac{2M}{r}} + r^2 d\Omega^2 - (1 - \frac{2M}{r})dt^2 \tag{29}$$

When $\gamma > 1$, the null energy condition $\rho + p_j > 0$ is violated [3] and a wormhole solution is obtained [6] with throat at

$$r_0 = \eta\left[1 + \gamma\sqrt{\frac{2}{1+\gamma}}\right] = M\left[\gamma + \sqrt{\frac{1+\gamma}{2}}\right] \tag{30}$$

to which corresponds the value R_0 given by equation (24a). In this last case beyond the throat, where $R > R_0$, we are faced with two possibilities according to the value of the radial coordinate r with respect to r_0.

1) If $r < r_0$, when $r \to 2\eta$ one has $R \to \infty$, $g_{tt} \to 0$, $g_{rr} \to 0$ and $G \to infty$. The singularity is beyond the throat and is smeared on a spherical surface, asymptotically large but not asymptotically flat.

2) If $r > r_0$, when $r \to \infty$ one has $R \to \infty$, $g_{tt} \to 1$, $g_{rr} \to 1$ and $G \to G_N$ (the Newton constant), so the space is asymptotically flat. In this case we have a two-way traversable wormhole which, more generally, will be a bridge connecting two quasi-universes.

5 Conclusions

The exterior and interior Schwarzschild solutions are rewritten replacing the usual radial variable with an angular one. This allows to obtain some results otherwise less apparent or even hidden in other coordinate systems. In particular we have proposed the concept of "quasi-universe" and described the Einstein-Rosen bridge as the extreme wormhole connecting two quasi-universes. Then we have employed the Brans-Dicke field to convert the non traversable Einstein-Rosen bridge into a traversable wormhole. There are however other possibilities to achieve this goal: the existence of exotic matter suffices for the violation of the null energy condition. Some other possibilities are:

Scalar fields acting in the low energy limit of string theories.

The Casimir energy.

Squeezed quantum states.

As a concluding remark, we have introduced wormholes connecting different universes; the possibility of wormholes connecting different regions of the same universe (called stargates in the fiction) does not seem too realistic, due to the large amount of exotic matter needed and the difficulty of its stabilization in time.

References

[1] R. C. Tolman, "Relativity, Thermodynamics and Cosmology", Clarendon Press, Oxford, 1950, p. 338.

[2] S. Weinberg, "Gravitation and Cosmology", John Wiley, New York, 1972, p.331.

[3] M. Visser, "Lorentzian Wormholes", Springer-Verlag, New York, 1996.

[4] C. W. Misner, K. S. Thorne and J. A. Wheeler, "Gravitation", W. H. Freeman and Company, San Francisco, 1973, p.834.

[5] M. S. Morris and K. S. Thorne, Am. J. Phys. **56**, 395, 1988.

[6] A. G. Agnese and M. La Camera, Phys. Rev. D, **51**, 2011, 1995.

[7] A. Einstein and E. G. Straus, Rev. Mod. Phys. **17**, 120, 1945.

[8] P. J. E. Peebles, "Principles of Physical Cosmology", Princeton University Press, Princeton, 1993, p.297.

[9] Ref. [4], p. 739.

[10] C. C. Dyer and C. Oliwa, E-print astro-ph/0004090.

[11] C. Brans and R. H. Dicke, Phys. Rev. **124**, 925, 1961.

[12] M. Visser and D. Hochberg, in "The Internal Structure of Black Holes and Spacetime Singularities", edited by L. M. Burko and A. Ori, Institute of Physics, Bristol, 1997.

[13] I. Quiros, Phys. Rev. D, **61**, 124026, 2000.

Normalized Weyl-type $\star$-product on Kähler manifolds

Takuya MASUDA
Department of Physics, Tokyo Metropolitan University,
Hachioji, Tokyo, 192-0397, Japan

1 Introduction

A new kind of mathematics is thought necessary for a non-perturbative description of the string theory just as Riemannian geometry is indespensable for the description of the theory of general relativity. Non-commutative geometry is one of the strong candidate for it.

We would like to construct a non-commutative manifold with a Kähler metric from this perspective. We take deformation quantization approach, which introduces a non-commutative product into the ring of functions on a commutative manifold. Kontsevich showed how to construct a non-commutative product on Poisson manifolds, which includes Kähler manifolds [1] [2], but the effect of a metric is implicit in their construction. Dependence of a non-commutativity on a metric is clear in our construction. We give an outline of our construction in the rest part of this section.

Among other approaches we focus our attention on two different types of non-commutative products, the Weyl-type and the Berezin-type. The ordinary Moyal product of functions f_1 and f_2 on C^1 [3] [4] can be written in an integral representation, as

$$(f_1 *_M f_2)(z,\overline{z}) = \int_{\mathrm{C}^1} f_1(w,\overline{v}) f_2(v,\overline{w}) \frac{e^{(z\overline{v}+v\overline{z}-z\overline{z}-v\overline{v})/\hbar}}{e^{(z\overline{w}+w\overline{z}-z\overline{z}-w\overline{w})/\hbar}} \frac{dv d\overline{v}}{2\pi i\hbar} \frac{dw d\overline{w}}{2\pi i\hbar},$$

which reduces to

$$f_1 f_2 + \frac{\hbar}{2\sqrt{-1}} \{f_1, f_2\}_P + O\left(\hbar^2\right)$$

in the small $\hbar$ limit and satisfies associativity. This approach of quantization is applicable only to a flat manifold.

There are papers on Berezin-type star-product on Kähler manifolds [5] [6]. The Berezin star-product is defined on certain Kähler manifolds by

$$(f_1 \bullet f_2)(z,\overline{z}) = \int_{\mathrm{C}^n} f_1(z,\overline{v}) f_2(v,\overline{z}) e^{(\Phi(z,\overline{v})+\Phi(v,\overline{z})-\Phi(z,\overline{z})-\Phi(v,\overline{v}))/\hbar} d\mu_\hbar(v,\overline{v}) \quad (1)$$

[0] tmasuda@phys.metro-u.ac.jp

where Φ is a Kähler potential of the manifold. This product also satisfies the associativity. The reduction to the Poisson bracket is achieved, however, only in the difference of two terms:

$$f_1 \bullet f_2 - f_2 \bullet f_1 = \frac{\hbar}{2\sqrt{-1}} \{f_1, f_2\}_P + O(\hbar^2).$$

Reshetikhin and Takhtajan showed (1) does not necessarily satisfy $f \bullet 1 = 1 \bullet f = f$ for the general Kähler potentials and defined a product which satisfies the above equality.

It was shown in [8] that there exists another possible approach

$$(f_1 \odot f_2)(z, \overline{z}) = \int_{\mathbf{C}^n} f_1(w, \overline{v}) f_2(v, \overline{w}) \frac{e^{(\Phi(z,\overline{v})+\Phi(v,\overline{z})-\Phi(z,\overline{z})-\Phi(v,\overline{v}))/\hbar}}{e^{(\Phi(z,\overline{w})+\Phi(w,\overline{z})-\Phi(z,\overline{z})-\Phi(w,\overline{w}))/\hbar}} d\mu_\hbar(v,\overline{v}) d\mu_\hbar(w,\overline{w}) \tag{2}$$

which interpolates between the Weyl-type and Berezin-type star-products. In the flat space this is the same as the Moyal product. The associativity, however, does not hold by itself, but is fulfilled in the functional integral limit of its multiple products.

Ref. [8] is insufficient in the sense that there is no consideration to the general Kähler manifolds for which $A = \frac{1}{2}\sum_{i,\bar{j}=1}^{n} h^{i\bar{j}} \partial_i \overline{\partial}_j \log \det H$ is not 0, where the matrix H is a metric.

We generalize the product defined in [8] so that it is applicable to general Kähler manifolds. Moreover we expand the product perturbatively and show that it really has a property characteristic of a Weyl-type product and that the normalization factor necesarry for a Berezin-type product introduced by Reshetikhin and Takhtajan is unnecessary for a Weyl-type product at least in the first order of $\hbar$.

2 Construction

Reshetikhin and Takhtajan has shown the Berezin-type product (1) does not necessarily satisfy $f \bullet 1 = 1 \bullet f = f$ for the general Kähler potentials and defined, as a product which satisfies the above equality, a normalized star-product

$$e_\hbar^{-1}(z, \overline{z}) \left((f_1 e_\hbar) \bullet (f_2 e_\hbar)\right)(z, \overline{z}), \tag{3}$$

where $e_\hbar$ is a normalization factor defined as $f \bullet e_\hbar = e_\hbar \bullet f = f$ [7].

As you can see, (3) consists of two kinds of Berezin-type products, an ordinary Berezin-type product

$$(f_1 \bullet f_2)(z, \overline{z}) = \int_{\mathbf{C}^n} f_1(z, \overline{v}) f_2(v, \overline{z}) e^{(\Phi(z,\overline{v})+\Phi(v,\overline{z})-\Phi(z,\overline{z})-\Phi(v,\overline{v}))/\hbar} d\mu_\hbar(v, \overline{v})$$

and a new kind of Berezin-type product

$$(f_1 \circ f_2)(z, \overline{z}) := \int_{\mathbf{C}^n} f_1(v, \overline{z}) f_2(z, \overline{v}) e^{-(\Phi(z,\overline{v})+\Phi(v,\overline{z})-\Phi(z,\overline{z})-\Phi(v,\overline{v}))/\hbar} d\mu_\hbar(v, \overline{v}).$$

A normalization factor is necessary for each kind of Berezin-type product, so a normalization factor is also necessary for a Weyl-type product.

We define a normalization factor $\hat{e}_\hbar$ for a new kind of Berezin-type product $f_1 \circ f_2$ so that $f \circ \hat{e}_\hbar = \hat{e}_\hbar \circ f = f$.

$$e_\hbar = 1 - \hbar A + O\left(\hbar^2\right)$$

whereas

$$\hat{e}_\hbar = 1 + \hbar A + O\left(\hbar^2\right).$$

We define a normalized star-product for $f_1 \circ f_2$ just as (3) for $f_1 \bullet f_2$ as

$$\hat{e}_\hbar^{-1}(z,\bar{z})\left((f_1\hat{e}_\hbar) \circ (f_2\hat{e}_\hbar)\right)(z,\bar{z}).$$

With two kinds of normalization factors $e_\hbar$ and $\hat{e}_\hbar$ we define a normalized Weyl-type prodct $f_1 * f_2$ as

$$\begin{aligned}
&(f_1 * f_2)(z,\bar{z}) \\
&:= e_\hbar^{-1}(z,\bar{z})\left((e_\hbar f_1 \hat{e}_\hbar) \odot (e_\hbar f_2 \hat{e}_\hbar)\right)(z,\bar{z})\hat{e}_\hbar^{-1}(z,\bar{z}) \\
&= \int_{\mathbf{C}^n} f_1(w,\bar{v}) f_2(v,\bar{w}) \frac{e_\hbar(w,\bar{v})e_\hbar(v,\bar{w})}{e_\hbar(z,\bar{z})} \frac{\hat{e}_\hbar(w,\bar{v})\hat{e}_\hbar(v,\bar{w})}{\hat{e}_\hbar(z,\bar{z})} \frac{e^{\Phi(z',\bar{v})+\Phi(v,\bar{z})-\Phi(z',\bar{z})-\Phi(v,\bar{v})}}{e^{\Phi(z,\bar{w})+\Phi(w,\bar{z}')-\Phi(z,\bar{z}')-\Phi(w,\bar{w})}} \\
&\qquad d\mu_\hbar(v,\bar{v})d\mu_\hbar(w,\bar{w})
\end{aligned}$$

This product satisfies

$$f * 1 = 1 * f = f$$

for general Kähler manifolds. It is clear from (3) generally $f_1 \odot f_2 \neq f_1 * f_2$ unless $\hat{e}_\hbar = e_\hbar^{-1}$.

As non-normalized star-product $f_1 \odot f_2$, normalized star-product $f_1 * f_2$ does not satisfy the associativity in this form, but the transition to the functional integral version goes as follows.

Multi-products from both sides $\left(f^{(0)} * \left(\cdots * \left(f^{(N-2)} * \left(f^{(N-1)} * f^{(N)}\right)\right)\cdots\right)\right)$ and $\left(\left(\cdots\left(\left(f^{(0)} * f^{(1)}\right) * f^{(2)}\right) * \cdots\right) * f^{(N)}\right)$ are respectively

$$\begin{aligned}
&\left(f^{(0)} * \left(\cdots * \left(f^{(N-2)} * \left(f^{(N-1)} * f^{(N)}\right)\right)\cdots\right)\right)(z,\bar{z}) \\
&= \int_{\mathbf{C}^n} \left(\prod_{j=1}^{N} d\mu_\hbar\left(z^{(j-1)},\bar{v}^{(j)}\right) d\mu_\hbar\left(v^{(j)},\bar{z}^{(j-1)}\right)\right. \\
&\quad \frac{e^{\phi\left(v^{(j-1)},\bar{v}^{(j-1)};v^{(j)},\bar{z}^{(j-1)}\right)}}{e^{\phi\left(v^{(j-1)},\bar{v}^{(j-1)};z^{(j-1)},\bar{v}^{(j)}\right)}} e_\hbar\left(z^{(j-1)},\bar{z}^{(j-1)}\right)\hat{e}_\hbar\left(z^{(j-1)},\bar{z}^{(j-1)}\right) \\
&\times \frac{e_\hbar\left(z^{(N)},\bar{z}^{(N)}\right)}{e_\hbar(z,\bar{z})} \frac{\hat{e}_\hbar\left(z^{(N)},\bar{z}^{(N)}\right)}{\hat{e}_\hbar(z,\bar{z})} \left(\prod_{j=0}^{N} f^{(j)}\left(z^{(j)},\bar{z}^{(j)}\right)\right)\left(v^{(N)} = z^{(N)}, \quad v^{(0)} = z\right),
\end{aligned}$$

$$\left(\left(\cdots\left(\left(f^{(0)} * f^{(1)}\right) * f^{(2)}\right) * \cdots\right) * f^{(N)}\right)(z,\bar{z})$$

$$= \int_{\mathbf{C}^n} \left(\prod_{j=1}^{N} d\mu_\hbar \left(z^{(j)}, \overline{v}^{(j-1)} \right) d\mu_\hbar \left(v^{(j-1)}, \overline{z}^{(j)} \right) \right.$$

$$= \frac{e^{\phi\left(v^{(j)}, \overline{v}^{(j)}; z^{(j)}, \overline{v}^{(j-1)}\right)}}{e^{\phi\left(v^{(j)}, \overline{v}^{(j)}; v^{(j-1)}, \overline{z}^{(j)}\right)}} e_\hbar \left(z^{(j)}, \overline{z}^{(j)} \right) \hat{e}_\hbar \left(z^{(j)}, \overline{z}^{(j)} \right)$$

$$\times \frac{e_\hbar \left(z^{(0)}, \overline{z}^{(0)} \right)}{e_\hbar (z, \overline{z})} \frac{\hat{e}_\hbar \left(z^{(0)}, \overline{z}^{(0)} \right)}{\hat{e}_\hbar (z, \overline{z})} \left(\prod_{j=0}^{N} f^{(j)} \left(z^{(j)}, \overline{z}^{(j)} \right) \right) \right) \left(v^{(0)} = z^{(0)}, \quad v^{(N)} = z \right).$$

Therefore they have the same functional integral limit,

$$\int \mathcal{D}\mu (z, \overline{v}) \mathcal{D}\mu (v, \overline{z})$$

$$\times \exp \left[\int d\tau \left\{ \frac{\partial \overline{v}}{\partial \tau} \left(\frac{\partial \Phi (z, \overline{v})}{\partial \overline{v}} - \frac{\partial \Phi (v, \overline{v})}{\partial \overline{v}} \right) - \frac{\partial v}{\partial \tau} \left(\frac{\partial \Phi (v, \overline{z})}{\partial v} - \frac{\partial \Phi (v, \overline{v})}{\partial v} \right) \right\} \right]$$

$$+ \log e_\hbar (z, \overline{z}) + \log \hat{e}_\hbar (z, \overline{z}) + \log f (z, \overline{z}) .$$

as in [8], where $f(z, \overline{z})$ is defined as

$$\int d\tau \ \log f (z, \overline{z}) = \lim_{N \to \infty} \sum_{j=0}^{N} \frac{1}{N} \log f^{(j)} \left(z^{(j)}, \overline{z}^{(j)} \right) \qquad \left(z^{(j)} - z^{(j-1)} = \frac{1}{N} \right).$$

From the same discussion as in [8], we define a normalized associative Weyl-type product as

$$f_1 \star f_2$$

$$= \int \mathcal{D}\mu (z, \overline{v}) \mathcal{D}\mu (v, \overline{z})$$

$$\times f_1 (z (\tau_1), \overline{z} (\tau_1)) f_2 (z (\tau_1), \overline{z} (\tau_2)) \exp \left[\int d\tau \left\{ \right. \right.$$

$$\left. \left. \frac{\partial \overline{v}}{\partial \tau} \left(\frac{\partial \Phi (z, \overline{v})}{\partial \overline{v}} - \frac{\partial \Phi (v, \overline{v})}{\partial \overline{v}} \right) - \frac{\partial v}{\partial \tau} \left(\frac{\partial \Phi (v, \overline{z})}{\partial v} - \frac{\partial \Phi (v, \overline{v})}{\partial v} \right) \right\} \right]$$

$$+ \log e_\hbar (z, \overline{z}) + \log \hat{e}_\hbar (z, \overline{z}) ,$$

where τ_1 and τ_2 are fixed points.

3 Perturbation

We expand $f_1 \odot f_2$ perturbatively in small $\hbar$. The perturbative expansion of one of the Berezin-type products is given by [7] :

$$f_1 \bullet f_2$$

$$= \quad \pi^{-n} \det H \int_{\mathbf{C}^n} e^{-(Hy, y)} \prod_{i=1}^{n} \frac{\left| dy^i \wedge d\overline{y}^i \right|}{2} \left[f_1 f_2 + \epsilon^2 \left\{ y^i \overline{y}^j \left(\overline{\partial}_j f_1 \right) \left(\partial_i f_2 \right) \right. \right.$$

$$+ \quad f_1 f_2 \left(-\frac{1}{4} y^i y^j \overline{y}^k \overline{y}^\ell \partial_i \overline{\partial}_k \Phi_{j\overline{\ell}} + y^i \overline{y}^j \frac{\left(\partial_i \overline{\partial}_j \det H\right) \det H - \left(\partial_i \det H\right) \left(\overline{\partial}_j \det H\right)}{\left(\det H\right)^2} \right)$$

$$+ \quad y^i \overline{y}^j \left(f_2 \overline{\partial}_j f_1 \frac{\partial_i \det H}{\det H} + f_1 \partial_i f_2 \frac{\overline{\partial}_j \det H}{\det H} \right)$$

$$+ \quad f_1 f_2 \left(\frac{1}{4} y^i y^j y^k \overline{y}^\ell \overline{y}^m \overline{y}^n \left(\partial_i \Phi_{j\overline{k}}\right) \left(\overline{\partial}_\ell \Phi_{m\overline{n}}\right) + y^i \overline{y}^j \frac{\left(\partial_i \det H\right) \left(\overline{\partial}_j \det H\right)}{\left(\det H\right)^2} \right.$$

$$\left. - \quad \frac{1}{2} y^i y^j \overline{y}^k \overline{y}^\ell \left\{ \frac{\left(\partial_i \Phi_{j\overline{k}}\right) \left(\overline{\partial}_\ell \det H\right)}{\det H} + \frac{\left(\partial_i \det H\right) \left(\overline{\partial}_k \Phi_{j\overline{\ell}}\right)}{\det H} \right\} \right) \Bigg\} + O\left(\epsilon^3\right) \Bigg]$$

$$= \quad f_1 f_2 + \hbar \left(A f_1 f_2 + \sum_{i\overline{j}} h^{i\overline{j}} \overline{\partial}_j f_1 \partial_i f_2 \right) + O\left(\hbar^2\right), \quad \left(\hbar = \epsilon^2\right) \tag{4}$$

We apply this method to perturbation expansion of the new Berezin-type product, $f_1 \circ f_2$ we have defined above :

$$f_1 \circ f_2$$

$$= \quad \pi^{-n} \det H \int_{\mathbf{C}^n} e^{(Hy,y)} \prod_{i=1}^{n} \frac{\left|dy^i \wedge d\overline{y}^i\right|}{2} \left[f_1 f_2 + \epsilon^2 \left\{ y^i \overline{y}^j \left(\partial_i f_1\right) \left(\overline{\partial}_j f_2\right) \right. \right.$$

$$+ \quad f_1 f_2 \left(\frac{1}{4} y^i y^j \overline{y}^k \overline{y}^\ell \partial_i \overline{\partial}_k \Phi_{j\overline{\ell}} + y^i \overline{y}^j \frac{\left(\partial_i \overline{\partial}_j \det H\right) \det H - \left(\partial_i \det H\right) \left(\overline{\partial}_j \det H\right)}{\left(\det H\right)^2} \right)$$

$$+ \quad y^i \overline{y}^j \left(f_1 \overline{\partial}_j f_2 \frac{\partial_i \det H}{\det H} + f_2 \partial_i f_1 \frac{\overline{\partial}_j \det H}{\det H} \right)$$

$$+ \quad f_1 f_2 \left(\frac{1}{4} y^i y^j y^k \overline{y}^\ell \overline{y}^m \overline{y}^n \left(\partial_i \Phi_{j\overline{k}}\right) \left(\overline{\partial}_\ell \Phi_{m\overline{n}}\right) + y^i \overline{y}^j \frac{\left(\partial_i \det H\right) \left(\overline{\partial}_j \det H\right)}{\left(\det H\right)^2} \right.$$

$$\left. + \quad \frac{1}{2} y^i y^j \overline{y}^k \overline{y}^\ell \left\{ \frac{\left(\partial_i \Phi_{j\overline{k}}\right) \left(\overline{\partial}_\ell \det H\right)}{\det H} + \frac{\left(\partial_i \det H\right) \left(\overline{\partial}_k \Phi_{j\overline{\ell}}\right)}{\det H} \right\} \right) \Bigg\} + O\left(\epsilon^3\right) \Bigg]$$

$$= \quad f_1 f_2 - \hbar \left(A f_1 f_2 + \sum_{i\overline{j}} h^{i\overline{j}} \partial_i f_1 \overline{\partial}_j f_2 \right) + O\left(\hbar^2\right) \quad \left(\hbar = \epsilon^2\right). \tag{5}$$

As a result we get

$$f_1 \odot f_2 = f_1 f_2 + \hbar \sum_{i,j=1}^{n} h^{i\overline{j}} \left(\overline{\partial}_j f_1 \partial_i f_2 - \partial_i f_1 \overline{\partial}_j f_2 \right) + O\left(\hbar^2\right). \tag{6}$$

A Poisson bracket appears in the first order of $\hbar$, which means the product $f_1 \odot f_2$ is really a Weyl-type product and which was not shown in [8].

What is surprising is the disapperance of a term $A = \frac{1}{2} h^{i\overline{j}} \partial_i \overline{\partial}_j \log \det H$ in the first order. Compare this result with the expansion of $f_1 \bullet f_2$ in [7], namely (4), and of $f_1 \circ f_2$. The cumbersome term A, which appears in a Berezin-type product, disappears in a Weyl-type product at least in the order of $\hbar$. This means, therefore, the normalization factor is unnecessary for the Weyl-type product at least in the first order of $\hbar$.

In conclusion we found that the Weyl-type product defined in [8] can be decomposed into two kinds of Berezin-type products and introduce a normalization factor into a Weyl-type product since a normalization factor is necessary for each kind of Berezin-type product. Note, in general, we must use $f_1 * f_2$ ($\neq f_1 \odot f_2$ unless $\hat{e}_\hbar = e_\hbar^{-1}$).

Moreover we perform a perturbative expansion of non-normalized Weyl-type product. In the first order of $\hbar$ a Poisson bracket appears and A does not appear. Therefore the normalization factor which is necessary for a Berezin-type product is unnecessary for a Weyl-type product at least in the first order of $\hbar$. The normalization factor is unnecessary for Weyl-type product non-perturbatively if $\hat{e}_\hbar = e_\hbar^{-1}$ is shown.

Acknowledgement

I would like to thank Dr. Saito for useful discussions and his correction and Dr. Iso for his introduction and explanation of [7] and Mr. Wakatsuki for his cooperation in reading [7].

References

[1] Maxim Kontsevich, "Deformation quantization of Poisson manifolds", q-alg/9709040.

[2] Alberto S. Cattaneo and Giovanni Felder, "A path integral approach to the Kontsevich quantization formula, Commun".Math.Phys. 212, 2000, 591-611.

[3] J. E. Moyal, Proc. Cambridge. Phil. Soc. 45, 1949, 99.

[4] H. J. Groenewold, Physica, 12, 1946, 405-60.

[5] Michel Cahen, Simone Gutt and John Rawnsley, Trans. AMS, 337, 1993, 73-98

[6] Alexander V. Karabegov, Comm. Math. Phys., 180, 1996, 745-755

[7] Nicolai Reshetikhin and Leon A. Takhtajan, "Deformation Quantization of Kähler Manifolds", math.QA/9907171.

[8] Satoru Saito and Kazunori Wakatsuki, "Symmetrization of Berezin Star Product and Path Integral Quantization", hep-th/9912265, Prog. Theor. Phys. 104 No.5, 2000.

Non-dopplerian cosmological redshift parameters in a model of graviton-dusty universe

M.A. Ivanov
Chair of Physics, Belarus State University of Informatics and Radioelectronics, 6 P. Brovka Street, BY 220027, Minsk, Belarus.

1 Introduction

In this paper, possible manifestations of the graviton background in a case of hypothetical superstrong gravitational quantum interaction are considered. On one hand the author gives the reasons that a quantum interaction of photons with the graviton background would lead to redshifts of remote objects too. The author considers a hypothesis on the existence of the graviton background to be independent from the standard cosmological model. One cannot affirm such an interaction is the only cause of redshifts. It is possible, that one gives a small contribution to an effect magnitude only. But we cannot exclude that such an interaction with the graviton background would be enough to explain the effect without an attraction of the big bang hypothesis. Comparing the own model predictions with supernova cosmology data by Riess et al [1], the author finds a good accordance between the redshift model and the observations.

On the other hand, it is shown, that every massive body, with a non-zero velocity v (relatively to the isotropic graviton background), should experience a constant acceleration. If one assumes that the full observable redshift magnitude is caused by such a quantum interaction with single gravitons, then this acceleration will have the same order of magnitude as a small additional acceleration of NASA deep-space probes (Pioneer 10/11, Galileo, and Ulysses), reported by Anderson's team [2].

For more details, one can find a preprint of my full original paper [3] in the e-print archive [4].

2 An interaction of photons and of massive bodies with the graviton background

If the isotropic graviton background exists, then, due to photon scattering on gravitons, the average energy losses of a photon with an energy E on a path dr will be equal to $dE = -aEdr$, where a is a constant [5, 4, 3]. It is shown here, that such an interaction with single gravitons should be superstrong to provide a full redshift magnitude. As it was reported by Anderson's team [2], NASA deep-space probes

[0] E-mail: ivanovma@gw.bsuir.unibel.by

experience a small additional constant acceleration, directed towards the Sun. It follows from an universality of gravitational interaction, that not only photons, but all other objects, moving relatively to the background, should loss their energy too due to such a quantum interaction with gravitons. If $a = H/c$, massive bodies or particles must experience a deceleration w of the same order as an additional acceleration of cosmic probes:

$$w = -ac^2(1 - v^2/c^2).$$

The acceleration w is directed against the body velocity in a particular system of reference, in which the graviton background is isotropic. It is for small velocities:

$$w \simeq -Hc \simeq -4.8 \cdot 10^{-10} m/s^2,$$

if Hubble's constant $H = 1.6 \cdot 10^{-18} s^{-1}$, that corresponds roughly to one half of the observed acceleration for NASA probes.

It is possible, that the annual periodic term in the residuals of the both Pioneers (see plot B in the Figure 1 [6]) may be caused by own additional acceleration of the Earth under its motion relatively to the graviton background.

3 Comparison of the redshift model with supernova cosmology data

In case of flat non-expanding universe, a photon flux relaxation due to non-forehead collisions of photons with gravitons, can be characterized by a factor b, so that in our model the luminosity distance D_L [1] is equal to:

$$D_L = a^{-1} \ln(1 + z) \cdot (1 + z)^{(1+b)/2} \equiv a^{-1} f_1(z; b),$$

where z is a redshift. The theoretical estimation for b is:

$$b = 3/2 + 2/\pi = 2.137$$

[4, 3]. Thus, the redshift

$$z = \exp(ar) - 1$$

and the luminosity distance D_L are characterized in the model by two parameters: H and b (r is a geometrical distance). One can introduce an effective Hubble constant

$$H_{eff} \equiv cdz/dr = H(z + 1);$$

in a language of expansion it can be interpreted as "a current deceleration of the expansion".

High-z Supernova Search Team data [1] give us a possibility to estimate H based on our model. We can use one of the best fits of the function $D_L(z; H_0, \Omega_M, \Omega_\Lambda)$ from [1] (see Eq.2 in [1]) with $\Omega_M = -0.5$ and $\Omega_\Lambda = 0$, which is unphysical in the original work. For $1 - \Omega_M > 0$ and $1 + \Omega_M z > 0$, the function $D_L(z; H_0, \Omega_M, \Omega_\Lambda)$ is equal to:

$$D_L = a^{-1}(1+z)m^{-1} \sinh(\ln|(k - m)/(k + m)| - \ln|(1 - m)/(1 + m)|)$$
$$\equiv a^{-1} f_2(z; H_0, \Omega_M, \Omega_\Lambda),$$

where $m \equiv (1 - \Omega_M)^{1/2}, k \equiv (1 + \Omega_M z)^{1/2}$. Assuming $b = 2.137$, we can find H from the relation:

$$HD_L/H_0D_L = f_1(z;b)/f_2(z;H_0,\Omega_M,\Omega_\Lambda)$$

(see Table). We see that $H/H_0 \simeq const$, a deviation from an average value $< H > \simeq 1.09H_0$ is less than 5%. If one would suggest that $f_1(z;b)$ describes results of

z	0	0.1	0.2	0.3	0.4	0.5	0.6	0.7	0.8	0.9	1.0
f_1	0	0.110	0.242	0.396	0.570	0.765	0.983	1.222	1.480	1.759	2.058
f_2	0	0.103	0.219	0,359	0.511	0.677	0.863	1.074	1.301	1.565	1.854
H/H_0	-	1.068	1.105	1.103	1.115	1.130	1.139	1.138	1.138	1.124	1.110

observations in an expanding universe, one could conclude that it is "an accelerating one". But a true conclusion may be strange: our universe is not expanding, and redshifts have the non-dopplerian nature.

4 Conclusion

If further investigations display that an anomalous NASA probes' acceleration cannot be explained by some technical causes, left out of account today, it will give a big push to a further development of physics of particles. Both supernova cosmology data and the Anderson's team discovery may change a gravity position in a hierarchy of known interactions, and, possibly, give us a new chance to unify their description.

References

[1] Riess, A.G., et al, AJ **116**, 1009, 1998.

[2] Anderson, J.D. et al, Phys. Rev. Lett. **81**, 2858, 1998.

[3] Ivanov, M.A., "General Relativity and Gravitation" (the paper is accepted by the editor and is scheduled for the March, 2001 issue of the journal).

[4] Ivanov, M.A., http://arXiv.org/astro-ph/0005084.

[5] Ivanov, M.A., "Quantum Electronics and Laser Science Conference (QELS'95)", May 21-26, 1995, Baltimore, USA; paper number: QThG1.

[6] Turyshev, S.G. et al, http://arXiv.org/gr-qc/9903024 v2.

Deformed Algebras, q-Hermite Polynomials and q-Bessel Functions

V. Srinivasan and S.Chaturvedi
School of Physics
University of Hyderabad, Hyderabad - 500 046
Email:vssp@uohyd.ernet.in, scsp@uohyd.ernet.in

1 Introduction

The q-calculus was invented by Jackson in 1908 [1]. In the last twenty years this area has received considerable attention both from physicists and mathematicians. Ramanujam's work extended by Andrews and Askey, the q-deformation of Lie-groups and Lie-algebras through the corresponding Hopf algebras which lead to non trivial R-matrices obeying Yang-Baxter equation [2] are some of the activities in this area. The various special functions and their q-deformations have also been studied[3]In this work we look at the q-deformed canonical comutation relations[4] of Bosons and show how some q-deformed functions naturally arise as eigenfunctions of certain operators.

2 q-exponential function

Consider the Heisenberg-Weyl algebra

$$[b, b^{\dagger}] = 1 \tag{1}$$

Here $b|0>= 0$ and $|0>$ is the vacuum state. The Fock states are

$$|n\rangle = \frac{(b^{\dagger})^n}{\sqrt{n!}}|0\rangle \tag{2}$$

$$b|n>= \sqrt{n}\ |n> \quad b^{\dagger}|n>= \sqrt{n+1}\ |n> \tag{3}$$

The deformed canonical commutation relations are [4]

$$aa^{\dagger} - qa^{\dagger}a = 1 \qquad |q| \leq 1 \tag{4}$$

The Fock states in this case are

$$|n>= \frac{(a^{\dagger})^n}{\sqrt{(n_q)!}}|0> \tag{5}$$

Frontiers of Fundamental Physics 4, Edited by Sidharth and Altaisky,
Kluwer Academic/Plenum Publishers, New York, 2001

where the vacuum state satisfies $a|0>=0$ and

$$n_q = \frac{1-q^n}{1-q} \quad ; \quad n_q! = n_q(n-1)_q \cdots (1)_q \tag{6}$$

The Fock states satisfy

$$a|n> \;=\; \sqrt{n_q}|n+1> \tag{7}$$

$$a^\dagger|n> \;=\; \sqrt{(n+1)_q}|n+1> \tag{8}$$

The number operator N satisfying the coomutation relations

$$[N,a] = -a \;\; , \;\; [N,a^\dagger] = a^\dagger \tag{9}$$

in this case turns out to be

$$N = \sum_n \frac{(1-q)^n}{1-q} a^{\dagger n} a^n \tag{10}$$

More general deformations of the form

$$[a,a^\dagger] = f(N+1) - f(N) \tag{11}$$

have been considered [5]
Consider the eigenfunctions of a

$$a|\alpha>=\alpha|\alpha> \tag{12}$$

Expanding $|\alpha>$ in terms of $|n>$

$$|\alpha>=\sum_{n=0}^{\infty} c_n|n> \tag{13}$$

one finds that $c_{n+1} = (\alpha/\sqrt{(n+1)_q})c_n$ and hence

$$\begin{aligned} |\alpha> &= \sum_{n=0}^{\infty} \frac{\alpha^n}{\sqrt{(n_q)!\; a^{\dagger n}|0>}} \\ &= \exp_q(\alpha a^\dagger)|0> \end{aligned} \tag{14}$$

where

$$\exp_q(\alpha x) = \frac{x^n}{(n_q)!}$$

Note that

$$\exp_q(x)\exp_q(y) \neq \exp_q(x+y)$$

Also, with the q-derivative defined as

$$\frac{d}{dx_q} f(x) = \frac{f(x) - f(qx)}{(1-q)x}, \tag{15}$$

the q-exponential satisfies

$$\frac{d}{dx_q} \exp_q(\alpha x) = \alpha \exp_q(x) \tag{16}$$

3 The $q-$ Hermite Polynomials

Define

$$\hat{x} = \frac{a + a^\dagger}{\sqrt{2}} \tag{17}$$

Consider the eigenstates of $\hat{x}$

$$\hat{x}|x> = x|x> \tag{18}$$

with

$$|x> = \sum_{n=0}^{\infty} C_n(x)|n> \tag{19}$$

The C_n's satisfy the recursion relation

$$\begin{aligned} C_q &= \sqrt{2}xC_0 \\ C_{n+1}\sqrt{(n+1)_q} + \sqrt{[n]_q}C_{n-1} - \sqrt{2}xC_n &= 0 \end{aligned} \tag{20}$$

Define

$$C_n(x) = \frac{H_n^q(x)}{(\sqrt{2})^n\sqrt{(n_q)!}} \tag{21}$$

then we are led to the following recursion relations for $H_n^q(x)$.

$$H_{n+1}^q(x) + 2[n]_q H_{n-1}^q - 2xH_n^q = 0 \tag{22}$$

As $q \to 1$ this becomes the recursion relation for the Hermite polynomials. As $q \to 0$, one obtains the recursion relations for the Chebyshev polynomials of the second kind $S_n(\sqrt{2}x)$. The orthognality relations are also obtained nicely.

$$\int_{-a}^{+a} dx H_n^q(x) H_m^q(x) \frac{1}{\exp_q(x^2)} = 0, \tag{23}$$

when $n \neq m$ and the limits of integration a and $-a$ are the zeros of the weight function in the integral.

As $q \to 1$ $\frac{1}{\exp_q(x^2) \longrightarrow \exp(-x^2)}$ with $a \to \infty$
and as $q \to 0$ $\frac{1}{\exp_q(x^2) \longrightarrow (1-x^2)}$ with $a \to 1$.

We therefore recover the conventional orthonormal relations of Hermite and Chebyshev. Thus we see that $H_n^q(x)$ are an orthonormal set that interpolates beautifully between Hermite and Chebyshev polynomial systems.

4 q modified Bessel function

We now consider a family of two mode state, the pair coherent states [6]. These are the eigenstates of ab and $a^\dagger a - b^\dagger b$, where a, b are two commuting bosonic oscillators.

$$\begin{aligned} ab|\xi, m> &= \xi|\xi, m> \\ a^\dagger a - b^\dagger b|\xi, m> &= m|\xi, m> \end{aligned} \tag{24}$$

Here m is an integer.
The normalised eigenstates are

$$|\xi, m> = N \sum_{n=0}^{\infty} \frac{\xi^k}{\sqrt{n+m!}\sqrt{n!}} |n+m, n> \tag{25}$$

with

$$N = [\sum_n \frac{|\xi|^{2k}}{(n+m)!n!} \quad I_m(2|\xi|)]^{-1}$$

where I_n is the modified Bessel function.
The q pair coherent states may be defined as the eigenstates of ab and $a^\dagger a - b^\dagger b$ with a and b obeying the following relations

$$\begin{aligned} aa^\dagger - qa^\dagger a &= 1 \\ bb^\dagger - qb^\dagger b &= 1 \\ [a, b] &= 0 \end{aligned} \tag{26}$$

This in turn leads to a normalization factor for the q-pair coherent states and the function that appears there may be taken as the definition of the q-modified Bessel function of the first kind of integer order

$$I_m(z, q) = \sum_{n=0}^{\infty} \frac{(\frac{z}{2})^{m+2n}}{n_q!(n+m)_q!} \tag{27}$$

We extend this definition to modified q-Bessel functions of non-integer order

$$I_\nu(z, q) = \sum_{n=0}^{\infty} \frac{(\frac{z}{2})^{\nu+2n}}{\Gamma_q(n+1)\Gamma_q(n+\nu+1)} \tag{28}$$

where the q-gamma function is defined as

$$\Gamma_q(z) = \frac{(q,q)_\infty (1-q)^{1-z}}{(q^z, q)_\infty} \tag{29}$$

Here

$$(a, q)_n = (1-a)(1-aq)\cdots(1-aq^{n+1}) \tag{30}$$

Some properties of the q-Gamma functions are listed below

(i) $\Gamma_q(z+1) = (z)_q \Gamma_q(z)$

(ii) $\frac{1}{\Gamma_q(-n)} = 0 \qquad n = 0, 1, \cdots$

(iii) The function

$$\psi_q(z) = \frac{d}{dz} \ln \Gamma_q(z)$$

obeys the recursion relation

$$\psi_q(z+1) = \psi_q(z) = \frac{q \ \ln q}{(1-q)} \frac{1}{(z)_q}$$

(iv) As $q \to 1$

$$I_n(z,q) = I_{-n}(z,q) \tag{31}$$

(v) Also

$$\frac{d}{dz}(z^{\nu/2}I_\nu(2\sqrt{z},q) = z^{(\nu-1)/2}I_{\nu-1}(2\sqrt{z},q) \tag{32}$$

One may also define $q-$ modified Bessel functions of the second kind

$$K_\nu(z,q) = \frac{1}{2}\left[\frac{I-\nu(z,q) - I_\nu(z,q)}{\sin \pi\nu}\right] \tag{33}$$

These I_ν, K_ν are different from that of in [7]. The way we have defined the modified q-Bessel functions preserves a large number of suitably modified properties of the ordinary modfied Bessel functions.

We thank Prof. Sidharth for the invitation.

References

[1] F.H. Jackson, Royal Soc, **46**, 253 , Quarterly J, 1908. Maths. **41**, 193 (1910); Quat. J. Math **2**, 1, 1951.

[2] G. Gasper and M. Rahman, "Basic Hypergeometric Series", Cambridge University Press, 1990.

[3] V. Chari and A. N. Pressley, "A Guide to Quantum Groups", Cambridge University Press, 1995.

[4] L. C. Biedenharn, J. Phys. A **22**, L873, 1989; A. J. Mcfarlane, J. Phys. A **22**, 4581, 1989; S. Chaturvedi, V. Srinivasan, Phys. Rev. A**44**, 8020, 1991.

[5] S. Chaturvedi, V. Srinivasan, Phys. Rev. A**44**, 8024, 1991.

[6] D. Bhaumik, K. Bhaumik and B. Dutta Roy, J. Phys A**9**, 1507, 1976; G. S. Agarwal, J. Opt. Soc. Am., **B5**, 1940, 1988.

[7] M. E. H. Ismail, SIAM J. Math. Anal. **12**, 454, 1981.

Quantum Hamilton-Jacobi Formalism and Broken Supersymmetric WKB Approximation Scheme.

Ramandip S. Bhalla, Ashok K. Kapoor

1 Introduction

Semi-classical quantization schemes have been quite useful since the early days of quantum mechanics, in estimating the approximate eigenspectra of various potentials. Interestingly, for certain potentials, the lowest order quantization condition yield the *exact* eigenvalues, e.g., the WKB scheme reproduces the spectra of the harmonic oscillator [1] and the Morse potentials [2]. Recently, in the context of supersymmetric (SUSY) quantum mechanics [3], it was observed that a semi-classical WKB-type approximation.

$$\frac{1}{\pi}\int_{x_1}^{x_2}\sqrt{E-\omega^2(x)}dx = n\hbar \tag{1}$$

gave the exact energy eigenvalues for many well-known potentials [4]. Here, $\omega^2(x)$ is the average of the supersymmetric partner potentials given by,

$$V_{\pm} = \omega^2(x) \pm \hbar\frac{\partial\omega(x)}{\partial x} \tag{2}$$

and where $\omega(x)$ is the super potential ($2m = 1$). For unbroken SUSY, these potentials have identical spectra, except for the unpaired zero energy groundstate. However, for some of these cases that exhibit broken SUSY, it is known that Eq. (1) does not give the exact answer; however this approximation, known as SUSY WKB approximation, gives better answers than the standard WKB approximation. It was also noticed that a new quantization condition given by broken SUSY WKB(BSWKB) quantization formula [5]

$$\frac{1}{\pi}\int_{x_1}^{x_2}\sqrt{E-\omega^2(x)}dx = (n+\frac{1}{2})\hbar \tag{3}$$

reproduces the energy spectra exactly for known cases of broken SUSY. Considering the significance of these quantization schemes in the potential problems, one is naturally led to investigate the origin of this exactness. We do so here using the

[0]akksp@uohyd.ernet.in
and Prasanta K. Panigrahi, School of Physics, University of Hyderabad, Hyderabad-500 046, India.
panisp@uohyd.ernet.in

Frontiers of Fundamental Physics 4, Edited by Sidharth and Altaisky,
Kluwer Academic/Plenum Publishers, New York, 2001

quantum Hamilton-Jacobi (QHJ) formalism [6, 7]. The motivation to use the QHJ formalism lies in the fact that, the quantization condition in this scheme, as will be seen later, is remarkably similar to the semi-classical formula Eq. (1).

The integral appearing in Eq. (1) can be evaluated using the techniques of complex integration. The integral

$$J_{SWKB} \equiv \frac{1}{2\pi} \oint_c \sqrt{E - \omega^2(x)} dx \tag{4}$$

can be easily seen to be equal to the integral in Eq. (1). Here C is a contour enclosing the branch cut on the positive x axis. Using the method of contour integration the SWKB integral can be evaluated.

In a previous paper [8], we have studied the reasons for exactness of Eq. (1) for the unbroken case using the QHJ method. We compared the singularities of the integrand in the action integral [8, 9],

$$J(E) = \frac{1}{2\pi} \oint_c p(x, E) dx \tag{5}$$

originating from the QHJ formalism with those of the integrand in the J_{SWKB}. This revealed the following interesting results. The location of the poles and the corresponding residues of these two integrals matched identically in the complex x plane. Since these poles completely determine the values of the two integrals, J_{SWKB} and $J(E)$, turn out to be equal. The exactness of quantization condition $J(E) = n\hbar$, in the QHJ formalism, then explains the success of SWKB approximation scheme. The integral J_{SWKB} may be regarded as semi classical type approximation to the action integral $J(E)$.

Recently, several Hamiltonians with spontaneously broken SUSY have been s-tudied in the literature [5, 10, 11]. Interestingly, these Hamiltonians exhibit broken SUSY for certain ranges of the parameters A and B appearing in the superpotential and for the other values, SUSY remains intact. It is seen that for ranges of parameters which gives unbroken SUSY, Eq. (1) gives the exact answers. For the ranges of parameters corresponding to the broken SUSY, $J_{SWKB} = (n + \frac{1}{2})\hbar$ gives the exact energy spectra.*However, apriori, it is not clear as to which of the two formulae is to be used to arrive at the exact energy eigenvalues.*

In our previous work [11], we show how the boundry conditions in the QHJ formalism, distinguishes between the broken and unbroken phases of SUSY. We have investigated Pöschl-Teller and the Scraf I potentials for the full ranges of parameters and determined ranges of A and B for phases of unbroken and broken SUSY. It is now natural to ask if we can gain an understanding of the exactness of the BSWKB quantization condition. Since, we will not be able to elaborate on the QHJ formalism here, we refer the interested readers to the references [6, 7] for the details.

The plan of the paper is as follows. In the following section, we explicitly work out the J_{SWKB} for the Pöschl-Teller and the Scraf I potentials, using contour integration method. Comparision with the quantum action variable, satisfying $J(E) = n\hbar$, then clearly reveals the origin of the exactness of the semi-classical quantization schemes. In §3, we study a one parameter dependent potentials, which breaks SUSY explicitly. Although the eigenvalues for these potentials can be computed following QHJ method, neither SWKB nor BSWKB, yields exact eigenvalues.

Interestingly, only for the unbroken or the spontaneously broken cases, relevant approximation methods give exact eigenvalues. We conclude in §4 with some points about the future directions of work.

2 Comparison of J_{SWKB} and $J(E)$

The integral J_{SWKB}, also called here as the approximate action integral, will be evaluated using the techniques of complex integration. The answer will then be compared with the exact answer for the action integral for, $J(E)$, in QHJ. We will at first take up the Pöschl-Teller potential whose superpotential is given by,

$$\omega(x) = A \coth \alpha x - B \text{cosech} \alpha x \qquad x \geq 0 \quad , \tag{6}$$

and its partner potentials $V_{\pm}(x)$ are,

$$\begin{aligned} V_{\pm}(x) &= A^2 + (A^2 + B^2 \pm A\alpha\hbar)\text{cosech}^2 \alpha x \\ &- B(2A \pm \alpha\hbar) \coth \alpha x \text{cosech} \alpha x \end{aligned} \tag{7}$$

using the mapping

$$\coth \alpha x + \text{cosech} \alpha x = y \quad . \tag{8}$$

the approximate action integral (see Eq. 3), in terms of the variable y is given by,

$$J_{SWKB} = -\frac{1}{\pi} \oint_{c_1} \frac{\sqrt{E - \tilde{\omega}^2(y)} dy}{\alpha(y^2 - 1)} \tag{9}$$

Here in Eq. (9) we have

$$\tilde{\omega}(y) = \frac{A(y^2 + 1)}{2y} - \frac{B(y^2 - 1)}{2y} \tag{10}$$

where $\tilde{\omega}(y)$ equals $\omega(x)$ expressed in terms of y. Also here C_1 is the contour in the complex y plane, enclosing the classical branch points; these are defined as the roots of equation$E - \tilde{\omega}^2(y) = 0$ lying in the classical region. There are two other roots present in the nonclassical ($y < 0$) region of the complex y plane. The above contour integral can be evaluated in terms of the other singularities of $\sqrt{E - \tilde{\omega}^2(y)}$, in the complex y plane.

We now evaluate this contour integral for the range $A - B < 0$ and $A + B > 0$; $A + B < 0$, and $A - B > 0$ for which SUSY is known to be broken [11]. It is clear from Eq. (9) that the integrand has singularities, at $y = 0$ and $y = \pm 1$. To evaluate the above integral, consider a counterclockwise contour integral J_{Γ_R} for a circle Γ_R of radius R, which is such that it encloses all the singular points of the above integrand (see Fig.1). Then,

$$J_{\Gamma_R} = J_{SWKB} + J_{c_2} + J_{\gamma_1} + J_{\gamma_2} + J_{\gamma_3}. \tag{11}$$

Here, J_{C_2} is the integral along the counterclockwise contour C_2 enclosing the branch cut in the real $y < 0$ region. J_{γ_1}, J_{γ_2}, and J_{γ_3} are the integrals along contours

γ_1, γ_2, γ_3 enclosing the singular points at $y = 1$, $y = -1$ and 0 respectively. The integral J_{Γ_R} is given by,

$$J_{\Gamma_R} = \frac{-1}{\pi} \oint_{\Gamma_R} \frac{\sqrt{E - \tilde{\omega}^2(y)}dy}{\alpha(y^2 - 1)}. \tag{12}$$

The integrals J_{γ_1},J_{γ_2}, and J_{γ_3} are similarly defined. It may be noticed that the symmetry $y \to -y$ of $\tilde{\omega}^2(y)$ implies,

$$J_{SWKB} = J_{C_2}. \tag{13}$$

Now for the calculation of J_{Γ_R} one more change of variable $z = 1/y$ is sought; the singularity at $y \to \infty$ is then mapped to the singularity at $z = 0$:

$$J_{\Gamma_R} = \frac{-1}{\pi} \oint_{\gamma_0} \frac{\sqrt{E - \tilde{\tilde{\omega}}^2(z)}dz}{\alpha(1 - z^2)} \equiv J_{\gamma_0}, \tag{14}$$

where γ_0 is a small circle in the complex z plane that encloses the singular point at $z = 0$, and $\tilde{\tilde{\omega}}(z)$ is equal to $\omega(x)$ expressed in terms of z. This, together with Eq. (11) gives

$$J_{SWKB} = \frac{1}{2}(J_{\gamma_0} - J_{\gamma_1} - J_{\gamma_2} - J_{\gamma_3}) \quad . \tag{15}$$

It is straightforward to compute the contour integrals appearing in Eq. (15) except the $\sqrt{E - \omega^2(x)}$, as a function of complex variable x, is a multiple valued function and therefore this square root function is to be defined carefully. It has a branch cut in the classically accessible region between the turning points and another one on the negative real axis. We demand that $\sqrt{E - \omega^2(x)}$ be positive for real x in the classical region as one approaches the real axis from the lower half plane, to fix its values uniquely. In terms of the variable y, we have $\sqrt{E - \tilde{\omega}^2(y)}$ given by,

$$\tilde{p}_o(y) \equiv \sqrt{E - \tilde{\omega}^2(y)} = \frac{i\xi\sqrt{(y - y_1)(y - y_2)(y + y_1)(y + y_2)}}{2y} \tag{16}$$

where y_1 and y_2 are the images of x_1 and x_2, given x_1 and x_2 are the two turning points in the classical region, of the complex x plane and $\xi = \pm \mid A - B \mid$. It is easy to check that a point lying just below the real axis between x_1 and x_2 in the complex x plane, is mapped to point just above the cut between y_1 and y_2 in the complex y plane. We thus choose the negative sign of ξ and this enables us to fix the sign of $\tilde{p}_o(y)$ in the cut complex y plane. The various integral J_{γ_K} in the right hand side of Eq. (15) can now be evaluated and are tabulated in the column four of the table.

We now compare the above computation with similar procedure applied to the action integral $J(E)$ of Eq. (5), where $p(x)$is the quantum momentum function satisfying the QHJ equation,

$$p^2 - i\hbar\frac{dp(x)}{dx} = E - V(x). \tag{17}$$

The exact action integral has been computed within the QHJ formalism in our previous paper.[11] We find that locations of singularities of $\sqrt{E - \omega^2}$, other than

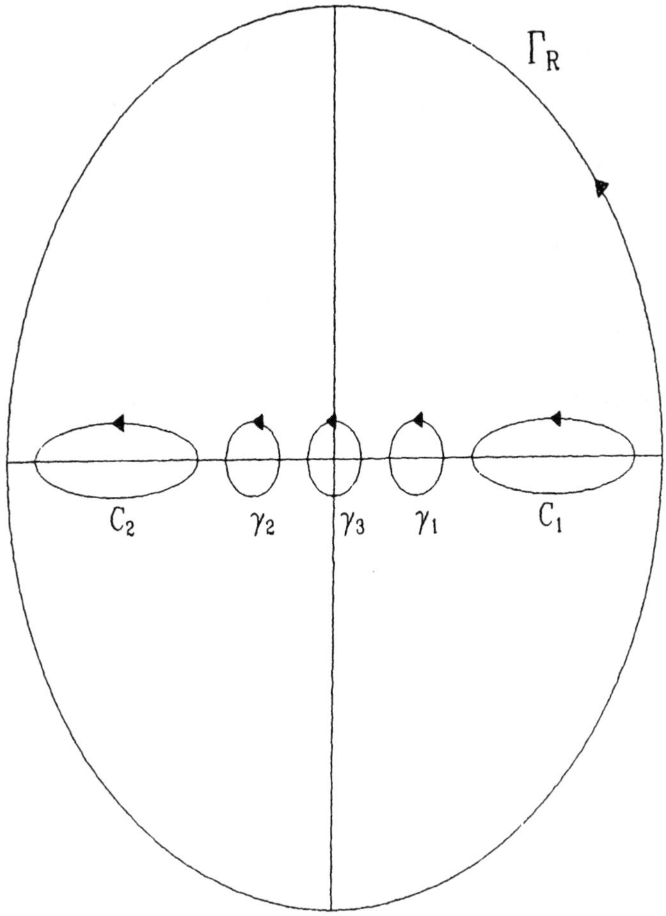

Figure 1. Contours for the generalized Pöschl-Teller potential.

the branch cuts, match exactly with that of the fixed poles of QMF $p(x, E)$. The value of the exact action integral $J(E)$ is again given by contour integrals as in Eq. (15) :

$$J(E) = \frac{1}{2}(I_{\gamma_0} - I_{\gamma_1} - I_{\gamma_2} - I_{\gamma_3}), \tag{18}$$

except that the integrals $I_{\gamma 0}$,$I_{\gamma 1}$,$I_{\gamma 2}$ and $I_{\gamma 3}$ are now given by expressions of the form

$$I_{\gamma\kappa} = -\frac{1}{\pi}\oint_{\gamma\kappa} \frac{\tilde{p}(y)dy}{\alpha(y^2-1)}, \tag{19}$$

and the contours are same as appearing in Fig.1. The values of the integrals $I_{\gamma\kappa}$ are given in the table.

It is seen that the value of the action integral in the SWKB approximation differ from the exact value by $\hbar/2$. This explains appearance of the $(n+\frac{1}{2})\hbar$ in the quantization condition in the SWKB approximation.

We repeat the above investigation for the Scarf I(trigonometric)case and we will highlight only the new points here. The superpotential in this case is given by

$$\omega(x) = -A \tan \alpha x + B \sec \alpha x \qquad (-\pi/2 \leq \alpha x \leq \pi/2). \tag{20}$$

Using the transformation

$$y = \sec \alpha x + \tan \alpha x \quad , \tag{21}$$

the quantization condition in the SWKB approximation is given by

$$J_{SWKB} = \frac{1}{2\pi\alpha} \oint_{c_1} \frac{\sqrt{E - \tilde{\omega}^2(y)}dy}{(y^2+1)} = (n+1/2)\hbar \quad . \tag{22}$$

The mapping from x to y gives rise to singularities in the integrand at $y = \pm i$ and at $y = 0$ and in order to calculate J_{SWKB}, we once again deform the contour so as to enclose all the singularities inside a large circle Γ_R or radious R (see Fig.2). We

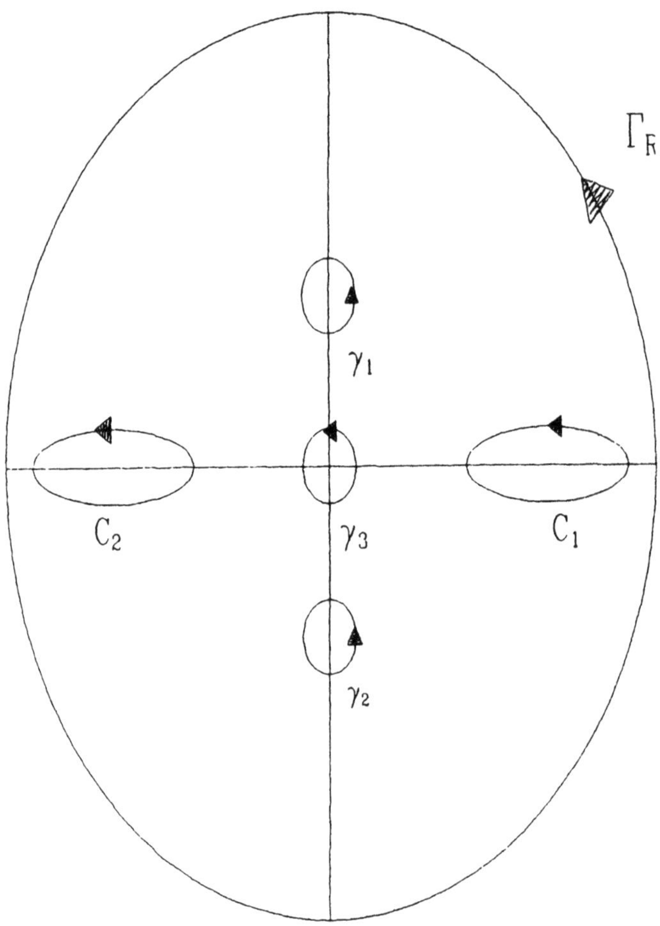

Figure 2. Contours for the Scarf I potential.

thus have

$$J_{\Gamma_R} = J_{SWKB} + J_{C_2} + J_{\gamma_1} + J_{\gamma_2} + J_{\gamma_3} \quad , \tag{23}$$

where J_{Γ_R} is the contour integral for the circle Γ_R;J_{γ_1},J_{γ_2} and J_{γ_3} are the contour integrals evaluated along the contours γ_1,γ_2 and γ_3 enclosing the poles at $y = i$, $-i$ and 0 respectively, and J_{C_2} is the integral along the counterclockwise contour C_2 enclosing the branch cut in the real $y < 0$ region. These contour integrals are given by

$$J_{\Gamma_R} = \frac{1}{\pi} \oint_{\Gamma_R} \frac{\sqrt{E - \tilde{\omega}^2(y)}}{\alpha(y^2+1)} dy \quad , \tag{24}$$

and similar equations for the other ones.

The contour integral J_{Γ_R} is once again calculated by the change of variable $z \to 1/y$ and we have,

$$J_{\Gamma_R} = \frac{1}{\pi} \oint_{\gamma_o} \frac{\sqrt{E - \tilde{\tilde{\omega}}^2(z)}}{\alpha(1+z^2)} dz \equiv J_{\gamma_o} \quad . \tag{25}$$

Table : Values of contour integrals contributing to J_{SWKB} and $J(E)$.

Potential	Range of Parameters	Contour γ_κ	$J_{\gamma_\kappa} \times \alpha$ (SWKB)	$I_{\gamma_\kappa} \times \alpha$ (QHJ)
Generalized	$A - B > 0$;	γ_1	$\sqrt{\lvert E - A^2\rvert}$	$\sqrt{\lvert E - A^2\rvert}$
Pöschl-	and	γ_2	$\sqrt{\lvert E - A^2\rvert}$	$\sqrt{\lvert E - A^2\rvert}$
Teller	$A + B > 0$	γ_3	$-(A+B)/2$	$-(A+B)/2$
		γ_0	$(B-A)/2$	$(B - A - \alpha\hbar)/2$
Generalized	$A - B < 0$;	γ_1	$-\sqrt{\lvert E - A^2\rvert}$	$-\sqrt{\lvert E - A^2\rvert}$
Pöschl-	and	γ_2	$-\sqrt{\lvert E - A^2\rvert}$	$-\sqrt{\lvert E - A^2\rvert}$
Teller	$A + B < 0$	γ_3	$(A+B)/2$	$(A + B + \alpha\hbar)/2$
		γ_0	$(A-B)/2$	$(A-B)/2$
Scarf I	$A - B < 0$	γ_1	$-\sqrt{E + A^2}$	$-\sqrt{E + A^2}$
Trigono-	and	γ_2	$-\sqrt{E + A^2}$	$-\sqrt{E + A^2}$
metric	$A + B > 0$	γ_3	$(A+B)/2$	$(A + B + \alpha\hbar)/2$
		γ_0	$(A-B)/2$	$(A-B)/2$
Scarf I	$A - B > 0$	γ_1	$-\sqrt{E + A^2}$	$-\sqrt{E + A^2}$
Trigono-	and	γ_2	$-\sqrt{E + A^2}$	$-\sqrt{E + A^2}$
metric	$A + B < 0$	γ_3	$-(A+B)/2$	$-(A+B)/2$
		γ_0	$-(A-B)/2$	$-(A - B + \alpha\hbar)/2$

Here again, due to the symmetry under $y \to -y$ of $\tilde{\omega}(y)$ the turning points on the real axis in the classical resion, $y > 0$, are interchanged with those in $y < 0$ region, and we have that J_{SWKB} and J_{C_2} are equal (see Fig.2). Therefore,

$$J_{SWKB} = \frac{1}{2}(I_{\gamma_0} - I_{\gamma_1} - I_{\gamma_2} - I_{\gamma_3}) \quad . \tag{26}$$

The various integrals appearing in the righthand side of the above equation can be computed in a straightforward fashion except that once again care is needed to define the multiple valued function $\sqrt{E - \tilde{\omega}^2(y)}$. This is done following a procedure similar to that used for the generalised Pöschl-Teller potential. The values of the integrals obtained are listed in the table. The exact action integral for the Scarf-I potential can be evaluated in the QHJ formalism and is given by an expression similar to Eq. (26) :

$$J(E) = \frac{1}{2}(I_{\gamma_0} - I_{\gamma_1} - I_{\gamma_2} - I_{\gamma_3}), \tag{27}$$

where in the above equation the integrals are to be taken as

$$I_{\gamma\kappa} = -\frac{1}{\pi}\oint_{\gamma\kappa} \frac{\tilde{p}(y)dy}{\alpha(1+y^2)} \quad . \tag{28}$$

The values of these integrals are given in the table. It is now clear that, for the two potentials under study, we have the relation

$$J_{SWKB} = J(E) + \frac{\hbar}{2}, \tag{29}$$

and the exact quantization condition $J(E) = n\hbar$, valid for all potentials, becomes $J_{SWKB} = (n + \frac{1}{2})\hbar$ for the two potentials when the supersymmetry is broken.

3 Analysis of potentials of the type $V(x) = \omega^2(x) + \hbar\beta\omega'(x)$

We have applied the QHJ method to study a general, one parameter dependent Hamiltonian,

$$H = p^2 + \omega^2(x) + \hbar\beta\omega'(x) \quad , \tag{30}$$

which reproduces the partner Hamiltonians H_+ and H_- for specific values of this parameter. The poles and the corresponding residues of the QMF are evaluated after taking into account the boundry condition.

We briefly highlight the analysis for generalized Pöschl-Teller superpotential given by,

$$\omega(x) = A\coth\ \alpha x - B\text{cosech}\alpha x \quad . \tag{31}$$

We thus have

$$\begin{aligned} V(x) &= \omega^2(x) + \beta\omega'(x) \quad , & (32)\\ &= A^2 + (A^2 + B^2 - \beta A\alpha\hbar)\text{cosech}^2\alpha x \\ &- B(2A - \beta\alpha\hbar)\coth\alpha x\text{cosech}\alpha x, & (33)\end{aligned}$$

making use of the transformation Eq. (8) as done earlier, the QHJ equation is given by

$$\tilde{p}^2 - \frac{i\hbar\alpha}{2}\frac{\partial\tilde{p}}{\partial y}(1-y^2) = E - A^2 - (A^2 + B^2 - A\alpha\beta\hbar)$$
$$\frac{(y^2-1)^2}{4y^2} + B(2A - \alpha\hbar\beta)\frac{(y^4-1)}{4y^2} \quad . \tag{34}$$

The quantization condition remains the same as in the case of generalised Pöschl-Teller case. With this inclusion of β in the potential, the singularity structure of QMF, $\tilde{p}(y, E)$ as well as $p_c(x, E)$ remains unchanged and the integral $J(E)$ is evaluated. We give answers only for

$$J(E) = \frac{1}{2}(I_{\gamma_0} - I_{\gamma_1} - I_{\gamma_2} - I_{\gamma_3}) \quad . \tag{35}$$

The contours for the integrals appearing in the above equation are given in Fig.1. The residues at different poles are determined by substituting Laurent expansion in the QHJ equation. The QHJ equation, being non-linear gives quadratic equations for the residues which are required to compute various contour integrals I_{γ_K}. The procedure for picking the correct root for the residues remains exactly the same as carried out in our previous paper [11]. Knowing the residues various contour integrals appearing in Eq. (26) can be computed. We give the answers only for $A + B > 0$, $A - B > 0$, and $\beta \approx -1$.

$$\alpha I_{\gamma_1} = -\sqrt{|E - A^2|} \quad , \tag{36}$$

$$\alpha I_{\gamma_2} = -\sqrt{|E - A^2|} \quad , \tag{37}$$

$$\alpha I_{\gamma_3} = \frac{1}{2}\left(\alpha\hbar - \left|\sqrt{(2A + 2B - \alpha\hbar)^2 + 4(\beta+1)\alpha\hbar(A+B)}\right|\right) \quad , \tag{38}$$

$$\alpha I_{\gamma_0} = \frac{1}{2}\left(\alpha\hbar - \left|\sqrt{(2A - 2B + \alpha\hbar)^2 + 4(\beta+1)\alpha\hbar(B-A)}\right|\right) \quad . \tag{39}$$

The quantization condition $J(E) = n\hbar$ gives the energy eigenvalues. It is seen that the difference between the exact action integral, $J(E)$, and the approximate action integral, J_{SWKB}, cannot be expressed as a simple function of β. Except for the values $\beta = \pm 1$, no simple modification of SUSY WKB quantization condition can give exact eigenvalues for all β. In addition to the generalised Pöschl-Teller potential, we have also carried out calculations for the Eckart, Rosen-Morse II, Scarf II (all hyperbolic); and the Rosen-Morse I and Scarf I (trigonometric) superpotentials using the QHJ formalism for entire range of parameters, A and B. Conclusions similar to those for generalised Pöschl-Teller potential are found to be true for these other potentials modified by including the parameter β.

4 Conclusion

In conclusion, we have studied a class of Hamiltonians which exhibit both broken and unbroken SUSY for suitable values of the parameters appearing in the potentials. It is found that, both for the broken and unbroken phases of SUSY, the poles of the QMF match identically with that of the integral $\sqrt{E-\omega^2(x)}$ appearing in the approximate quantization formula. However for the broken case, there is a difference in the residues of the QMF and $\sqrt{E-\omega^2(x)}$ in the complex plane which gives rise to the modified BSWKB quantization condition. Since these poles determine the eigenvalues completely, the exactness of the semi-classical scheme is accounted for.

The finding that the residues of $\sqrt{E-\omega^2(x)}$ and QMF match for the unbroken SUSY, and do not match in the broken SUSY example, suggests that it is the quantum effect that breaks SUSY. It is amusing that the difference in residues for the above two cases is precisely $\hbar\pi$. Analysis of a general one parameter dependent Hamiltonian revealed that only for the broken and spontaneously broken cases of SUSY, the appropriate semi-classical quantization formulae can reproduce the exact spectra for the examples considered in the text. These type of potentials and their generalizations can be further studied, in the QHJ approach, to find better semi-classical approximation schemes.

There are potentials for which the singularity structure of $\sqrt{E-\omega^2(x)}$ and of QMF are very different. In these cases, no simple SWKB type approximation is known which may give exact answers [12]. The QHJ analyses of these potentials will be presented elsewhere [13].

Acknowledgements :
We acknowledge useful discussions with Profs. V. Srinivasan, Pankaj Sharan and Dr. C. Nagraja Kumar.

References

[1] C.M. Bender, K. Olauseen and P.S. Wang: Phys. Rev. **D 6**, 1977, 1740.

[2] K. Raghunathan, M. Seetharaman and S.S. Vasan: Phys. Lett. **B188**, 1987, 351.

[3] E. Witten: Nucl. Phys.**B 188**, 1981, 513; F. Cooper, A. Khare and U. Sukhatme: Phys. Rep.**251**, 1995, 2 and references therein.

[4] A. Comtet, A. Bandrauk and D. Campbell: Phys. Lett. **B150**, 1985, 159; A. Khare: Phys. Lett. B **161**, 1985, 131.

[5] A. Inomata and G. Junker, "Proceedings of Adriatico Research conference on path-integration and its application", ICTP, Italy, 3-6 Sept., 1991. R. Dutt, A. Gangopadhyaya, A. Khare, A. Pagnamenta and U. Sukhatme: Phys. Rev **A 48**, 1993, 1845.

[6] R.A. Leacock and M.J. Padgett: Phys. Rev. Lett.**50**, 1983, 3; R.A. Leacock and M.J. Padgett: Phys. Rev. **D28**, 1983, 2491.

[7] For a pedagogical review see: R.S. Bhalla, A.K. Kapoor and P.K.Panigrahi: Am. J. Phys. **65**, 1997, 1187.

[8] R.S. Bhalla, A.K. Kapoor and P.K. Panigrahi: Phys. Rev. **A 54**, 1996, 951.

[9] R.S. Bhalla, A.K. Kapoor and P.K. Panigrahi: Mod. Phys. Lett **A 12**, 1997, 295.

[10] A. Murali, T.R. Govindrajan and A. Khare: Phys. Rev. **A 52**, 1995, 4259.

[11] R.S. Bhalla, A.K. Kapoor and P.K. Panigrahi: Int. J. Mod. Phys. **A12**, 1997, 1875.

[12] A. Khare and Y.P. Varshini: Phys. Lett. **A 142**, 1989, 1; D. Delaney and M.M. Nieto: Phys. Lett. **B247**, 1990, 301.

[13] R.S. Bhalla, A.K. Kapoor and P.K. Panigrahi (under preparation).

Theory of Quantum Hall Effect: Effective Fractional Charge

Keshav N. Shrivastava
School of Physics, University of Hyderabad,
Hyderabad 500 046, India

1 Introduction

We assume that in a semiconductor if the electric field is applied along the x-direction and a magnetic field along the z-direction, then there is a Hall voltage along the y direction. The Hall resistivity is given by

$$\rho_H(x,y) = -\frac{B_z}{nec} \tag{1}$$

where n is the concentration of electrons, e the charge and c is the velocity of light. It may be observed that all of the three dimensions are equally important at this stage. Since the magnetic field is directed along the z-direction, the current is along x direction and the electric voltage is along y direction, the electrons are confined in the $x-y$ plane and hence may exhibit two-dimensional behaviour. It was found by v. Klitzing et al[1,2] that the above resistance is quantized as

$$\rho(x,y) = \frac{h}{\nu e^2} \tag{2}$$

where ν is an integer. It can be interpretted as a Landau level filling factor for electrons in two-dimensions. For $\nu = 1$, the quantity

$$\frac{h}{e^2} = 25812.8 \qquad \text{ohms} \tag{3}$$

is called "one Klitzing". Later on, it was found by Tsui, Stormer and Gossard[3] that ν need not be an integer. While measuring the transverse resistivity as a function of field, a plateau was found at $\nu = 1/3$. Later studies discovered many different values of the fraction, ν.

In this brief review, we describe our theory of the quantum Hall effect. In particular, effort is made to compare the theory with the experimental data.

2 Theory:

We consider the spin as well as the orbital motion so that

$$g_j j = g_s\vec{s} + g_l\vec{l} = \frac{1}{2}(g_l + g_s)\vec{j} + \frac{1}{2}(g_l - g_s)(\vec{l} - \vec{s}) \; . \tag{4}$$

Frontiers of Fundamental Physics 4, Edited by Sidharth and Altaisky,
Kluwer Academic/Plenum Publishers, New York, 2001

The conduction electrons have $l = 0$. However, we consider also the bound electrons which have finite orbital angular momentum quantum number. Multiplying both sides of the above equation by $\vec{j} = \vec{l} + \vec{s}$ and taking eigenvalues, we find,

$$g_j j(j+1) = \frac{1}{2}(g_l + g_s) j(j+1) + \frac{1}{2}(g_l - g_s)[l(l+1) - s(s+1)] , \tag{5}$$

which upon substituting $s = \frac{1}{2}$ gives,

$$g_j = g_l \pm \frac{g_s - g_l}{2l+1} \tag{6}$$

for $j = l \pm \frac{1}{2}$. For $g_s = 2, g_l = 1$ we find

$$g_\pm = 1 \pm \frac{1}{2l+1} . \tag{7}$$

The cyclotron frequency is defined in terms of the magnetic field as,

$$\omega = \frac{eB}{mc} . \tag{8}$$

Corresponding to this frequency, the voltage along y direction is

$$\hbar\omega = eV_y . \tag{9}$$

From (8) and (9)

$$\hbar\frac{eB}{mc} = eV_y . \tag{10}$$

We multiply the above by e/h from both sides so that,

$$\frac{e^2 B}{2\pi mc} = \frac{e^2}{h} V_y \tag{11}$$

which describes the current in the x-direction so that it is proved that

$$\rho_{xy} = \frac{h}{e^2} \tag{12}$$

which is the same as (2) for $\nu = 1$ because of the Ohm's law. We take into account, the gyromagnetic ratio so that (11) may be written as,

$$I_x = \frac{1}{2} g \frac{e^2 B}{2\pi mc} = \frac{1}{2} \frac{g e^2 V_y}{h} . \tag{13}$$

For $l = 0, g = 2$,

$$I_x = \frac{e^2}{h} V_y \tag{14}$$

which describes the correctly quantized current for $\nu = 1$. From (13) $\nu = \frac{1}{2} g_\pm$ which gives the filling factor, one value for + sign and the other value for − sign in (7). For $l = 0$, we obtain $\frac{1}{2} g_+ = 1$ and $\frac{1}{2} g_- = 0$, for $l = 1$, we get $\frac{1}{2} g_+ = 2/3$ and $\frac{1}{2} g_- = 1/3$. These values of $\nu = \frac{1}{2} g_\pm$ are given in Table 1 from the theory of Shrivastava,[4] which is also described in refs. 5 and 6.

Table 1. The calculated values of ν from ref. 4.

l	$\frac{g_-}{2} = \frac{l}{(2l+1)}$	$\frac{g_+}{2} = \frac{(l+1)}{(2l+1)}$
0	0	1
1	1/3	2/3
2	2/5	3/5
3	3/7	4/7
4	4/9	5/9
5	5/11	6/11
6	6/13	7/13

The Landau levels are introduced by projecting the above to two dimensions. We shall consider only the Landau level quantum number, n so that $\nu = n(\frac{1}{2}g_\pm)$. Thus we can multiply the above fractions by an integer n wheneven needed to interpret the data.

3 Supporting Data

We prove that this is the correct theory by means of comparing the above values with the experimental data. The values of ρ_{xx} and ρ_{xy} for a single interface of GaAs/AlGaAs have been measured by Willett et al[7] at a temperature of 150 mK and at high fields at 85 mK. These values are shown in Fig.1. The fractions occur in two sets, the values 2/5, 3/7, 4/9, 5/11, 6/13 etc., shown in the experimental data are exactly the same as predicted by column 2 of table 1. Similarly, the values 2/3, 3/5, 4/7, 5/7, 6/11, etc., seen in the experimental data are exactly the same as predicted by column 3 of table 1. When we mutliply the values by n we can interpret all of the experimentally measured values by use of table 1. The two columns of table 1 belong to two Kramers conjugates one belonging to $+\frac{1}{2}$ and the other to $-\frac{1}{2}$ spin and the experimental data also displays them in two groups. Thus this grouping aspect of the theory is also in agreement with the data.

Accurate measurements of the quantized Hall resistance upto a magnetic field of 30 Tesla have been performed by Eisenstein and Stormer[8]. The dip in ρ_{xx} at $\nu = 1/3$ is clearly seen in this measurement displayed here in Fig. 2. The value of 1/3 is in agreement with that predicted in Table 1. The entire series of fractions along with grouping in the experimental data is exactly the same as predicted. It was shown in ref. 4 that the cyclotron frequency is determined by the Bohr magneton,

$$\hbar\omega_c = g\mu_B B \tag{15}$$

where $\mu_B = e\hbar/2mc$. Therefore, the correction $(\frac{1}{2})g_\pm$ can be considered as a correction to the charge. Therefore, we can write the effective charge as

$$e_{eff} = (\frac{1}{2})ge = \nu e.$$

Therefore for $\nu = 1/3$, the effective charge becomes $(1/3)e$ which is in agreement with the ideas of a fractional charge.

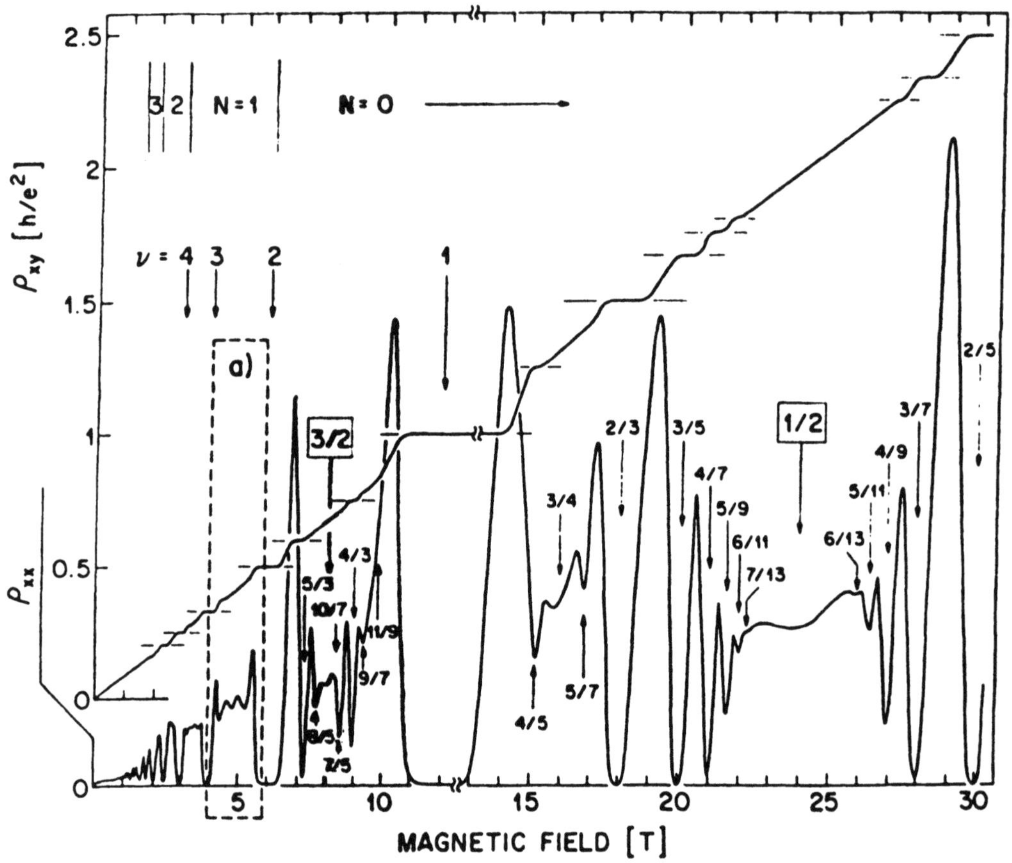

Fig.1: The experimental data of Willett et al [7] is shown. The series, on the right hand side of 1/2 as well as that on the left hand side are exactly the same as those predicted by Shrivastava.[4]

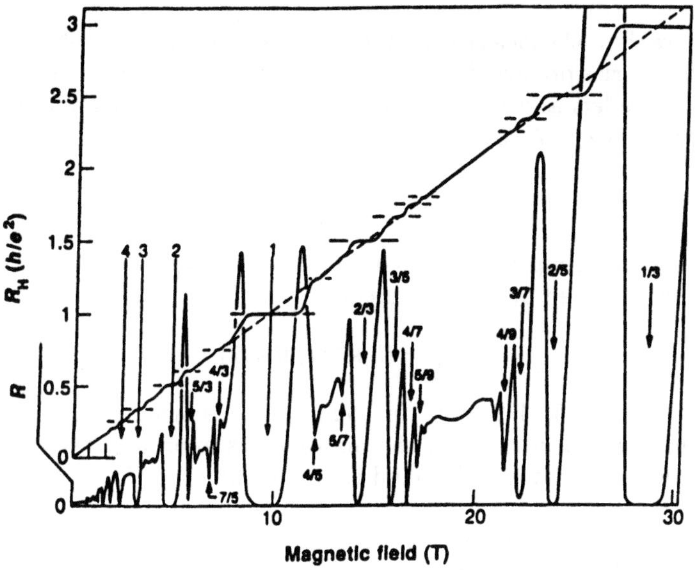

Fig.2: The experimental data of the Hall resistance as a function of magnetic field as measured by Eisenstein and Stormer.[8] All of the fractions between 15 and 30 T are exactly the same as those predicted by Shrivastava.[4]

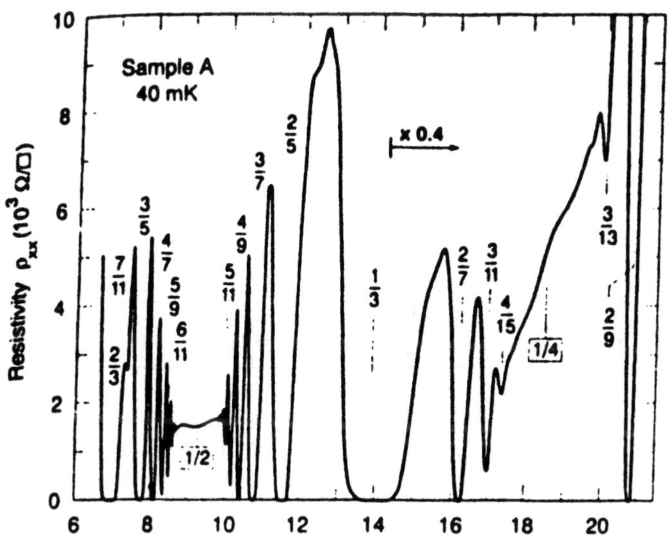

Fig. 3: The experimental data of the Hall resistance measured by Du et al[9] are shown. All the values on the left or right side of 1/2 are exactly the same as those predicted by Shrivastava.[4]

In Fig.3 we show the measurements of the fractions by Du et al[9] at a temperature of 40 mK. The series predicted by table 1 are in full agreement with the data. All of the values on the left as well as on the right hand side of 1/2 are in full agreement with our series. From (7), one of the series is

$$\nu_- = l/(2l+1) \tag{16}$$

which predicts one group of fractions 0, 1/3, 2/5, 3/7, 4/9, 5/11 etc. Another group of fractions is predicted by the expression,

$$\nu_+ = (l+1)/(2l+1) \tag{17}$$

which are 1, 2/3, 3/5, 4/7, etc. in complete agreement with the experimental measurements when $l = \infty$ both the above series approach 1/2 except that one series approaches from the right hand side and the other from the left hand side exactly as observed.

The series (16) and (17) can be used to explain the high Landau levels easily. Eisenstein et al[10] have found that at higher values of the Landau level quantum number, n, the number of fractions observed in much less than at the lowest Landau level. At the magnetic field of 4 to 5 Tesla only a small number of fractions are observed,[11] the strongest ones are at 8/3, 5/2 and 7/3. Since the effective charge uses the Landau level quantum number, it is not possible to distinguish large charge from a large Landau level quantum number. The series, $l/(2l+1)$ is the particle-hole conjugate of $(l+1)/(2l+1)$. For $l = 7$ two values, 7/15 and 8/15 are predicted and $l = \infty$ value is 1/2. When the same particle occurs in different levels its charge remains unchanged. We can multiply the values by $n = 5$ so that the predicted values of 1/2, 7/15 and 8/15 become 5/2, 7/3 and 8/3. These predicted values are exactly the same as those observed experimentally by Eisenstein et al[10] and shown in Fig. 4. Thus 7/3 is the particle-hole conjugate of 8/3 as seen in Table 2 for $n = 5$. The states at 7/3, 8/3 and 5/2 are also observed experimentally by Pan et al[12] to which our interpretation applies satisfactorily.

Table. 2: Derivation of 7/3 and 8/3 states for $n = 5$

l	$l/(2l+1)$	$(l+1)/(2l+1)$	$nl/(2l+1)$	$n(l+1)/(2l+1)$
∞	1/2	1/2	5/2	5/2
7	7/15	8/15	7/3	8/3

It may be noted that 5/2 obtained by this method which is the value of 1/2 multiplied by $n = 5$ is indistinguishable from the second level filled upto 1/2 which is $2 + \frac{1}{2}$. Since all values are emerging in a coherent manner, the $n = 5$ interpretation is correct.

The first fraction discovered was 1/3. Later on other fractions were seen. For a long time only odd denominators were reported which shows that even denominators are weak. For some time odd denominator rule was established but such a rule is not correct because even denominators occur as well as even numerators with odd denominators are found. Later on even denominators were reported. Therefore, it is observed that the fractions corresponding to even denominators are weak compared with those of odd denominators. There is a limiting

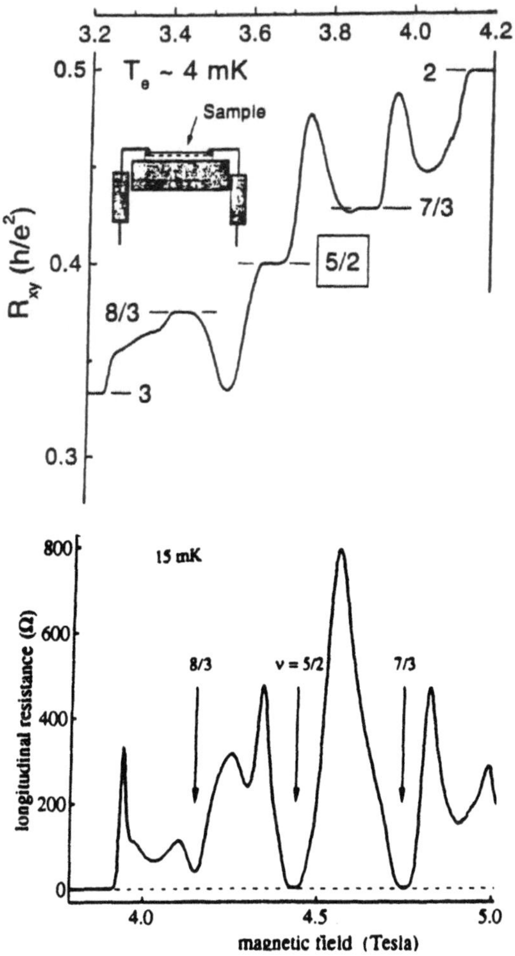

Fig. 4: The ρ_{xx} measured by Pan et al[12] is shown in the upper graph and those of Eisenstein et al[10] [cond-mat/9909238] are shown in the lower graph. The interpretation of 8/3 and 7/3 is described in the text and 5/2 is the $n = 5$ of $n/2$, also given in the text.

value of the series$l/(2l+1)$ which for $l \gg 1$, gives 1/2 which has an even denominator. Because of the limit, there is a Fermi surface at 1/2. When higher Landau levels are populated the Fermi surface will have to be shifted from 1/2 to $n/2$. Thus 1/2, 2/2, 3/2, 4/2, 5/2, 6/2, 7/2 are pedicted which are all observed by Yeh et al[13] as shown in Fig. 5. Thus at least 14 fractions given in Table 1 and 7 fractions in the limit $l \to \infty$ which adds upto 21 are correctly predicted by ref. 4. Besides nν are also predicted. Therefore, for n=2 the predicted number of fractions becomes 42, etc. which are predicted correctly. Yoshioka[14], has pointed out that some of the fractions claimed by the experimentalists are incorrect and have to be ignored. Hence the theory of ref. 4 is correct.

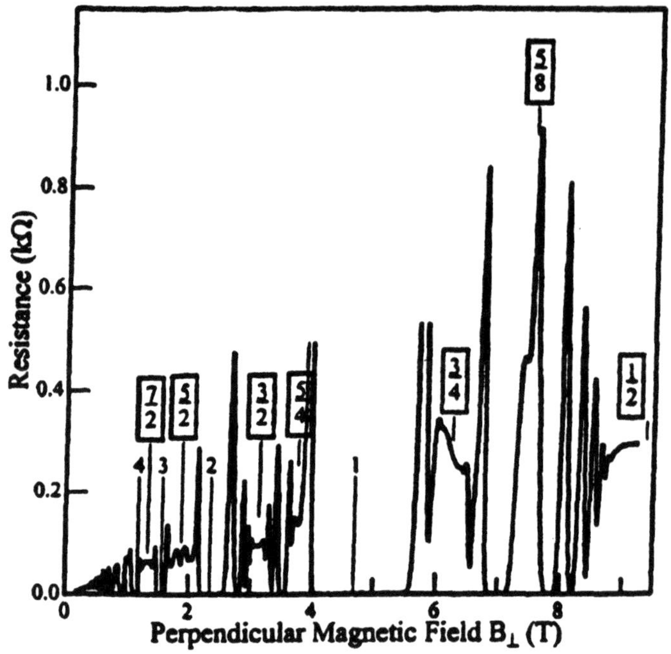

Fig. 5: Experimental data of Yeh et al[13] showing $n/2$ states exactly as predicted by Shrivastava's theory.

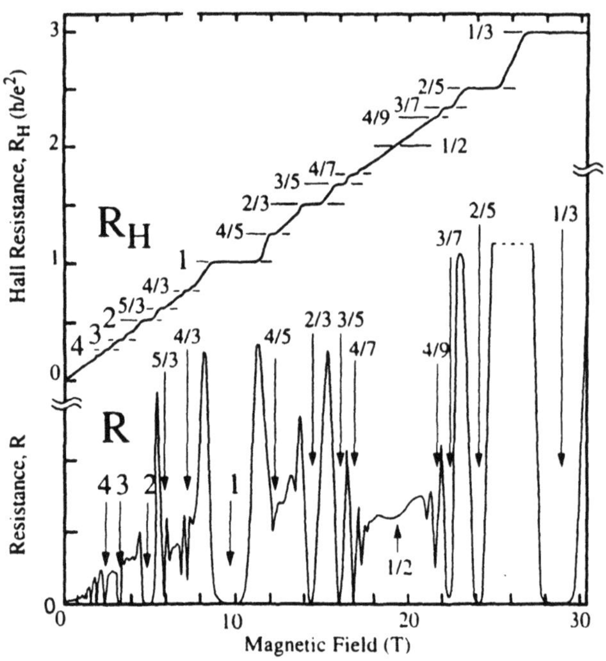

Fig. 6: Experimental data of Stormer from ref. 39 is exactly the same as predicted by us in ref. 4.

In Fig. 6, we show the experimental measurement by Stormer. The fractions found are 1/3, 2/5, 3/7, 4/9 and 1/2 which are correctly predicted by our Table 1. The values 2/3, 3/5, 4/7 and 1 are all the same as predicted by us. The values 4/3 and 5/3 are the $n(1/3)$ which are also predicted. The values n multiplied by 1 and predicted as observed. The value 2 multiplied by 2/5 is seen. Thus all of the observed values are predicted.

It has been reported by Yeh et al[13] that the effective mass and g factor of some of the fractions are equal to those of others. We have shown[5] that the effective masses can be equal only when the two quasiparticles are particle-hole conjugates. The particle-hole conjugates should obey the following relation,

$$\nu_p + \nu_h = 1 \ . \tag{18}$$

For a given value of l, the two fractions given in table 1 always obey this relation and hence the fractions of the second column have the same masses as those in the corresponding row of the third column. For example $1/3 + 2/3 = 1, 2/5 + 3/5 = 1$, etc.

4 Half-filled Landau level

In the expressions (16) and (17), we take the $l = \infty$ limit so that

$$\begin{aligned} \lim_{l\to\infty} \nu_+ &= 2/2l = \frac{1}{2}(+) \\ \lim_{l\to\infty} \nu_- &= \frac{1}{2}(-) \ . \end{aligned} \tag{19}$$

Since these are obtained from different series we use an additional index A for + and B for −. In this way 1/2 comes from the right hand side as well as from the left hand side such as in Fig. 1. Various fractions from the ν_- series start from 1/3 at the highest magnetic field and move to lower fields (right to left) as the value of l increases. Similarly, the fractions from the series ν_+ start from a fixed value 2/3 and move to higher fields (left to right) as l increases. Both the series merge in the centre of 1/3 and 2/3. The remaining pattern of fractions at which there are minima in ρ_{xx} are generated by multiplying $\nu_\pm$ by n, the Landau level quantum number. The series represented by ν_+ and ν_- are also conjugate states representing two spin orientations. Therefore, one of the 1/2 values is like a hole, $\frac{1}{2}(B)$ and the other half is like an electron $\frac{1}{2}(A)$. Since the electron and the hole are separated by a distance, this state is compressible. When we multiply this result by n, the Landau level quantum number, we obtain,

$$\frac{1}{2}, \frac{2}{2}, \frac{3}{2}, \frac{4}{2}, \frac{5}{2}, \frac{6}{2}, \frac{7}{2}, \frac{8}{2}, \ \text{etc.}$$

which are in agreement with the experimental data. The predicted values $n/2$ are exactly as arranged in the data of Yeh et al[13]. Our result that the state is compressible is in agreement with the theoretical work of Halperin et al.[15,16] We now determine the wave vector dependance of the response function of the 1/2 filled Landau state.

We write the Fourier transform of the Coulomb interaction as

$$v(q) = \frac{4\pi e^2}{q^2} \tag{20}$$

so that the relation between the dielectric function, $\epsilon(q,\omega)$ and the response function becomes,

$$\frac{1}{\epsilon(q,\omega)} = 1 + v(q)\chi(q,\omega) \; . \tag{21}$$

The well known dielectric constant has an anomaly at $2k_F$ which is called as the Kohn anomaly.[17,18] The dielectric constant showing logarithmic divergence is given by

$$\epsilon(q,\omega) = 1 + \frac{k_s^2}{2q^2}\left\{1 + \frac{k_F}{q}\left(1 - \frac{q^2}{4k_F^2}\right)\ln\frac{q+2k_F}{q-2k_F}\right\} \tag{22}$$

where k_F is the Fermi wave vector and k_s is the screening.
We write eq.(19) as,

$$\chi(q,\omega) = \left(\frac{1}{\epsilon(q,\omega)} - 1\right)\frac{q^2}{4\pi e^2} \tag{23}$$

which does not vary as q^2 because of the wave vector dependence of the dielectric constant. The q^2 dependence is associated with the incompressibility but clearly the response function given by (23) is compressible. For

$$\epsilon(q,0) \;=\; 1 + \frac{k_s^2}{q^2} \tag{24}$$

$$\chi(q,0) \;=\; \left(\frac{q^2}{q^2 + k_s^2} - 1\right)\frac{q^2}{4\pi e^2} \; . \tag{25}$$

For small $q(q^2 \ll k_s^2)$,

$$\begin{aligned}\chi(q,0) &= \left(\frac{q^2}{k_s^2} - 1\right)\frac{q^2}{4\pi e^2} \\ &\simeq \frac{q^4}{4\pi e^2 k_s^2} - \frac{q^2}{4\pi e^2} \qquad (26)\\ &\qquad\qquad\qquad\qquad\qquad\; (27)\end{aligned}$$

there is a correction term which varies as q^4. For large q

$$\chi(q,0) = \left[\left\{1 + \frac{k_s^2}{q^2}\right\}^{-1} - 1\right]\frac{q^2}{4\pi e^2} \simeq -\frac{k_s^2}{4\pi e^2} \tag{28}$$

which is independent of q. Thus the response function varies as q^2 at small q with a q^4 term as q increases and becomes constant at large q.

According to eq.(19) there are two types of liquids, one is the + or A type and the other a − or B type. Therefore, we suggest a two-component theory for the half filled Landau level. We define the induced density

$$\delta n_A(r,t) = n_A(r,t) - \bar{n}_A \tag{29}$$

for $\frac{1}{2}(+)$ and

$$\delta n_B(r,t) = n_B(r,t) - \bar{n}_B \tag{30}$$

for $\frac{1}{2}(-)$ type quasiparticles. Here $\bar{n}_A$ and $\bar{n}_B$ correspond to homogeneous values and $v_A(q,\omega)$ and $v_B(q,\omega)$ are the Fourier transforms of $v_A(r,t)$ and $v_B(r,t)$ respectively. In the generalized RPA, we write

$$\begin{aligned}
\delta n_A(q,\omega)[1-\chi_o^A(q,\omega)\psi_{AA}(q)] &= \chi_o^A(q,\omega)v_A(q,\omega) + \chi_o^A(q,\omega)\psi_{AB}(q) \\
&\times \ < \delta n_B(q,\omega) > \\
\delta n_B(q,\omega)[1-\chi_o^B(q,\omega)\psi_{BB}(q)] &= \chi_o^B(q,\omega)v_B(q,\omega) + \chi_o^B(q,\omega)\psi_{BA}(q) \\
&\times \ < \delta n_A(q,\omega) >
\end{aligned} \tag{31}$$

where $\psi_{AA}, \psi_{BB}, \psi_{BA}$ and ψ_{AB} represent the effective static interaction between particles belonging to components A and B with 1/2 field Landau level as in eq.(19). $\chi_o^A(q,\omega)$ is the noninteracting response function of the A component and $\chi_o^B(q,\omega)$ is that of the B component. The above two relations may be combined with $\psi_{AB} = \psi_{BA}$ so that,

$$\begin{aligned}
\delta n_A(q,\omega) &= \chi_o^A(q,\omega)[1-\chi_o^B(q,\omega)\psi_{BB}(q)]v_A(q,\omega)/\Delta(q,\omega) \\
&\quad +\chi_o^A(q,\omega)\chi_o^B(q,\omega)\psi_{BA}(q)v_B(q,\omega)/\Delta(q,\omega) \\
\Delta(q,\omega) &= \{1-\chi_o^A(q,\omega)\psi_{AA}(q)\}\{1-\chi_o^B(q,\omega)\psi_{BB}(q)\} \\
&\quad -\chi_o^A(q,\omega)\chi_o^B(q,\omega)\psi_{AB}(q)\psi_{BA}(q).
\end{aligned} \tag{32}$$

The equation for $\delta n_B(q,\omega)$ can be written from the above by interchanging A and B. From (28) and (31) we can write the response function as a symmetric matrix with matrix elements,

$$\begin{aligned}
\chi_{AA}(q,\omega) &= \chi_o^A(q,\omega)[1-\chi_o^B(q,\omega)\psi_{BB}(q)]/\Delta(q,\omega) \\
\chi_{AB}(q,\omega) &= \chi_o^A(q,\omega)\chi_o^B(q,\omega)\psi_{AB}(q)/\Delta(q,\omega) \\
\chi_{BB}(q,\omega) &= \chi_o^B(q,\omega)[1-\chi_o^A(q,\omega)\psi_{AA}(q)]/\Delta(q,\omega)
\end{aligned} \tag{33}$$

Thus the wave vector dependence of the response function in (23) is continued in the matrix elements and small corrections occur over the unperturbed values. There is enhancement in the response function at $q = 2k_F$. The half filled Landau level consists of A and B type liquids which are coupled by spin-orbit interaction. For small wave vectors there is a correction in the response function varying as the fourth power of the wave vector over and above quadratic behaviour. For large wave vectors, the response function is independent of the wave vector. The compressibility arises from the filling structure of the Landau levels.

Halperin et al[15,16] have reported that the quasiparticles at the half filled Landau levels are compressible and their response function varies linearly as a function of wave vector going to a finite value as the wave vector approaches zero, $\chi(q) =$ const. + $q/(2\pi e^2)$. In our theory, eq.(26-27), the wave vector dependence is more complicated but does not contradict Halperin.

5 Diamagnetic points.

We assume that magnetic induction is independent of time so that $dB/dt = 0$. According to a Maxwell equation,

$$\nabla \times E = -\frac{\partial B}{\partial t} \tag{34}$$

so that $\nabla \times E = 0$ gives $E = 0$. Since the current $j = \left(\frac{1}{\rho}\right) E$, the value $E = 0$ is consistent with $\rho = 0$. In quantum Hall effect when $\rho_{xx} = 0$ there is a large magnetic field along the z direction. Therefore, we replace B_z by $B_z - B_o$ so that

$$\frac{d}{dt}(B_z - B_o) = 0 \tag{35}$$

which means that the Maxwell equation is satisfied with no field inside the semiconductor only when the magnetic field is shifted by B_o.
Using the Maxwell equation, we can write,

$$\nabla \times (E_x - E_o) = 0 \tag{36}$$

so that $E_x - E_o = 0$. The Ohm's law is

$$\rho_{xx} j = E_x - E_o \tag{37}$$

which is consistant with $\rho_{xx} = 0$ for $E_x - E_o = 0$. This means that $\rho_{xx} = 0$ points are consistent with shifts in both the electric as well as the magnetic field. The magnetic field of 8.38 T may be written as energy $H^2/8\pi = 1.59 \times 10^{20}$ eV/cm^3 which for a volume of 80 μm$\times$ 80μm$\times$15 nm, where the dimensions of the quantum well are used, is equal to 0.1598 meV. Thus the field shift is associated with an energy of the order of 0.1598 meV. As the x-component of the resistivity approaches zero, the energy of about 0.1598 meV is emitted.

6 Laughlin's wave function and ground state[19-21]

Laughlin's prototype wave function for the state with fractional charge is described by

$$\psi_m(z_1, z_2, \cdots, z_n) = \prod_{j<k}^{N} (z_j - z_k)^m \exp\left(-\frac{1}{4\pi l^2} \sum_j^N |z_j|^2\right) \tag{38}$$

where m is an odd integer, in this case 3 and $z_j = x_j + iy_j$ is the complex coordinate. For large N, the electron density is $\rho = 1/2\pi m l^2$ where l is the magnetic length, $l = (\hbar c/eB)^{1/2}$. The repulsive Coulomb interactions by themselves can not produce a state with quasiparticle charge 1/3. Therefore, it is first assumed on the basis of experimental data that the quasi particle charge is 1/3 and then the wave function is written down. Similarly, for Coulombian interactions the wave function for the quasi particle charge 1/5 may be written but the correct series can not be obtained.

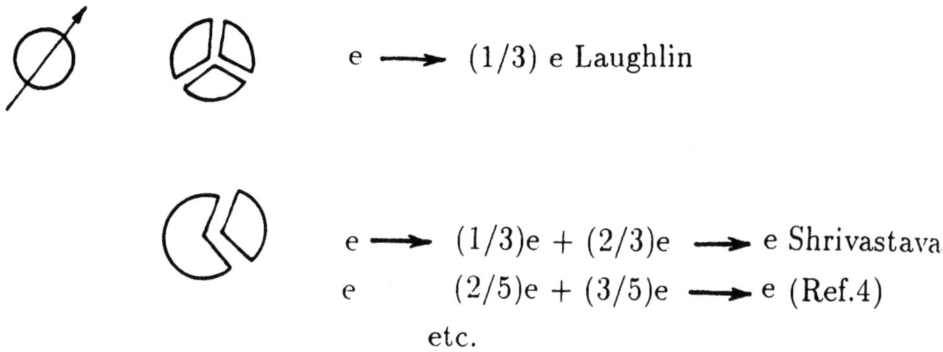

Fig. 7. Pictorial representation of Laughlin's fractional charge and Shrivastava's adding up to unity. In our theory every fragment has spin whereas in Laughlin's theory only charge is fractionalized.

Lam and Girvin[22] find that ψ_m is unstable to crystallization for $m > 7$. Haldane and Rezayi[23] and Morf and Halperin[24] have calculated the electron charge densities and the creation energies associated with the wave functions. By means of eq.(15) and the subsequent discussion, we have no difficulty in producing an effective charge of 1/3 except that mapping has to be done from Laughlin's wave functions of fractional charge of 1/3 to the spin, ordbit and Landau state, $\psi_m \rightarrow \psi_s \psi_L |n>$. Laughlin's charge is independent of spin so that there is complete decoupling between spin and charge. Thus charge can be fixed without any knowledge of spin. Besides. Laughlin's wave function can not get equal mass for unequal charge. On the other hand, the theory of ref. 4 which agrees with the data fixes the charge as well as the spin. Wilczek[25,26] has given the ideas of attaching magnetic flux tubes to particle and Jain[27] has suggested a composite fermion model with even number of flux quanta attached to each electron. The success of Jain's paper is the result of the correct series, but there is a prior dated paper[4],. The idea of attaching flux tubes to particles[25-27] is not in agreement with the experimental data and as we have seen above, both the electric as well as the magnetic fields have to be shifted. Accordingly, both the vector potential as well as the scalar potential of the electromagnetic field requires corrections, which are called Chern-Simons fields.[28,29] There are several experiments which show that the effective fractional charge is observable.[30-35].

7 Conclusions

The fractional charges found in the quantum Hall effect can be well explained by the angular momentum theory of this author.[4] Usually, the electrons are in the conduction band and hence the orbital angular momentum quantum number is zero, $l = 0$. However, when finite values of l are considered, we can define an effective fractional charge. Laughlin's theory of fractional charge defines the wave function and the energies. When the energy computations are performed by others[23,24] good agreement is found with Laughlin's calculations. However, Laughlin's wave functions do not use the spin or the orbit and hence there is spin-charge decoupling. Our theory of ref. 4 makes use of the spin and orbit to predict the effective charge and the various Hall steps correctly and in agreement with the experimental data. We also study the wave vector dependence of the response function from which we conclude that the quasi particles are compressible. Störmer[37] and Tsui[38] have obviously performed very ingenious measurements of the Hall resistance, and Laughlin has fractionalized the charge. In our theory, the fractions add up to unity as in eq.(18) and exhibited in the pictorial form in Fig.7.

References

[1] K.v. Klitzing, G. Dorda and M. Pepper, Phys. Rev. Lett. **45**, 494, 1980.

[2] K.v. Klitzing, Rev. Mod. Phys. **58**, 519, 1986.

[3] D.C. Tsui, H.L. Störmer and A.C. Gossard, Phys. Rev. Lett. **48**, 1559, 1982.

[4] K.N. Shrivastava, Phys. Lett. A**113**, 435, 1986; **115**, 459(E), 1986.

[5] K.N. Shrivastava, Mod. Phys. Lett. **13**, 1087, 1999;

[6] K.N. Shrivastava, "Superconductivity: Elementary Topics", World Scientific, Singapore, 2000.

[7] R. Willett, J.P. Eisenstein, H.L. Stormer, D.C. Tsui, A.C. Gossard and J.H. English, Phys. Rev. Lett. **59**, 1776, 1987.

[8] J.P. Eisenstein and H.L. Stormer, Science **248**, 1510, 1990.

[9] R.R. Du, H.L. Stormer, D.C. Tsui, L.N. Pfeiffer and K.W. West, Phys. Rev. Lett. **70**, 2944, 1993.

[10] J.P. Eisenstein, M.P. Lilly, K.B. Cooper, L.N. Pfeiffer and K.W. West Cond-mat/9909238; 13th International Conference on the electronic properties of two dimensional systems, Ottawa, Canada, 1999; Physica E**6**, 29, 2000.

[11] M.P. Lilly, K.B. Cooper, J.P. Eisenstein, L.N. Pfeiffer and K.W. West, Phys. Rev. Lett. **82**, 394, 1999.

[12] W. Pan, J.S. Xia, V. Shvarts, D.E. Adams, H.L. Stormer, D.C. Tsui, L.N. Pfeiffer, K.W. Baldwin and K.W. West, Phys. Rev. Lett. **83**, 3530, 1999.

[13] A.S. Yeh, H.L. Stormer, D.C. Tsui, L.N. Pfeiffer, K.W. Baldwin and K.W. West, Phys. Rev. Lett. **62**, 592, 1999.

[14] D. Yoshioka, Phys. Rev. Lett. **83**, 886, 1999.

[15] B.I. Halperin and A. Stern, Phys. Rev. Lett. **80**, 5457, 1998.

[16] A. Stern, B.I. Halperin, F.v. Oppen and S.H. Simon, Phys. Rev. **B59**, 12547, 1999.

[17] C. Kittel, Quantum theory of solids, Wiley, New York, 1963.

[18] E.J. Woll, Jr. and W. Kohn, Phys. Rev. **126**, 1693, 1962.

[19] R.B. Laughlin, Rev. Mod. Phys. **71**, 863, 1999.

[20] R.B. Laughlin, Phys. Rev. Lett.**50**, 1395, 1983.

[21] R.B. Laughlin, in "The Quantum Hall Effect", edited by R.E. Prange and S.M. Girvin, Springer Verlag, New York, 1987.

[22] P.K. Lam and S.M. Girvin, Phys. Rev. **B30**, 473, 1984; **31**, 613(E), 1985.

[23] F.D.M. Haldane and E.H. Rezayi, Phys. Rev. Lett. **54**, 237, 1985.

[24] R. Morf and B.I. Halperin, Phys. Rev. **B33**, 2221, 1986.

[25] F. Wilczek, Phys. Rev. Lett. **49**, 957, 1982.

[26] S.M. Girvin and A.H. MacDonald, Phys. Rev. Lett. **58**, 1252, 1987.

[27] J.K. Jain, Phys. Rev. Lett. **63**, 199, 1989.

[28] S.S. Chern and J. Simons, Proc. Nat. Acad. Sci. **68**, 791, 1971.

[29] M. Stone, "Quantum Hall Effect", World Scientific, Singapore, 1992.

[30] L. Saminadayar, D.C. Glattli, Y. Jin and B. Etienne, Phys. Rev. Lett. **79**, 2526, 1997.

[31] R. de Picciotto, M. Reznickov, M. Heiblum, V. Umansky, G. Bunin and D. Mahalu, Nature **389**, 162, 1997.

[32] R. de Picciotto, M. Reznikov, M. Heiblum, V. Umansky, G. Bunin, D. Mahalu, Physica **B249-251**, 395, 1998.

[33] V.J. Goldman and B. Su, Science **267**, 1010, 1995.

[34] V.J. Goldman, Physica **E1**, 15, 1997.

[35] V.J. Goldman, Cond-mat/9708041.

[36] T.G. Griffiths, E. Comforti, M. Heiblum, A. Stern and V. Umansky, Phys. Rev. Lett. **85**, 3918, 2000.

[37] H.L. Stormer, Rev. Mod. Phys. **71**, 875, 1999.

[38] D.C. Tsui, Rev. Mod. Phys. **71**, 891, 1999.

Large Fractals in Condensed Matter Physics

Vipin Srivastava
School of Physics, University of Hyderabad,
Hyderabad 500046 (India)
Email: vpssp@uohyd.ernet.in

1 Introduction

The concept of fractal[1] entered condensed matter physics in the 1970's through a problem in electrical conduction in disordered solids (with random potential) that eluded solution for quite sometime. Since then it has set a new trend in condensed matter research and scenarios ranging from percolation to quasi crystals, to Penrose tiles to subtle objects like electron wavefunctions and electron-state distributions in disordered and quasi-disordered systems, have received new insight by the use of the concept of fractal. We will briefly touch upon some of the prominent examples and develop the case of self-similar wavefunctions in some detail to investigate the famous problem of metal-insulator transition driven entirely by disordered potential. The examples of application of fractals fall in two categories: classical and quantum mechanical. The examples of classical systems are infinite clusters formed at the percolation threshold, diffusion limited aggregates, colloidal emulsions and many from the realm of chaos. The prominent candidates in the domain of quantum systems are electron wavefunctions and energy state distributions (density of states) in random potentials, quasi crystals and Penrose tiles. The latter two systems offer almost periodic potentials that can be commensurate or incommensurate with the lattice periodicity. In either situation one learns new physics[2].

2 Electrical Conduction in Disordered Systems

The difficult problem of electrical conduction in a random potential was first studied in the simple framework of classical percolation[3]. It is simple because it involves a binary system where on a lattice the sites are either occupied with probability p or empty with probability $1-p$. The onset of conduction is marked by the formation of an infinite connected network of nearest neighbour occupied sites at a *critical* concentration p_c. The conductivity in this network at a concentration p of occupied sites goes as

$$\sigma(p) = \sigma(1)P(p) \propto (p-p_c)^t, \tag{1}$$

where t is the conductivity exponent, typically > 1 (~ 1.1 for two dimensions, $d = 2$) in experimental measurements, and $P(p)$ represents the percolation probability, which is the ratio of number of sites in the infinite cluster and total number of sites in the system,

$$P(p) \propto (p-p_c)^{\beta_P}. \tag{2}$$

Frontiers of Fundamental Physics 4, Edited by Sidharth and Altaisky,
Kluwer Academic/Plenum Publishers, New York, 2001

Calculations gave very small values for β_P – just about 0.14 for $d = 2$ and 0.4 for $d = 3$. The vast disagreement between calculations and measurements was puzzling. An initial suspecion was that an spurious negative contribution might come from the dangling bonds in the percolating cluster which form a large fraction of occupied sites. These bonds were cut off and the resulting "backbone" of the percolating cluster was considered. It made only marginal improvement to the value of β.
It was felt that the coarse grained channel picture was not good enough and that one needed to consider microscopic details like voids etc. This brought in the so-called "network picture"[4]. However, this also did not improve β considerably. At this stage a closer inspection of the backbone revealed that it was highly ramified and had holes or gaps (of unoccupied sites) of all sizes and shapes. It was treated as a fractal starting from a length scale $\xi(p)$ on which the network looked homogeneous down to the scale of the lattice constant. The *matter* in the backbone inside a cell of side ξ was computed to be ξ^{d_B}, with the fractal dimension of the backbone measuring about 1.6 in $d = 2$ and approximately 2 in $d = 3$, very close to the values for 2 and 3 dimensional Sierpinsky carpets [1]. The Sierpinski carpet model for the percolating backbone raised the values of β considerably – to 0.9 in $d = 2$ and 1.6 in $d = 3$ – in reasonably good agreement with experimental values. It also became clear that the agreement with experiments would improve further if one took the backbone as a multifractal, and also took into account geometrical and topological parameters like *lacunarity*, which measures deviations from translational invariance, and *soccularity*, which measures the filamentary nature of the backbone[1].
Thus it was learnt that the fractal texture of the backbone is not only a natural hypothesis but is quantitatively justified too. This insight about the percolation network was naturally extended to studies on diffusion limited aggregates and collodal emulsions[5], and chaos[2]. We will now move on to the quantum mechanical case.

3 Eigenstates in Random Potentials

Quasi Random Potentials: The nature of wavefunction and the eigenenergy distribution have been studied extensively in Fibonacci lattices and Penrose tilings [6]. The essential findings for the quasi-random potentials are that both, the energy distribution (or the density of states) and the wavefunction possess fractal and sometimes multifractal character. The wavefunction can be spatially localized or extended depending on the eigenenergy. Unlike the case of random potential (discussed below) where the localized and extended states are sharply separated in energy, here both are distributed over the entire spectrum although *they are never degenerate*. In Penrose tiling and quasi random potentials that are commensurate with the lattice the spectral measure is singular continuous (say a Cantor set) and the eigenstates are of critical nature in that they are *neither extended nor localized*. We will discuss such states at length in connection with random potentials later on. When the quasi random potential is not commensurate with the lattice several exotic types of localization become possible and one gets glimpses of chaos though limited by quantum interference[2].
Random Potentials: The case of random potentials refers to the famous Anderson localization problem[7]. An electron in an environment of a random potential

undergoes multiple scattering which gives rise to quantum interference. As a result the possibility of an electron getting localized in space arises. The disordered system can now have two types of states – extended and localized – separated in energy space sharply by an energy called "mobility edge". If the Fermi energy E_F is made to travel through the mobility edge a metal-insulator transition (MIT) occurs from the metallic behaviour when the E_F is amongst the extended states to the insulating behaviour for the E_F in the localised states' regime. The nature of the MIT is of great interest due to its numerous practical applications. An important question is: is the MIT continuous or abrupt? There have been theoretical as well as experimental claims on either type of MIT and it is necessary to resolve this question.

The abrupt or discontinuous transition requires a "minimum metallic condivity" originally postulated and calculated by Mott[8] and supported by several numerical and experimental estimates. The claim is that before the conductivity on the metallic side attains a universal value σ_{min} just before becoming zero at the mobility edge. On the other hand the continuous-transition picture claims that $\sigma_{min} = 0$ and there is a gradual and smooth transition to the insulating state. Lately there have been experimental evidences to this effect also.

If one subscribes to the continuous-transition picture one has to investigate carefully the nature of the states in the vicinity of the mobility egde on the extended states' side of the energy spectrum. We suggest that one should recall the fact that the single particle Greens function has a branch-cut (i.e. an absolutely continuous spectrum) in the extended states' regime and a dense distribution of poles (i.e. a point spectrum) on the localized side of the spectrum[7]. The branch-cut and the poles are sharply separated by an energy called the mobility edge. We ask the question: what if the two regimes overlap such that some of the poles near the mobility edge go inside the branch-cut? This situation would amount to certain poles appearing in the higher Riemann sheets. The nature of the states represented by such poles can be studied by investigating Anderson's "stay-put" probability in this scenario. Recall the stay-put probability, $\lim_{t\to\infty} P_{oo}(0,t)$, which is the probability of finding an electron at an arbitrary site 'o' at time $t = \infty$ if it was there with probability one at $t = 0$, ought to be zero for the extended states and non-zero for localized states[7]. The calculations for the poles inside the Riemann surface reveals something very interesting. While it vanishes in the limit $t = \infty$ indicating delocalization in the strict sense, its time-integral is found to diverge [9]:

$$\lim_{t\to\infty} P_{oo}(0,t) = 0, \tag{3}$$

such that

$$\int_o^{\infty} P_{oo}(0,t)dt = \infty. \tag{4}$$

This indicates that the electrons in the states represented by such poles do diffuse through the system (in the thermodynamic limit) but take infinite time to do so. That is the conduction in this regime of the spectrum is very sluggish.

Sluggish conduction is a clear indication that the conduction path is tortuous. Such a situation arises in the percolating cluster when it is a fractal just above the percolation threshold. We therefore propose that in the vicinity of the mobility edge, on the extended side, the wavefunction should have a highly fragmented character like a fractal. Such a wavefunction can be viewed as a confluence or a mixture of degenerate localized and extended states[9] – an infinitely connected main path with a number of tortuous loops of all shapes and sizes weakly connected to it all along. While an electron can sail through the former more or less smoothly like in an extended state, inevitable excursions or side walks into the tortuous loops connected to it would make the electron relax into these localized states for varying lengths of time. The net effect would be a sluggish conduction. Essentially, we are suggesting that the conduction in the vicinity of mobility edge can be described as "quantum percolation"[10, 11]. The fractal dimension has been computed by us for these states[12]. It turns out to be 1.58 in $d = 2$ and 2.20 in $d = 3$ which are quite close to the fractal dimensions of the percolating clusters in $d = 2$ and 3 respectively[12]. The calculations of quantum percolation and fractal dimension of wavefunction involve computation of "inverse participation ratio" which measures the number of sites under a wavefunction [13].

With the insight that the states in the vicinity of the mobility edge have fractal wavefunctions which quantum percolate along the infinite fractal regions occupied by electrons, it appears plausible to propose that the fractal nature of the wavefunction should vary as the energy approaches the band centre in a systematic manner, both qualitatively and quantitatively: qualitatively, in terms of decreasing lacunarity and soccularity, and quantitatively, in terms of increasing fractal dimension (which will at some energy become 2 in $d = 2$ and 3 in $d = 3$). We highlight below the topological aspects of random potential and its implications on the shape of the wavefunction.

4 New Insight

The fractal nature of the percolating cluster and that of the wavefunction have vital implications on the nature of localized states and therefore on the details of MIT and phenomena related to it. For instance, it can lead to "superlocalization" where the wavefunction may decay faster than exponential, and give rise to new types of hopping conduction [14]. We summarise in the following how the results obtained by us so far in this connection when combined with some new ideas can lead to new insight.

We suggest that the disordered potential in a real lattice can possess fractal nature if it is truly disordered, free from any short- and long-range orders. At different values of the Fermi energy, E_F, the contours formed by equipotential lines will change their shapes. There will be a critical energy at which infinite contours will appear. These will enclose an infinte percolating cluster. This will have the fractal character which will affect the spread of the wavefunction in the manner described above. The wavefunctions at this energy will more or less mimic the tortous shape of the cluster though it will lie *well within* the confines of the cluster because of rapid decay of amplitude near the edges of the cluster. If the E_F is made to move towards the band centre the fractal nature of the infinite cluster will diminish by and by and that will be reflected in the shape of the wavefunction of the corresponding energy. The changes in the shape of the wavefunction as a function of energy, we suggest, can be measured in terms of the changes in the fractal dimension of the infinite cluster that supports the wavefunction. In this way the approach to the

mobility edge from the band centre and the transition to the insulating state at the mobility edge can be quantified in terms of the fractal dimension of the percolating cluster. The critical lacunarity corresponding to the *smallest* value of the fractal dimension will correspond to the critical disorder for the MIT. One thus has the correspondence between the MIT caused by changing energy at a fixed disorder and that caused by changing disorder at a fixed energy.
Following[12] if it is reaffirmed by computer experiments that the suggestion that the wavefunction of an extended state mimics (even approximately) the shape of the cluster that holds it then we can map the states starting from the mobility edge down to the band centre by a series of Seirpinsky carpets of increasing fractal dimension. This can be easily achieved by constructing carpets of steadily decreasing lacunarity. With this construction study of not only the details of MIT will become easier but the study of any related critical phenomenon will be facilitated.

Acknowledgements: Discussions with R. Krishnan and financial support from the Ministry of Information Technology, Government of India are acknowledged.

References

[1] B.B. Mandelbrot, "The Fractal Geometry of Nature", W. H. Freeman and Co., New York, 1983.

[2] V. Srivastava, "Some quasi-random remarks on Anderson localization and quantum limitation of chaos" in Non-Linear Dynamical Systems Eds. V. S-rinivasan, A. K. Kapoor and P. K. Panigrahi, Allied Publishers, Hyderabad, 2000, p 61-70.

[3] D. Stauffer, "Introduction to Percolation Theory", Taylor and Francis, London, 1985.

[4] S. Kirkpatrick, in "Ill-Condensed Matter", Eds. Balian, Maynard and Toulouse, Noth Holland, Amsterdam, 1979.

[5] See e.g. "Physics of Biomaterials: Fluctuations, Selfassembly and Evolution", Eds. Riste, T. and Sherrington, D., NATO ASI Series E 322, Kluwer, 1995.

[6] T.Fujiwara, M. Kohomoto, and T. Tokihira, Phys. Rev. B **40**, (Rapid Communications), 7413, 1989 and the references therein.

[7] P.W. Anderson, Phys. Rev. **109**, 1472, 1958.

[8] N.F. Mott, and E.A. Davis, "Electronic Processes in Non-Crystalline Materials", Clarendon Press, Oxford, 1979.

[9] V. Srivastava, Phys. Rev. **41**, 5667, 1990.

[10] V. Srivastava, and D. Weaire, **18**, 6635, 1978.

[11] V. Srivastava, and Chaturvedi, Meena, Phys. Rev. **30** (Rapid Communications), 2238, 1984.

[12] V. Srivastava, and E. Granato, Phys. Rev. **46**, 14 893, 1992.

[13] D. Weaire, and V. Srivastava, J. Phys. C.**10**, 4309, 1977.

[14] See D. van der Putten, et al., Phys. Rev. Lett. **69**, 494 1992 and references therein.

Canonical Forms of Positive Definite Matrices under Congruence: Extensions of the Schweinler-Wigner Extremum Principle

S. Chaturvedi and V. Srinivasan
School of Physics
University of Hyderabad, Hyderabad - 500 046
Email:scsp@uohyd.ernet.in, vssp@uohyd.ernet.in
R. Simon
Institute of Mathematical Sciences
C.I.T. Campus, Chennai - 600-013
Email:simon@imsc.ernet.in

1 Introduction

It is well known that a N-dimensional real symmetric [complex hermitian] matrix V is congruent to a diagonal matrix modulo an orthogonal [unitary] matrix[1]. That is, $V = S^{\dagger}DS$ where D is diagonal and $S \in SO(N)$ $[S \in SU(N)]$. It is not so well known that if, in addition, V is also positive definite, new possibilities exist for establishing its congruence to a diagonal matrix [2].

- For N even, such a V is also congruent to a diagonal matrix modulo a symplectic matrix in $Sp(N, \mathcal{R})$ $[Sp(N, \mathcal{C})]$ (Williamson's theorem[3],[4],[5]). That is, $V > 0$ implies $V = S^{\dagger}D'S$ where D' is diagonal and $S \in Sp(N, \mathcal{R})$ $[S \in Sp(N, \mathcal{C})]$. (This theorem has been exploited in defining quadrature squeezing and symplectically covariant formulation of the uncertainty principle for multimode states[6]).
- For any N and for any partition $N = n + m$ such a matrix is congruent to a diagonal matrix modulo a pseudo-orthogonal [pseudo-unitary] matrix. That is, $V > 0$ implies $V = S^{\dagger}D''S$ where D'' is diagonal and $S \in SO(m, n)$ $[S \in SU(m, n)]$, for any choice of partition $N = m + n$.

It may be noted that in neither of the last two cases are the diagonal entries the eigenvalues of V.

In this work, we outline the proofs of the last two statements. The strategy for proving the second statement is easily adapted to offer a simple proof of the Williamson result. We also show that an extremum principle akin to the Schweinler-Wigner extremum principle[7] for the orthogonal case can also be formulated for the other two cases as well.

Frontiers of Fundamental Physics 4, Edited by Sidharth and Altaisky,
Kluwer Academic/Plenum Publishers, New York, 2001

2 Congruence of a positive matrix under pseudo-orthogonal [pseudo-unitary] transformations

The fact that a real symmetric [complex hermitian] matrix is congruent to a diagonal matrix modulo an orthogonal [unitary] matrix crucially depends on the circumstance that congruence coincides with conjugation in the real orthogonal and complex unitary cases, they become distinct when more general sets of transformations are involved. A question which naturally arises is whether congruence to a diagonal form can also be achieved through a pseudo orthogonal [pseudo-unitary] transformation. The answer to this question turns out to be in the affirmative with the caveat that the matrix in question be positive definite, and can be formulated as the following theorem:

Theorem 1: Let V be a real symmetric positive definite matrix of dimension N. Then, for any choice of partition $N = m + n$, there exists an $S \in SO(m, n)$ such that

$$S^T V S = D^2 = \text{diagonal (and} > 0). \tag{1}$$

Proof: Recall that the group $SO(m,n)$ consists of all real matrices which satisfy $S^T g S = g$, $\det S = 1$, where $g = \text{diag}(\underbrace{1, 1, \cdots, 1}_{m}, \underbrace{-1, \cdots, -1}_{n})$. Consider the matrix $V^{-1/2} g V^{-1/2}$ constructed from the given matrix V. Since $V^{-1/2} g V^{-1/2}$ is real symmetric, there exists a rotation matrix $R \in SO(N)$ which diagonalizes $V^{-1/2} g V^{-1/2}$:

$$R^T V^{-1/2} g V^{-1/2} R = \text{diagonal} \equiv \Lambda\,. \tag{2}$$

This may be viewed also as a congruence of g using $V^{-1/2}R$, and signatures are preserved under congruence. (Indeed, signatures are the only invariants if we allow congruence over the full linear group $GL(N, \mathcal{R})$). As a consequence, the diagonal matrix Λ can be expressed as the product of a positive diagonal matrix and g :

$$R^T V^{-1/2} g V^{-1/2} R = D^{-2} g = D^{-1} g D^{-1}\,. \tag{3}$$

Here D is diagonal and positive definite. Define $S = V^{-1/2} R D$. It may be verified that S satisfies the following two equations :

$$\begin{aligned} S^T g S &= g\,, \\ S^T V S &= D^2 = \text{diagonal}\,. \end{aligned} \tag{4}$$

The first equation says that $S \in SO(m,n)$ and the second says that V is diagonalized through congruence by S. Hence the proof.

By replacing the superscript ^{T} by †, the group $SO(m,n)$ by $SU(m,n)$, and $R \in SO(N)$ by $U \in SU(N)$ in the statement and proof of the above theorem, we have the following theorem which applies to the complex case.

Theorem 2: Let V be a hermitian positive definite matrix of dimension N. Then, for any partition $N = m + n$, there exists an $S \in SU(m,n)$ such that

$$S^\dagger V S = D^2 = \text{ diagonal (and } > 0). \tag{5}$$

3 Congruence of an even dimensional positive matrix under $Sp(2n,\mathcal{R})$ $[Sp(2n,\mathcal{C})]$ transformations

The strategy outlined above, when applied to the real symplectic group of linear canonical transformations, leads a particularly simple proof of Williamsons's theorem.

Theorem 3: Let V be a $2n$-dimensional real symmetric positive definite matrix. Then there exists an $S \in Sp(2n,\mathcal{R})$ such that

$$\begin{aligned} S^TVS &= D^2 > 0\,, \\ D^2 &= \mathrm{diag}(\kappa_1,\kappa_2,\cdots,\kappa_n,\kappa_1,\kappa_2,\cdots,\kappa_n). \end{aligned} \tag{6}$$

Proof: Note that the $2n$-dimensional diagonal matrix D has only n independent entries. The group $Sp(2n,\mathcal{R})$ consits of all real matrices S which obey the condition

$$S^T\beta S = \beta\,, \quad \beta = \begin{pmatrix} 0 & 1 \\ -1 & 0 \end{pmatrix}, \tag{7}$$

with 1 and 0 denoting the $n \times n$ unit and zero matrices respectively. Even though $S^T\beta S = \beta$ may appear to suggest that $\mathrm{det}S = \pm 1$, it turns out that $\det S = 1$. In other words, $Sp(2n,\mathcal{R})$ consists of just one connected (though not simply connected) piece. Indeed, for every $n \geq 1$ the connectivity property of $Sp(2n,\mathcal{R})$ is the same as that of the circle.

The most general $S \in GL(2n,\mathcal{R})$ which solves $S^TVS = D^2$ is $S = V^{-1/2}RD$, where $R \in O(2n)$. Note that none of the factors D, R or $V^{-1/2}$ is an element of $Sp(2n,\mathcal{R})$. However, a V-dependent choice of D, R can be so made that the product $V^{-1/2}RD$ is an element of $Sp(2n,\mathcal{R})$ as we shall now show.

Since $\beta^T = -\beta$, it follows that $\mathcal{M} = V^{-1/2}\beta V^{-1/2}$ is antisymmetric. Hence there exists an $R \in SO(2n)$ such that[8]

$$R^TV^{-1/2}\beta V^{-1/2}R = \begin{pmatrix} 0 & \Omega \\ -\Omega & 0 \end{pmatrix}, \quad \Omega = \text{diagonal} > 0\,. \tag{8}$$

Define a diagonal positive definite matrix

$$D = \begin{pmatrix} \Omega^{-1/2} & 0 \\ 0 & \Omega^{-1/2} \end{pmatrix}. \tag{9}$$

Then we have

$$DR^TV^{-1/2}\beta V^{-1/2}RD = \beta\,. \tag{10}$$

Now define $S = V^{-1/2}RD$. It may be verified that S enjoys the following properties:

$$\begin{aligned} S^T\beta S &= \beta\,, \\ S^TVS &= D^2 = \text{diagonal}. \end{aligned} \tag{11}$$

The first equation says that $S \in Sp(2n,,\mathcal{R})$ and the second one says that V is diagonalized by congruence through S. This completes the proof of the Willianson theorem. To appreciate the simplicity of the present the reader may like to compare it with two recently published proofs of the Williamson theorem[5]. As in the pseudo-orthogonal case, by replacing the supercript ^{T} by † in the statement and proof of Theorem 3, one obtains the following result.

Theorem 4: Let V be a $2n$-dimensional hermitian positive definite matrix. Then there exists an $S \in Sp(2n, C)$ such that

$$S^\dagger V S = D^2 > 0\,, \tag{12}$$

The diagonal matrix in both the cases can be shown to have the following structure

$$D^2 = \mathrm{diag}(\kappa_1, \kappa_2, \cdots, \kappa_n, \kappa_1, \kappa_2, \cdots, \kappa_n). \tag{13}$$

From the theorems stated above it is evident that for a real symmetric [complex hermitian] positive definite matrix we can not talk about *the* canonical form under congruence, for there are $m + n$ possible choices of $SO(m, n)$ [$SU(m, n)$], and in the case of even dimension one more choice coming from Williamson's theorem. Needless to add that for the same matrix V, the diagonal matrix D will be different for different choices.

4 Orthogonalization Procedures: Schweinler-Wigner Extremum Principle

The age old orthogonalization problem consists in constructing a set of orthonormal vectors $(z_1, z_2, \cdots, z_N)$ out of a given set of linearly independent N dimensional vectors $\mathbf{v} = (v_1, v_2, \cdots, v_N)$. The orthogonalization problem can also be rephrased as follows: Let G denote the associated Gram matrix of pairwise inner products: $G_{ij} = (v_i, v_j)$. The Gram matrix is hermitian by construction, and positive definite by virtue of the linear independence of the given vectors. The orthogonalization problem, i.e., constructing a set of orthonormal vectors out of the given set of linearly independent vectors, amounts to finding a matrix S that solves

$$S^\dagger G S = 1, \;\; i.e., \;\; G^{-1} = S S^\dagger\,. \tag{14}$$

Each such S defines an orthogonalization procedure. The freedom available for the solution of the orthonormalization problem is exactly as large as the unitary group $U(N)$. Of the infinitely many possible orthogonalization procedures, the most well known is the Gram-Schmidt basis.

In a significant work Schweinler and Wigner[7] posed and answered the following questions

- Is it possible to construct an orthonormal basis which, unlike the familiar Gram-Schmidt basis (which depends on the particular initial order in which the given linearly independent vectors are listed), treats all the linearly independent vectors on an equal footing ?
- Is there a way of discriminating between various choices of S that solve (13) and hence between various orthogonalization procedures? They argued that a particular choice of orthogonalization procedure should correspond ultimately to the extremization of a suitable scalar function over the manifold of all orthonormal bases, with the given linearly independent vectors appearing as parameters in the function. Different choices of onthonormal bases will then correspond to different functions to be extremized.

The answers to these questions found by Schweinler and Wigner are as follows:

- A democratic orthonormal basis i.e. which treats all the v's on an equal footing is given by $\mathbf{z} = \mathbf{v}U_0P^{-1/2}$, where U_0 is the unitary matrix which diagonalizes G: $U_0{}^{\dagger}GU_0 = P$. We may refer to this as the Schweinler-Wigner basis,
- The function $m(\mathbf{z})$ which the Schweinler-Wigner basis extremizes is the following permutation symmetric quartic expression in the given vectors

$$gm(\mathbf{z}) = \sum_k \left(\sum_l |(z_k, v_l)|^2 \right)^2 . \tag{15}$$

They showed that the extremum (maximum in this case) value of $m(\mathbf{z})$ is given by $\mathrm{tr}(G^2)$, and that it attains this value for the the Schweinler-Wigner basis. Further, U_0 and hence the Schweinler- Wigner basis is essentially unique if the eigenvalues of the Gram matrix G are all distinct.

5 A reformulation Schweinler-Wigner work

The content of the work of Schweinler and Wigner can be reformulated[8] in a manner that offers a clearer and more general picture of the Schweinler-Wigner quartic form $m(\mathbf{z})$ and of the orthonormal basis which maximizes it. This perspective on the orthogonalization problem plays an important role in our generalizations of the Schweinler-Wigner extremum principle, and hence we summarise it briefly.

Since every orthonormal basis is the eigenbasis of a suitable hermitian operator, it is of interest to characterize the Schweinler-Wigner basis in terms of such an operator. Given linearly independent N-dimensional vectors $\mathbf{v} = (v_1, v_2, \cdots, v_N)$, the operator $\hat{M} = \sum_j v_j v_j^{\dagger}$ is hermitian positive definite. In a *generic orthonormal* basis $\mathbf{z}$, it is represented by a hermitian positive definite matrix $M(\mathbf{z}) : M(\mathbf{z})_{ij} = (z_i, \hat{M} z_j)$. Under a change of orthonarmal basis $\mathbf{z} \to \mathbf{z}' = \mathbf{z}S$, $M(\mathbf{z})$ transforms as follows

$$M(\mathbf{z}) \to M(\mathbf{z}') = S^{\dagger} M(\mathbf{z}) S \,, \quad S \in U(N) \,. \tag{16}$$

Recall that $U(N)$ acts transitively on the set of all orthonormal bases and that $\mathrm{tr}(M(\mathbf{z})^2) = \sum_{j,k} |M(\mathbf{z})_{jk}|^2$ is invariant under such a change of basis, and hence is endependent of $\mathbf{z}$. The Schweinler-Wigner quartic form $m(\mathbf{z})$ can easily be identified as $\sum_k (M(\mathbf{z})_{kk})^2$. In view of the above invariance, maximization of $\sum_k (M(\mathbf{z})_{kk})^2$ is the same as minimization of $\sum_{j \neq k} |M(\mathbf{z})_{jk}|^2$. The absolute minimum of $\sum_{j \neq k} |M(\mathbf{z})_{jk}|^2$ equals zero, and obtains when $M(\mathbf{z})$ is diagonal. Thus, the orthonormal basis which maximizes $\sum_k (M(\mathbf{z})_{kk})^2$ is the same as the one in which $\hat{M}$ is diagonal, and we arrive at the following important conclusion of ref.[9]:

Theorem 5: The distinquished orthonormal basis which extremizes the Schweinler-Wigner quartic form $m(\mathbf{z})$ over the manifold of all orthonormal bases is the same as the orthonormal basis in which the positive definite matrix $M(\mathbf{z})$ becomes diagonal.

Important for the above structure is the fact that the invariant $\mathrm{tr}(M(\mathbf{z})^2)$ is the sum of non-negative quantities, and therefore a part of it is necessarily bounded. It is precisely this property, which can be traced to the underlying unitary symmetry, that is not available when we try to generalize the Schweinler-Wigner procedure to construct pseudo-orthonormal and symplectic bases wherein the underlying symmetries are the noncompact groups $SO(m,n)$ and $Sp(2n,\mathcal{R})$ respectively..

6 Lorentz basis with an extremum property

In this Section we show how the Schweinler-Wigner procedure can be generalized to construct pseudo-orthonormal basis based on an extremum principle. We begin with the case of real vectors.

We are given a set of linearly independent real N-dimensional vectors $\mathbf{v} = (v_1, \cdots, v_N)$ and we want to construct out of it a pseudo-orthonormal basis [$SO(m,n)$ Lorentz basis with $N = m + n$], i.e., a set of vectors $\mathbf{z} = (z_1, \cdots, z_N)$ satisfying

$$(z_k, g z_l) = g_{kl}\,, \quad g = \mathrm{diag}(\underbrace{1, 1, \cdots, 1}_{m}, \underbrace{-1, \cdots, -1}_{n}). \tag{17}$$

Let $\hat{M} = \sum_j v_j v_j^T$, and let the symmetric positive definite matrix $M(\mathbf{z}):\ M(\mathbf{z})_{ij} = (z_i, \hat{M} z_j)$ represent $\hat{M}$ in a *generic pseudo-orthonormal* basis $\mathbf{z}$. Under a pseudo-orthogonal change of basis $\mathbf{z} \to \mathbf{z}' = \mathbf{z}S$, the matrix $M(\mathbf{z})$ transforms as follows:

$$M(\mathbf{z}) \to M(\mathbf{z}') = S^T M(\mathbf{z}) S\,, \quad S \in SO(m,n)\,. \tag{18}$$

Since $S^T g S = g$ (or $g S^T = S^{-1} g$) by definition, we have

$$S:\ \ gM(\mathbf{z}) \to gM(\mathbf{z}') = S^{-1} g M(\mathbf{z}) S. \tag{19}$$

That is, as $M(\mathbf{z})$ undergoes congruence, $gM(\mathbf{z})$ undergoes conjugation. Thus, $\mathrm{tr}(gM(\mathbf{z}))^l$, $l = 1, 2, \cdots$, are invariant. In what follows we shall often leave implicit the dependence of M on the generic pseudo-orthonormal basis $\mathbf{z}$.

Consider the invariant $\mathrm{tr}(gM(\mathbf{z})gM(\mathbf{z}))$ corresponding to $l = 2$. Write $M = M^{\mathrm{even}} + M^{\mathrm{odd}}$ where

$$M^{\mathrm{even}} = \frac{1}{2}(M + gMg)\,, \ M^{\mathrm{odd}} = \frac{1}{2}(M - gMg)\,. \tag{20}$$

In the above decomposition we have exploited the fact that g is, like parity, an *involution*.

With M expressed in the (m,n) block form

$$M = \begin{pmatrix} A & C \\ C^T & B \end{pmatrix}, \ A^T = A\,, \ B^T = B\,, \tag{21}$$

we have

$$M^{\mathrm{even}} = \begin{pmatrix} A & 0 \\ 0 & B \end{pmatrix}, \ M^{\mathrm{odd}} = \begin{pmatrix} 0 & C \\ C^T & 0 \end{pmatrix}. \tag{22}$$

Symmetry of M implies that M^{odd} and M^{even} are symmetric. Further, M^{odd} and M^{even} are trace orthogonal: $\text{tr}(M^{\text{odd}}M^{\text{even}}) = 0$. Thus,

$$\text{tr}(gMgM) = \text{tr}(M^{\text{even}})^2 - \text{tr}(M^{\text{odd}})^2 , \tag{23}$$

which can also be written as

$$\text{tr}(MgMg) = \text{tr}(M^2) - 2\text{tr}(M^{\text{odd}})^2 . \tag{24}$$

A few observations are in order:

- In contradistinction to the original unitary case, the invariant in the present case is no more a sum of squares. This can be traced to the non-compactness of the underlying $SO(m,n)$ symmetry. As one consequence, $\sum_k (M_{kk})^2$ is not bounded. As an example, consider the simplest case $m = 1$, $n = 1$ and let

$$M = \begin{pmatrix} a & 0 \\ 0 & b \end{pmatrix}, \quad a, b > 0. \tag{25}$$

Under congruence by the $SO(1,1)$ element

$$S = \begin{pmatrix} \cosh\mu & \sinh\mu \\ \sinh\mu & \cosh\mu \end{pmatrix}, \tag{26}$$

the value of $\sum_k (M_{kk})^2$ changes from $a^2 + b^2$ to $a^2 + b^2 + 2ab\sinh^2\mu\cosh^2\mu$, which grows with μ without bounds, showing that $\sum_k (M_{kk})^2$ and hence $\text{tr}(M^2)$ is not bounded. Thus, in contrast to the unitary case, extremization of the Schweinler-Wigner quartic form $\sum_k (M_{kk})^2$ will make no sense in the absence of further restrictions.

- The structure of the invariant $\text{tr}(gMgM)$ in (30) suggests the further restriction needed to be imposed: within the submanifold of pseudo-orthogonal bases $\mathbf{z}$ which keep $\text{tr}(M(\mathbf{z})^{\text{odd}})^2$ (and hence $\text{tr}(M(\mathbf{z})^2)$) at a fixed value we can maximize $\sum_k M(\mathbf{z})^2_{kk}$. In particular we can do this within the submanifold which minimizes $\text{tr}(M(\mathbf{z})^{\text{odd}})^2$, and hence $\text{tr}(M(\mathbf{z})^2)$. Clearly, zero is the absolute minimum of the nonnegative object $\text{tr}(M(\mathbf{z})^{\text{odd}})^2$. But by theorem 1 there exists a Lorentz basis $\mathbf{z}$ in which $M(\mathbf{z})$ is diagonal and hence $M(\mathbf{z})^{\text{odd}} = 0$. Thus the minimum $\text{tr}(M(\mathbf{z})^{\text{odd}})^2 = 0$, and hence the minimum of $\text{tr}(M(\mathbf{z})^2)$, namely $\text{tr}(gM(\mathbf{z})gM(\mathbf{z}))$, is attainable.

The above observations suggest the following *two step analogue of the Schweinler-Wigner extremum principle for Lorentz bases.* Choose the submanifold of Lorentz bases which minimize the quartic form $\text{tr}(M(\mathbf{z})^{\text{odd}})^2$, and maximize the Schweinler-Wigner quartic form $m(\mathbf{z}) = \sum_k (M(\mathbf{z})_{kk})^2$ within this submanifold. Clearly, the first step takes M to a block-diagonal form, and the second one diagonalizes it.

Thus we have established the following generalization of Theorem 5 to the pseudo-orthonormal case:

Theorem 6: The distinquished pseudo-orthonormal basis which extremizes the quartic form $m(\mathbf{z})$ over the submanifold of pseudo-orthonormal bases which minimize the quartic form $\mathrm{tr}(M(\mathbf{z})^2)$ is the same as the pseudo-orthonormal basis in which the positive definite matrix $M(\mathbf{z})$ becomes diagonal.

The submanifold under reference consists of Lorentz bases which are related to one another through the maximal compact (connected) subgroup of $SO(m,n)$, namely $SO(m) \times SO(n)$. This subgroup consists of matrices of the block-diagonal form

$$\begin{pmatrix} R_1 & 0 \\ 0 & R_2 \end{pmatrix}, \quad R_1 \in SO(m)\,, \quad R_2 \in SO(n)\,, \tag{27}$$

and this is precisely the subgroup of $SO(m,n)$ transformations that do not mix the even and odd parts of $M(\mathbf{z})$.

To conclude, we may note that the above construction carries over to the complex case, with obvious changes like replacing ^{T} by † and $SO(m,n)$ by $SU(m,n)$.

7 Sympletic Basis with an Extremal Property

Our construction in the pseudo-orthogonal case suggests a scheme by which the Schweinler-Wigner extremum principle principle can be generalized to construct a symplectic basis. Suppose that we are given a set of linearly independent vectors $\mathbf{v} = (v_1, v_2, \cdots, v_{2n})$ in $\mathcal{R}^{2n}$. The natural symplectic structure in $\mathcal{R}^{2n}$ is specified by the standard symplectic "metric" β defined in (7). Let $\mathbf{z} = (z_1, z_2, \cdots, z_{2n})$ denote a generic symplectic basis. That is, $(z_j, \beta z_k) = \beta_{jk}\,,\ j,k = 1,2,\cdots,2n$. The real symlectic group $Sp(2n,R)$ acts transitively on the set of all symplectic bases.

To generalize the Schweinler-Wigner principle to the symplectic case, we begin be defining $\hat{M} = \sum_{j=1}^{2n} v_j v_j^T$. Let $M(\mathbf{z}) : \ M(\mathbf{z})_{ij} = (z_i, \hat{M} z_j)$ be the symmetric positive definite matrix representing the operator $\hat{M}$ in a ***generic symplectic*** basis $\mathbf{z}$. Under a symplectic change of basis $\mathbf{z} \to \mathbf{z}' = \mathbf{z}S$, $S \in Sp(2n,\mathcal{R})$, the matrix $M(\mathbf{z})$ undergoes the following transformation:

$$M(\mathbf{z}) \to M(\mathbf{z}') = S^T M(\mathbf{z}) S\,, \quad S \in Sp(2n,R)\,. \tag{28}$$

Since $S^T\beta S = \beta$ implies $\beta S^T = S^{-1}\beta$, we have

$$S: \ \beta M(\mathbf{z}) \to \beta M(\mathbf{z}') = S^{-1}\beta M(\mathbf{z}) S. \tag{29}$$

That is, under a symplectic change of basis $M(\mathbf{z})$ undergoes congruence, but $\beta M(\mathbf{z})$ undergoes conjugation. Hence $\mathrm{tr}(\beta M(\mathbf{z}))^{2l}$, $l = 1,2,\cdots,n$ are invariant (Note that $\mathrm{tr}(\beta M(\mathbf{z}))^{2l+1} = 0$ in view of $\beta^T = -\beta$, $M(\mathbf{z})^T = M(\mathbf{z})$).

Since $i\beta$ is an ***involution*** we can use it to separate $M(\mathbf{z})$ into even and odd parts :

$$\begin{aligned} M(\mathbf{z}) &= M(\mathbf{z})^{\text{even}} + M(\mathbf{z})^{\text{odd}}\,, \\ M(\mathbf{z})^{\text{even}} &= \frac{1}{2}(M(\mathbf{z}) + \beta M(\mathbf{z})\beta^T)\,, \\ M(\mathbf{z})^{\text{odd}} &= \frac{1}{2}(M(\mathbf{z}) - \beta M(\mathbf{z})\beta^T)\,. \end{aligned} \tag{30}$$

The even and odd parts of $M(\mathbf{z})$ satisfy the symmetry properties

$$\beta M(\mathbf{z})^{\text{even}}\beta^T = M(\mathbf{z})^{\text{even}}\,, \quad \beta M(\mathbf{z})^{\text{odd}}\beta^T = -M(\mathbf{z})^{\text{odd}}\,. \tag{31}$$

Further, $M(\mathbf{z})^{\text{odd}}$ and $M(\mathbf{z})^{\text{even}}$ are trace orthogonal: $\text{tr}\left(M(\mathbf{z})^{\text{odd}}M(\mathbf{z})^{\text{even}}\right) = 0$.

The structure of the even and odd parts of $M(\mathbf{z})$ may be appreciated by writing $M(\mathbf{z})$ in the block form

$$M(\mathbf{z}) = \begin{pmatrix} A & C \\ C^T & B \end{pmatrix}, \quad A^T = A\,, \; B^T = B\,. \tag{32}$$

We have

$$\begin{aligned} M(\mathbf{z})^{\text{even}} &= \begin{pmatrix} \frac{1}{2}(A+B) & \frac{1}{2}(C-C^T) \\ -\frac{1}{2}(C-C^T) & \frac{1}{2}(A+B) \end{pmatrix}, \\ M(\mathbf{z})^{\text{odd}} &= \begin{pmatrix} \frac{1}{2}(A-B) & \frac{1}{2}(C+C^T) \\ \frac{1}{2}(C+C^T) & \frac{1}{2}(B-A) \end{pmatrix}. \end{aligned} \tag{33}$$

Now consider the invariant $-\text{tr}(\beta M(\mathbf{z})\beta M(\mathbf{z})) = \text{tr}(\beta^T M(\mathbf{z})\beta M(\mathbf{z}))$. We have

$$\text{tr}(\beta^T M(\mathbf{z})\beta M(\mathbf{z})) = \text{tr}(M(\mathbf{z})^{\text{even}})^2 - \text{tr}(M(\mathbf{z})^{\text{odd}})^2\,, \tag{34}$$

which can also be written as

$$\text{tr}(\beta^T M(\mathbf{z})\beta M(\mathbf{z})) = \text{tr}(M(\mathbf{z})^2) - 2\text{tr}(M(\mathbf{z})^{odd})^2\,. \tag{35}$$

The structural similarity of this invariant to that in the pseudo-orthogonal case should be appreciated.

Now, by an argument similar to the pseudo-orthogonal case one finds that, owing to the noncompactness of $Sp(2n,\mathcal{R})$, the function $\text{tr}(M(\mathbf{z})^2)$ and hence the Schweinler-Wigner quartic form $\sum_{k=1}^{2n}(M(\mathbf{z})_{kk})^2$ is unbounded if $\mathbf{z}$ is allowed to run over the entire manifold of all symplectic bases. For instance, in the lowest dimensional case $n = 1$ with M chosen to be

$$M = \begin{pmatrix} a & u \\ d & b \end{pmatrix}, \quad a,b > 0, \;\; ab - ud > 0, \tag{36}$$

under congruence by the $Sp(2,,\mathcal{R})$ matrix

$$S = \begin{pmatrix} \mu & 0 \\ 0 & 1/\mu \end{pmatrix}, \tag{37}$$

the value of $\sum_k (M_{kk})^2$ changes from $a^2 + b^2$ to $\mu^2 a^2 + (1/\mu^2)b^2$ which, by an appropriate choice of μ, can be made as large as one wishes.

However, it follows from (41) that over the submanifold of symplectic bases which leave $\text{tr}(M(\mathbf{z})^{\text{odd}})^2$ fixed, the function $\text{tr}(M(\mathbf{z})^2)$ remains invariant and so the quartic form $\sum(M(\mathbf{z})_{kk})^2$ is bounded within this restricted class of symplectic bases and hence can be maximised. In particular the nonnegative $\text{tr}(M(\mathbf{z})^{\text{odd}})^2$ can be chosen to take its minimum value. Williamson theorem implies that there are symplectic bases which realize the absolute mimumum $\text{tr}(M(\mathbf{z})^{\text{odd}})^2 = 0$.

We can now formulate the ***analogue of the Scweinler-Wigner extremum principle for symplectic bases*** in the following way: Take the subfamily of symplectic bases in which $\text{tr}(M(\mathbf{z})^{\text{odd}})^2$ and hence $\text{tr}(M(\mathbf{z})^2)$is minimum. [This minimum of $\text{tr}(M(\mathbf{z})^2)$ equals the invariant $\text{tr}(\beta^T M(\mathbf{z})\beta M(\mathbf{z}))$]. Then maximise the Schweinler-Wigner quartic form $m(\mathbf{z}) = \sum_k (M(\mathbf{z})_{kk})^2$ within this submanifold of symplectic bases. This will lead, not just to a basis in which $M(\mathbf{z})$ is diagonal, but to one where $M(\mathbf{z})$ has the Williamson canonical form $M(\mathbf{z}) = \text{diag}(\kappa_1, \cdots, \kappa_n; \kappa_1, \cdots, \kappa_n)$. We have thus established the following generalization of the Schweinler-Wigner extremum principle to the symplectic case.

Theorem 7: The distinquished symplectic basis which extremizes the "Schweinler-Wigner" quartic form $m(\mathbf{z})$ over the submanifold of symplectic bases which minimize the quartic form $\text{tr}(M(\mathbf{z})^2)$ is the same as the symplectic basis in which the positive definite matrix $M(\mathbf{z})$ assumes the Williamson canonical diagonal.

Note that once $M(\mathbf{z})^{\text{odd}} = 0$ is reached, as implied by $\text{tr}(M(\mathbf{z})^{\text{odd}})^2 = 0$, $M(\mathbf{z})$ has the special even form

$$\begin{pmatrix} A & C \\ -C & A \end{pmatrix}, \quad A^T = A, \; C^T = -C, \tag{38}$$

so that $A + iC$ is hermitian. The subgroup of symplectic transformations which do not mix $M(\mathbf{z})^{\text{even}}$ with $M(\mathbf{z})^{\text{odd}}$, and hence maintain the property $M(\mathbf{z})^{\text{odd}} = 0$ have the special form

$$S = \begin{pmatrix} X & Y \\ -Y & X \end{pmatrix}, \quad X + iY \in U(n). \tag{39}$$

This subgroup, isomorphic to the unitary group $U(n)$, is the maximal compact subgroup[10] of $Sp(2n, \mathcal{R})$. Thus, diagonalizing $M(\mathbf{z})$ using symplectic change of basis, after it has reached the even form, is the same as diagonalizing an n-dimensional hermitian matrix using unitary transformations.

8 Concluding Remarks

To conclude, we have shown that an $N \times N$ real symmetric [complex hermitian] positive definite matrix is congruent to a diagonal form modulo a pseudo-orthogonal [pseudo-unitary] matrix belonging to $SO(m, n)$ [$SU(m, n)$], for any choice of partition $N = m + n$. The method of proof of this result is adapted to provide a simple proof of Williamson's theorem. An important consequence of these theorems is that while a real-symmetric [complex-hermitian] positive definite matrix has a unique diagonal form under conjugation, it has several different canonical diagnal forms under congruence. The theorems developed here are used to formulate an extremum principle a lá Schweinler and Wigner for constructing pseudo-orthonormal[pseudo-unitary] and symplectic bases from a given set of linearly independent vectors. Conversely, the extremum principle thus formulated can be used for finding the congruence transformation which brings about the desired diagonalization.

References

[1] See, for instance, F. C. Gantmacher *The Theory of Matrices*, Vol 1 (Chelsea, New York, 1960).

[2] R. Simon, S. Chaturvedi and V. Srinivasan, J. Math. Phys. **40**, 3632 (1999).

[3] J. Williamson, Am. J. of Math. **58**, 141 (1936); **59**, 599 (1936); **61**, 897 (1936). Williamson's results are more general than the theorem quoted, and obtain all the different canonical forms a real symmetric (not necessarily positive definite) matrix can take under congruence by the real symplectic group. The results of Williamson are summarized in a manner that should appeal to physicists in V. I. Arnold, *Mathematical Methods of Classical Mechanics* (Springer-Verlag, New York, 1978), Appendix 6.

[4] J. Moser, Comm. Pure Appl. Math. **11**, 81 (1958); A. Weinstein, Bull. Am. Math. Soc. **75**, 814 (1971); N. Burgoyne and R. Cushman, Celes. Mech. **8**, 435 (1974); J. Laub and K. Meyer, Celes. Mech. **9**, 213 (1974).

[5] A. J. Dragt, F. Neri, and G. Rangarajan, Phys. Rev. **A45**, 2572 (1992); E. C. G. Sudarshan, C. B. Chiu, and G. Bhamathi, Phys. Rev. **A52**, 43 (1995).

[6] R. Simon, E. C. G. Sudarshan, and N. Mukunda, Phys. Rev. **A36**, 3668 (1987); R. Simon, N. Mukunda, and B. Dutta, Phys. Rev. **A49** 1567 (1994); Arvind, B. Dutta, N. Mukunda, and R. Simon, Pramana J. Phys. **45** 471 (1995); Arvind, B. Dutta, N. Mukunda, and R. Simon, Phys. Rev. **A52**, 1609 (1995).

[7] H. C. Schweinler and E. P. Wigner, J. Math. Phys. **11** 1693 (1970).

[8] Just as diagonal form is the canonical form for real symmetric matrices under rotation, $i\sigma_2 \otimes K$, with K diagonal, is the canonical form for a real antisymmetric matrix under rotation. Further K can be chosen to be non-negative, in general, and positive definite when the antisymmetric matrix is nonsingular.

[9] S. Chaturvedi, A. K. Kapoor, and V. Srinivasan, J. Phys. **A 31**, L367 (1998).

[10] R. Simon, N. Mukunda, and B. Dutta, Phys. Rev. **A49** 1567 (1994).

A Novel Method to Solve Familiar Differential Equations and its Applications

* N. Gurappa, Prasanta K. Panigrahi, T. Shreecharan, S. Sree Ranjani
*Institute of Mathematical Sciences, Taramani,
Chennai 600 113, INDIA.
School of Physics, University of Hyderabad, Hyderabad,
Andhra Pradesh 500 046, INDIA.

1 Introduction

Text books on mathematical methods [1] and non-relativistic quantum mechanics [2] routinely take recourse to the method of series solution for solving linear differential equations encountered in various physical problems. The tediousness of the above procedure has often led to the search for alternate methods; the elegant raising and lowering operator approach for solving the harmonic oscillator problem serves as a classic example of these attempts. The other algebraic approach is group theoretical in nature and takes advantage of the symmetries of the problem at hand [2, 3]. The text book example of the Coulomb problem, where the angular momentum and the Rungé-Lenz vectors combine to yield a complete algebraic description of the Hilbert space, demonstrates the efficacy of this procedure. Most of these methods fail to generalize to many-variable interacting systems, a field lately attracting considerable attention because of its relevance to many areas of physics [4, 5].

In this paper, the usefulness of a recently developed method for solving linear differential equations (DE) of arbitrary order [6] is demonstrated, by applying it to a number of examples encountered in physics and mathematics literature. This simple and elegant method allows an unified treatment of both the single and many-variable interacting systems. It is worth emphasizing that, the method presented does not, *a priori* assume any symmetry about the equation under study. The material presented here is at a pedagogical level, so that any student of physics or mathematics will be able to follow and appreciate the utility of the approach. In section 2, we develop the method to solve single, as well as, many variable linear DE of arbitrary order. For the purpose of illustration, the example of Laguerre DE is worked out explicitly and the novel expression for the solution of hypergeometric equation is displayed. Using this technique in section 3, we illustrate the utility of the approach in finding the ladder operators for the DE and physical systems, e.g., the Coulomb problem. This procedure strightforwardly generalizes to other examples [7]. Section 4 is devoted to the many-body case and we conclude in the

[0] gurappa@imsc.ernet.in, panisp@uohyd.ernet.in,
panisprs@uohyd.ernet.in and akksprs@uohyd.ernet.in

subsequent section pointing out a host of other problems, where this method can be employed profitably.

2 Solution for Linear Differential Equations of Arbitrary Order

A single variable linear differential equation, after suitable manipulations (shown later), can be written in the form,

$$[F(D) + P(x, d/dx)]\, y(x) = 0 \quad , \tag{1}$$

where, $D \equiv x\frac{d}{dx}$ is the Euler operator, $F(D) \equiv \sum_{n=-\infty}^{n=\infty} a_n D^n$ and a_n's are some parameters; $P(x, d/dx)$ can be an arbitrary polynomial function of x and $\frac{d}{dx}$. Using the property of the Euler operator, $Dx^n = nx^n$, it is easy to see that $F(D)$ is a diagonal operator in the space of monomials spanned by x^n. Here, x can take values on the full real line or the half line depending on the equation under consideration.

The solution to Eq. (1) can be written in the form [6],

$$\begin{aligned} y(x) &= C_\lambda \left\{ \sum_{m=0}^{\infty} (-1)^m \left[\frac{1}{F(D)} P(x, d/dx) \right]^m \right\} x^\lambda \\ &\equiv C_\lambda \hat{G}_\lambda x^\lambda \quad , \end{aligned} \tag{2}$$

provided, $F(D)x^\lambda = 0$ and the coefficient of x^λ in $y(x) - C_\lambda x^\lambda$ is zero (no summation over λ); here, C_λ is a constant. The case, when the equation $F(D)x^\lambda = 0$ does not have distinct roots, is not considered here. It is necessary to point out that, since $F(D)$ is diagonal in the space of monomials the inverse of this operator, in the above expression, is well defined. The proof of Eq. (2) is straightforward. Substituting Eq. (2), modulo C_λ, in Eq. (1),

$$\begin{aligned} &[F(D) + P(x, d/dx)] \\ &\left\{ \sum_{m=0}^{\infty} (-1)^m \left[\frac{1}{F(D)} P(x, d/dx) \right]^m \right\} x^\lambda \\ &= F(D) \left[1 + \frac{1}{F(D)} P(x, d/dx) \right] \\ &\quad \left\{ \sum_{m=0}^{\infty} (-1)^m \left[\frac{1}{F(D)} P(x, d/dx) \right]^m \right\} x^\lambda \\ &= F(D) \sum_{m=0}^{\infty} (-1)^m \left[\frac{1}{F(D)} P(x, d/dx) \right]^m x^\lambda \\ &+ F(D) \sum_{m=0}^{\infty} (-1)^m \left[\frac{1}{F(D)} P(x, d/dx) \right]^{m+1} x^\lambda \end{aligned}$$

$$
\begin{aligned}
&= F(D)x^\lambda \\
&- F(D)\sum_{m=0}^{\infty}(-1)^m\left[\frac{1}{F(D)}P(x,d/dx)\right]^{m+1}x^\lambda \\
&+ F(D)\sum_{m=0}^{\infty}(-1)^m\left[\frac{1}{F(D)}P(x,d/dx)\right]^{m+1}x^\lambda \\
&= 0 \quad , \qquad (3)
\end{aligned}
$$

Eq. (2) connects the space of solutions to the monomials and provides a *single expression* for the solutions of the DE of arbitrary order. Apart from yielding all the known polynomials and functions [8], this method allows one to construct the raising and lowering operators in a rather simple manner. These operators, which generate new solutions from a given one, are straightforwardly introduced at the level of monomials and the corresponding operators for the solution of the DE can then be easily obtained by similarity transformations. Generalization to the many-variable case is immediate; writing $\bar{D} \equiv \sum_i D_i$ where $D_i \equiv x_i \frac{d}{dx_i}$, one finds that $F(\bar{D})X^\lambda = 0$ has solutions, for arbitrary $F(\bar{D})$, when X^λ are monomial symmetric functions. Monomial symmetric function is a homogeneous symmetric polynomial in the particle cordinates $x_1, x_2, ... x_N$ (for the N particle case) with the degree of homogenity λ i.e., $\bar{D}m_\lambda = \lambda m_\lambda$. It is defined as, $m_\lambda = \sum_{\lambda_1,\lambda_2,...} x_1^{\lambda_1} x_2^{\lambda_2}...$, where the summation is over all distinct permutations of λ_i and $\lambda = \sum_i \lambda_i$. Though there are different basis sets for the symmetric polynomials [9], only the monomial symmetric functions are useful in the present context.

To illustrate the working of this method, we explicitly solve the Laguerree differential equation with the method just proposed. The solution for the Laguerre DE [1],

$$
\left[x\frac{d^2}{dx^2} + (\alpha - x + 1)\frac{d}{dx} + n\right] L_n^\alpha(x) = 0 \quad , \qquad (4)
$$

for which $F(D) = (n - D)$ and $P(x, d/dx) = (xd^2/dx^2 + (\alpha + 1)d/dx)$ is given by [10](with the chosen normalization $(-1)^n/n!$),

$$
\begin{aligned}
&L_n^\alpha(x) \\
&= \frac{(-1)^n}{n!}\left\{\sum_{m=0}^{\infty}(-1)^m\left[\frac{1}{D-n}\left(x\frac{d^2}{dx^2} + (\alpha+1)\frac{d}{dx}\right)\right]^m\right\}x^n \\
&= \frac{(-1)^n}{n!}\exp\left[-x\frac{d^2}{dx^2} - (\alpha+1)\frac{d}{dx}\right]x^n \quad . \qquad (5)
\end{aligned}
$$

The hypergeometric differential equation, manifesting in the context of various quantum mechanical and other problems,

$$
\begin{aligned}
&\left[x^2\frac{d^2}{dx^2} + (\alpha+\beta+1)x\frac{d}{dx} + \alpha\beta - x\frac{d^2}{dx^2} - \gamma\frac{d}{dx}\right] \\
&F(\alpha,\beta,\gamma,x) = 0 \quad , \qquad (6)
\end{aligned}
$$

[1] Note that different books follow different nomenclature. Throughout this paper, we consistently follow the nomenclature adopted in reference [8].

can be written as,

$$\left[(D+\alpha)(D+\beta) - x\frac{d^2}{dx^2} - \gamma\frac{d}{dx}\right] F(\alpha,\beta,\gamma,x) = 0 \quad . \tag{7}$$

Using the ansatz of Eq. (2) the solution is given by,

$$F(\alpha,\beta,\gamma,x) =$$

$$C_\lambda \left\{ \sum_{m=0}^{\infty} \left[\frac{1}{(D+\alpha)(D+\beta)} \left(x\frac{d^2}{dx^2} + \gamma\frac{d}{dx} \right) \right]^m \right\} x^\lambda \quad . \tag{8}$$

Unlike the previous case, here $F(D)x^\lambda$ gives two solutions for λ i.e., $\lambda = -\alpha$ and $\lambda = -\beta$. The normalized solution for $F(\alpha,\beta,\gamma,x)$ (with $\lambda = -\beta$) is,

$$F(\alpha,\beta,\gamma,x) = (-1)^{-\beta} \frac{\Gamma(\alpha-\beta)\,\Gamma(\gamma)}{\Gamma(\gamma-\beta)\,\Gamma(\alpha)}$$

$$\exp\left[-\frac{1}{(D+\alpha)} \left(x\frac{d^2}{dx^2} + \gamma\frac{d}{dx} \right) \right] x^{-\beta} \quad . \tag{9}$$

We observe from the above expression that, only when β is a negative integer does the series converge. It may also be noticed that, our series solution has decreasing powers in x compared to the standard solution [8], where the powers of x are in the increasing order. To the best of the authors' knowledge, the above is a novel expression for the solution of the hypergeometric differential equation.

It is worth emphasizing that, once $F(D)$ and $P(x, d/dx)$ are identified, we immediately have the solution in the form of an infinite series. Although it is not essential to express the series as an exponential, the exponential form for the solution does have many advantages as will be seen later. As is evident, if $P(x, d/dx)$ has a definite degree d, i.e., $[D, P^d] = dP^d$, the series will not produce singular expressions.

3 Applications

Apart from finding novel, compact expressions for the solutions of DE, the advantage of the present approach, as metioned earlier, lies in the fact that, the ladder operators can be straightforwardly obtained, leading to the familiar recurrence relations of the orthogonal polynomials. We will illustrate this by deriving the operators for the Laguerre DE, from which the construction of ladder operators for the Coulomb problem and the general DE will become clear. The method also yields the ladder operators for the other quantum numbers present in the polynomials e.g., α in L_n^α.

At the level of the monomials x^n, the simplest raising and lowering operators are x and d/dx respectively.

Defining $B = [xd^2/dx^2 + (\alpha+1)d/dx]$, the raising operator for the Laguerre polynomials follows from,

$$e^{-B} x e^{B} e^{-B} x^n = e^{-B} x^{n+1} \quad , \tag{10}$$

or

$$\left[x - 2x\frac{d}{dx} - (\alpha+1) + x\frac{d^2}{dx^2} + (\alpha+1)\frac{d}{dx}\right] L_n^\alpha(x)$$
$$= \quad -(n+1)L_{n+1}^\alpha(x) \quad . \tag{11}$$

In deriving the above relation, we have made use of the Baker-Campbell-Hausdorff (BCH) formula,

$$e^{-A}Be^{A} = B + [B,A] + \frac{1}{2!}[[B,A],A] + ... \quad . \tag{12}$$

Before constructing the lowering operator, it is important to mention that, the operator d/dx, is not the correct choice here at the monomial level. The reason being, the similarity transformation on this operator, for finding the lowering operator at the level of the polynomial, does not give a convergent answer. It is easy to see that B is an appropriate choice, yielding

$$\left[x\frac{d^2}{dx^2} + (\alpha+1)\frac{d}{dx}\right] x^n = n(n+\alpha)x^{n-1} \quad , \tag{13}$$

or

$$e^{-B}\left[x\frac{d^2}{dx^2} + (\alpha+1)\frac{d}{dx}\right] e^{B}e^{-B}x^n = n(n+\alpha)e^{-B}x^{n-1} \quad , \tag{14}$$

and hence

$$\left[x\frac{d^2}{dx^2} + (\alpha+1)\frac{d}{dx}\right] L_n^\alpha(x) = -(n+\alpha)L_{n-1}^\alpha(x) \quad . \tag{15}$$

The above results can be used to generate the ladder operators for physical systems associated with the Laguerre polynomials, e.g., the Coulomb problem. The knowledge of the unnormalized wave function, $R_n^\alpha(x) = x^l e^{-\frac{x}{2}} L_n^\alpha(x)$ with $\alpha = 2l+1$, leads to

$$x^l e^{-\frac{x}{2}}\left[x - 2x\frac{d}{dx} - (\alpha+1) + x\frac{d^2}{dx^2} + (\alpha+1)\frac{d}{dx}\right]$$
$$e^{\frac{x}{2}}x^{-l}R_n^\alpha(x) = -(n+1)R_{n+1}^\alpha(x) \quad , \tag{16}$$

and hence

$$\left[\frac{x}{2} - x\frac{d}{dx} + x\frac{d^2}{dx^2} + 2\frac{d}{dx} - \frac{l(l+1)}{x} - 1\right] R_n^\alpha(x)$$
$$= \quad -(n+1)R_{n+1}^\alpha(x) \quad . \tag{17}$$

The lowering operator follows similarly:

$$x^l e^{-\frac{x}{2}}\left[x\frac{d^2}{dx^2} + (\alpha+1)\frac{d}{dx}\right] e^{\frac{x}{2}}x^{-l}R_n^\alpha(x) = -(n+\alpha)R_{n-1}^\alpha(x) \quad , \tag{18}$$

implying

$$\left[\frac{x}{2} + x\frac{d}{dx} + x\frac{d^2}{dx^2} + 2\frac{d}{dx} - \frac{l(l+1)}{x} + 1\right] R_n^\alpha(x)$$
$$= \quad -(n+\alpha)R_{n-1}^\alpha(x) \quad . \tag{19}$$

The exponential forms for the solutions of the DE makes it easy to identify the operators which change other quantum numbers. For example, in the Laguerre case, the angular quantum number can be changed by a simple operator, using the fact $d/dxB(\alpha) = B(\alpha+1)d/dx$, we find,

$$\frac{d}{dx}L_n^\alpha(x) = (-1^n/n!)\exp\left[-\frac{d}{dx}x\frac{d}{dx} - (\alpha+1)\frac{d}{dx}\right]\frac{d}{dx}x^n \quad , \tag{20}$$

or

$$\frac{d}{dx}L_n^\alpha(x) = (-1^n)/(n-1)!\exp\left[-x\frac{d^2}{dx^2} - (\alpha+2)\frac{d}{dx}\right]x^{n-1} \quad , \tag{21}$$

The above is the standard result [8]

$$\frac{d}{dx}L_n^\alpha(x) = -L_{n-1}^{\alpha+1}(x) \quad . \tag{22}$$

Making use of the of the previous ladder operators which change n, it is clear that composite operators can be constructed which change only α. These operators naturally lead to the full symmetry algebra behind the Coulomb problem. It is easy to convince oneself that, the procedure yields the functions in the appropriate cases.

From the above, it is clear that, once the Schrödinger eigenvalue problem is brought to the level of an equation for an appropriate polynomial or function, our procedure is quite handy in, not only generating the solutions, but also yielding the raising and lowering operators for various quantum numbers. This obviously requires a prior knowledge of the ground-state wave function or a suitable measure. These inputs can be provided either from the asymptotic analysis of the DE or from the factorization method [11].

In the following, we illustrate how the present approach can be used for solving quantum mechanical problems directly. For the familiar harmonic oscillator ($\hbar=\omega=m=1$), the Schrödinger eigenvalue equation

$$\left[\frac{d^2}{dx^2} + (2E_n - x^2)\right]\psi_n = 0 \quad , \tag{23}$$

can be written, after multiplying by x^2, in the form,

$$\left[(D-1)D + x^2(2E_n - x^2)\right]\psi_n = 0 \quad . \tag{24}$$

With $F(D) = D(D-1)$, $F(D)x^\lambda = 0$, yields $\lambda = 0$ or 1. Using Eq. (2), the solution for $\lambda = 0$ is,

$$\begin{aligned}\psi_0 &= C_0\left\{\sum_{m=0}^{\infty}(-1)^m\left[\frac{1}{(D-1)D}(x^2(2E_0 - x^2))\right]^m\right\}x^0 \\ &= C_0\left[1 - \frac{[2E_0]}{2!}x^2 + \frac{(2! + [2E_0]^2)}{4!}x^4 - \cdots\right] \quad .\end{aligned} \tag{25}$$

One can notice that, ψ_0 is an expansion in the powers of x, whose coefficients are polynomials in E_0. Only when $E_0 = 1/2$, the above series can be written in the closed, square integrable form $C_0e^{-\frac{1}{2}x^2}$. Instead of proceeding by trial and error for finding an approprate E_0, one can make use of the asymptotic analysis of the DE as the input. As has been mentioned earlier, in most of the cases, once ψ_0 is known the Schrödinger equation can be cast in the form of confluent hypergeometric or hypergeometric equation. For finding the nth excited state in the above manner, one can differentiate the Schrödinger equation n times, multiply it by x^n and use $x^n \frac{d^n}{dx^n} = \prod_{l=0}^{n-1}(D-l) = F(D)$ and proceed along the lines as outlined above.

4 Many-Body Systems

In this section, we demonstrate the applicability of our method for diagonalizing interacting many-body Hamiltonians and finding out their eigenfunctions. In particular, we will concentrate on Calogero-Sutherland type models [12], which have attracted considerable attention in the recent literature. They find applications in diverse branches of physics, starting from fluid flow, random matrix models, to novel statistics and quantum Hall effect [4, 5]. Conventionally, one employs Lax-pair and Bethe-ansatz techniques [13, 14], for tackling these problems, tools which are not familiar to the uninitiated ones. I t will be shown below that, the same method as employed in the previous sections for the single variable cases, can be applied, with equal ease, to the many-body correlated systems.

Consider, the Calogero-Sutherland eigenvalue problem ($\hbar = \omega = m = 1$);

$$\left[-\frac{1}{2}\sum_{i=1}^{N}\frac{\partial^2}{\partial x_i^2} + \frac{1}{2}\sum_{i=1}^{N} x_i^2 + \frac{\alpha(\alpha-1)}{2}\sum_{\substack{i,j=1\\ i\neq j}}^{N}\frac{1}{(x_i - x_j)^2} - E_n\right] (\psi_0\phi_n) = 0 \quad , \tag{26}$$

for which the correlated ground-state wave function is given by, $\psi_0 = \exp\{-\frac{1}{2}\sum_i x_i^2\}\prod_{i<j}^{N}|x_i - x_j|^\alpha$. We have quantized the many-body system as bosons, without any loss of generality. Apart from the Gaussian, ψ_0 contains the Jastrow factor, because of which it cannot be constructed from the single particle wave functions, for $\alpha \neq 0, 1$. This indicates that the system is a correlated one. It is easy to see that the polynomial part of the wave function satisfies,

$$\left[\sum_i x_i \frac{\partial}{\partial x_i} + E_0 - E_n - \hat{A}\right]\phi_n = 0 \quad , \tag{27}$$

where, $E_0 = \frac{1}{2}N + \frac{1}{2}N(N-1)\alpha$ is the ground-state energy and $\hat{A} \equiv [\frac{1}{2}\sum_i \frac{\partial^2}{\partial x_i^2} + \alpha\sum_{i\neq j}\frac{1}{(x_i - x_j)}\frac{\partial}{\partial x_i}]$. It has been noticed quite some time back by Calogero that, only the ground state energy depends on the coupling parameter α, rest of the spectrum is unaffected by the interaction. Rearranging Eq. (27),

$$\left[\tilde{D} - n + n + E_0 - E_n - \hat{A}\right]\phi_n = 0 \quad , \tag{28}$$

the solution can be written as,

$$\phi_n = C_n\left\{\sum_{m=0}^{\infty}(-1)^m\left[\frac{1}{(\sum_i x_i \frac{\partial}{\partial x_i} - n)}(n + E_0 - E_n - \hat{A})\right]^m\right\} S_n(\{x_i\}) \quad . \tag{29}$$

The purpose of adding and subtracting n is to satisfy the requirement that, $F(\bar{D})S_n(\{x_i\}) = (\sum_i x_i \frac{\partial}{\partial x_i} - n)S_n(\{x_i\}) = 0$, with $S_n(\{x_i\})$ being a polynomial. In the present case $S_n(\{x_i\})$ is a homogeneous polynomial of degree n. However, it is clear from the presence of the constant terms in the numerator of Eq. (29) that ϕ_n's are singular in nature. Noticing that $\hat{A}$ has degree -2, to avoid the singular solutions, when the inverse of $(\sum_i x_i \frac{\partial}{\partial x_i} - n)$ acts on $S_n(\{x_i\})$, we choose $n + E_0 - E_n = 0$. This yields the familiar energy spectrum of the Calogero-Sutherland model: $E_n = E_0 + n$. Further, one has to choose $S_n(\{x_i\})$ to be completely symmetric under the exchange of x_i's, such that, the action of $\hat{A}$ yields polynomial solutions, which are normalizable with respect to ψ_0 as the weight function. Similar to the case of the Hermite polynomial, one can easily prove that,

$$\phi_n = C_n e^{-\frac{1}{2}\hat{A}} S_n(\{x_i\}) \quad . \tag{30}$$

It can be further shown, following the method of similarity transformations of the previous sections, that the Calogero-Sutherland model can be made equivalent to N decoupled oscillators [5]. The above approach can be extended to many other interacting systems on the real line and the circle [15].

5 Conclusions

In conclusion, we have elaborated on a method of solving single, as well as, many variable linear differential equations and demonstrated its usefulness for various physical problems. The solutions can be straightforwardly obtained and the ladder operators, generating new solutions from a given one, also follow rather effortlessly. The possibility of generating new exactly or quasi-exactly solvable potentials, both for single and many-variable cases, also needs to be explored.

Acknowledgements

T.S thanks U.G.C. India, for providing financial support through the J.R.F fellowship scheme. S. S thanks Prof. A. K. Kapoor for useful discussions.

References

[1] P. M. Morse and H. Feschbach, "Methods of Theoretical Physics", Vol 1, McGraw-Hill, New-York, 1953.

[2] L. Schiff, "Quantum Mechanics", McGraw-Hill, New-York, 1968;
L. D. Landau and E. Lifshitz, "Quantum Mechanics", Pergamon, New-York, 1977.

[3] Y. Alhassid, F. Gürsey and F. Iachello, Phys. Rev. Lett., **50**, 873, 1983;
Y. Alhassid, F. Gürsey and F. Iachello, Ann. Phys., **148**, 346, 1983;
A. O. Barut, A. Inomata and R. Wilson, J. Phys., **A 20**, 4075, 1987; **20**, 4083, 1987;
A. Gangopadhyaya, Jeffrey V. Mallow and U. Sukhatme, Phys. Rev, **A 58**, 4287, 1998;
S. Chaturvedi, R. Dutt, A. Gangopadhyaya, P.K. Panigrahi, C. Rasinariu and U. Sukhatme, Phys. Lett., **A 248**, 2, 1998.

[4] B. D. Simons, P. A. Lee and B. L. Altshuler, Phys. Rev. Lett., **70**, 4122, 1993; **72**, 64, 1994 and references therein.

[5] N. Gurappa and P. K. Panigrahi, Phys. Rev, **B 59**, R2490, 1999 and references therein.

[6] N. Gurappa and P.K. Panigrahi, hep-th/9908127.

[7] N. Gurappa, P. K. Panigrahi and T. Shreecharan, manuscript under preparation.

[8] I. S. Gradshteyn and I. M. Ryzhik, "Tables of Integrals, Series and Products", Academic Press Inc., 1965.

[9] I. G. Macdonald, "Symmetric Functions and Hall Polynomials", 2nd edition, Oxford: Clarendon press, 1995.

[10] F. M. Fernández, E. A. Castro, "Algebraic Methods in Quantum Chemistry and Physics", CRC, Boca Raton, 1996;
N. Gurappa, A. Khare and P. K. Panigrahi, Phys. Lett, **A 224**, 467, 1998.

[11] E. Schrödinger, Proc. Roy. Irish. Acad, **46A**, 9, 1940 and **46A**, 183, 1941;
L. Infeld and T.E. Hull, Rev. Mod. Phys., **23**, 21, 1951.

[12] F. Calogero, J. Math. Phys, **12**, 419, 1971;
B. Sutherland, ibid., **12**, 246, 1971; **12**, 251, 1971.

[13] M. A. Olshanetsky and A. M. Perelomov, Phys. Rep., **71**, 313, 1981; **94**, 6, 1983.

[14] B. Sutherland and B. S. Shastry, Phys. Rev. Lett., **71**, 5, 1993.

[15] N. Gurappa and P. K. Panigrahi, Phys. Rev, **B 62**, 1943, 2000.

Towards a Landau–Ginzburg Theory for Granular Fluids

M.H. Ernst*, J. Wakou* and R. Brito**

* Institute for Theoretical Physics
University of Utrecht
3508 TA Utrecht, The Netherlands
E-mail: J.Wakou@phys.uu.nl

** Departamento Física Aplicada I
Universidad Complutense
28040 Madrid, Spain

Granular matter [1] consits of small or large macroscopic particles. When out of equilibrium, its dynamics is controlled by dissipative interactions, and distinguished in quasi-static flows or granular solids on the one hand, and rapid flows or granular fluids [2] on the other hand.

Typical realizations of granular solids are sand piles, avalanches, Saturn's rings, grain silos, stress distributions. Here particles remain essentially in contact, and the dynamics is controlled by gravity, friction and surface roughness. In this article we concentrate on granular fluids. Typical examples are driven granular flows, such as Poisseuille flow, vibrated bed with, for instance, the occurrence of oscillons, or rapid flows with some form of continuous energy input. Here the dynamics is controlled by inelastic binary collisions, separated by ballistic motion of the particles. The forces are of short range and repulsive, and frequently modelled as collection of rough or smooth inelastic hard spheres [3].

Our focus here is the idealized limiting case of a freely evolving rapid flow without energy input and with nearly elastic collisions, that is slowly cooling [4]. It is described by a system of smooth inelastic hard spheres (IHS), of diameter σ and mass m. Momentum is conserved during collisions, which makes the system a fluid, but energy is not conserved. In a collision, on average, a fraction ϵ of the relative kinetic energy of the colliding pair is lost, where $\epsilon = 1 - \alpha^2$ is referred to as the degree of inelasticity in the literature [2, 3, 4], usually expressed in terms of the coefficient of restitution α. Its detailed definition does not concerns us here.

By performing computer simulations, Goldhirsch and Zanetti [4] have discovered in this system an interesting instability. The system, when prepared initially in a spatially homogeneous equilibrium state, does not stay there, but slowly developes patterns, both in the flow field (vortices), and in the density (clusters), the so called clustering instability, as illustrated in Fig.1.

The question of interest is then: can these models be fitted in the generic classification of Landau–Ginzburg–type models, as given by Hohenberg and Halperin [5],

Frontiers of Fundamental Physics 4, Edited by Sidharth and Altaisky,
Kluwer Academic/Plenum Publishers, New York, 2001

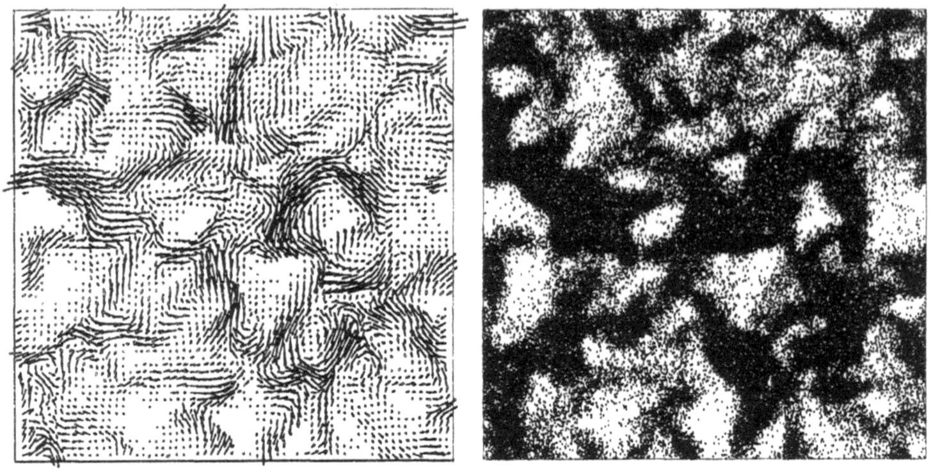

Figure 1: **Flow field (left) and density field (right) for a system of $N = 50\,000$ particles at area density of $\phi = 0.4$ and dissipation $\alpha = 0.85$, after $\tau = 160$ collisions per particle.**

to describe critical dynamics and hydrodynamic instabilities? The goal of this article is to illustrate how, under certain restrictions, the equations of motion for the IHS-fluid can be cast into Landau–Ginzburg–type equation of motion for the order parameter, which can be derived from an energy functional, and to verify a number of predictions of this theory by means of molecular dynamics (MD) computer simulations.

Instabilities: an intuitive picture

To facilitate the presentation of later arguments, we first give an intuitive explanation for the formation of vortices and clusters. In order to explain the large scale vortices in Fig.1, we note that a binary collision destroys a fraction of the kinetic energy of relative motion of the colliding pair. The cumulative effect of many successive collisions is that they make the particles locally move in a more parallel and coherent fashion. This creates the local patches of vorticity observed in Fig.1. These patches grow in size by selective supression (stronger damping) of short wavelengths modes, as explaind in Ref. [6]. Moreover, the vortex patches appear *long before* the density clusters start to appear.

Next consider the density clustering. Suppose there occurs locally a negative spontaneous pressure fluctuation ('depression'). The resulting particle flows from the surrounding try to compensate for the local depression. However the increased local density increases the collision frequency, and in turn decreases the temperature. This creates again a depression, and the process keeps repeating itself, thus creating cold dense clusters, surrounded by a hot dilute gas. Moreover, these arguments also suggest that the pressure fluctuations and the corresponding gradients remain substantially smaller than those in density and temperature. This is in fact a crucial assumption in this article, that remained to be tested by MD simulations. In a later publication [7], a more systematic and formal derivation will be presented.

Dynamic Equations for the IHS fluid

The macroscopic time evolution of the IHS fluid on large spatial and temporal scales [2, 4] may be described by the nonlinear hydrodynamic equations for the local density $n(\mathbf{r},t)$, the local flow field $\mathbf{u}(\mathbf{r},t)$ and the local temperature $T(\mathbf{r},t)$, supplemented with a sink term Γ accounting for the energy loss through inelastic collisions [4],

$$\begin{aligned} D_t n &= -n\nabla\cdot\mathbf{u}, \\ D_t\mathbf{u} &= -\frac{1}{mn}\nabla p + 2\nu\nabla\cdot\mathsf{D}, \\ D_t T &= -\frac{2p}{dn}\nabla\cdot\mathbf{u} + b_T\nabla^2 T + 2b_\perp \mathsf{D}:\mathsf{D} - \Gamma. \end{aligned} \tag{1}$$

These equations are given for a d-dimensional systems for later convenience. Here $D_t = \partial_t + \mathbf{u}\cdot\nabla$ is a material derivative, the local energy density of the IHS fluid is $e = \frac{1}{2}mnu^2 + \frac{d}{2}nT$, and p is the pressure. The shear rate $\mathsf{D} = \{D_{\alpha\beta}|\alpha,\beta = x, y, \ldots, d\}$ is a second rank symmetric traceless tensor,

$$D_{\alpha\beta} = \frac{1}{2}\left(\nabla_\alpha u_\beta + \nabla_\beta u_\alpha - \frac{2}{d}\delta_{\alpha\beta}\nabla\cdot\mathbf{u}\right). \tag{2}$$

The coefficient $b_T = 2\kappa/dn$ is proportional to the heat conductivity κ, and $b_\perp = 2m\nu/d$ to the shear viscotiy ν. For simplicity of presentation the bulk viscosity has been set equal to zero. The four terms on the right hand side of the energy balance or temperature equation acounts respectively for the work done by the pressure, heat conduction, viscous heating and collisional dissipation. Inhomogeneous flows slows down the collisional cooling process considerably through viscous heating.

On the basis of kinetic theory one finds that the rate of collisional energy loss is proportional to the collision frequency ω multiplied by the fraction of energy ϵT, lost per collision [8, 9],

$$\begin{aligned} \Gamma &= 2\gamma_0\omega T, \\ \gamma_0 &= \epsilon/2d = (1-\alpha^2)/2d, \end{aligned} \tag{3}$$

where the collision frequency for IHS fluids, $\omega(T)$, is proportional to the root mean square velocity, $v_0 = \sqrt{2T/m}$, and its explicit form for a fluid of hard spheres can be found in Refs. [11, 12]. When the system is evolving at short times in a spatially homogeneous cooling state (HCS), combination of Eqs. (1) and (3) yields $\partial_t T \sim -\gamma_0 T^{3/2}$. This leads to Haff's homogeneous cooling law [10] for the mean energy per particle in the IHS,

$$E(t) = \frac{d}{2}T(t) = \frac{E_0}{(1+\gamma_0\omega_0 t)^2}, \tag{4}$$

where $E_0 = (d/2)T_0$ is the initial energy per particle, and $t_0 = 1/\omega_0$ with $\omega_0 = \omega(T_0)$ is the mean free time in the initial state. In order to study fluctuations, noise terms of Langevin type may be added to these equations [11, 12].

Incompressible flows

The incompressible limit is an interesting limiting case of the full set of macroscopic equations (1), that greatly simplifies and elucidates the analytic structure of the full set of coupled NL-equations. As is well known from standard fluid dynamics and from the theory of turbulence [13, 14], elastic fluid flows are quite ***incompressible.*** This implies, $\nabla \cdot \mathbf{u} = 0$, or for the longitudinal Fourier mode,

$$\mathbf{u}_{\parallel}(\mathbf{k}) = \widehat{\mathbf{k}} \cdot \mathbf{u}(\mathbf{k}) = 0. \tag{5}$$

Moreover, both MD simulations and the theory of hydrodynamic fluctuations in granular flows [11, 12] show that the incompressibility asuumption, $\mathbf{u}_{\parallel}(\mathbf{k}) = 0$, remains valid for a large range of wavenumbers,

$$k\xi_{\parallel} \gtrsim 1, \tag{6}$$

and ultimately breaks down at the largest wavelengths. Here $\xi_{\parallel} \sim 1/\gamma_0$ is the largest intrinsic dynamic correlation length in IHS fluid, which satisfies the inequality, $\xi_{\parallel} \gg \xi_{\perp} \equiv (\nu/\omega\gamma_0)^{1/2}$ for nearly elastic systems. Both correlation lengths are defined in [12].

Therefore, as a zeroth approximation to our nonlinear theory, we make the incompressibility assumption (5), $\nabla \cdot \mathbf{u} = 0$, following Refs. [11, 12], and the macroscopic equations become,

$$\begin{aligned} D_t n &= 0, \\ D_t \mathbf{u} &= -\frac{1}{mn}\nabla p + \nu \nabla^2 \mathbf{u}, \\ D_t T &= b_T \nabla^2 T + b_{\perp}\left[\nabla \mathbf{u} + (\nabla \mathbf{u})^{\dagger}\right] : \nabla \mathbf{u} - 2\gamma_0 \omega T. \end{aligned} \tag{7}$$

Some comments are in order here. The incompressibility assumption keeps $n(\mathbf{r}, t)$ constant in a comoving frame. So the set of nonlinear equations (7) cannot describe the growth of inhomogeneities in the density field, but would describe time evolution where density fluctuations are small. This may happen under two circumstances where density fluctuations are strongly constrained: high densities or small systems:

(i) At ***high packing*** fraction: for instance at coverage density $\phi \gtrsim 0.4$ in two dimensions, a local density fluctuations can barely grow a factor of 2 before close packed configurations are reached.

(ii) In ***small systems***: To explain this we consider the spectra of exponential growth rates $z_{\lambda}(k)$ of the Fourier modes in the IHS fluid, as shown in Fig.2, and calculated from linear stability analysis [15, 16, 12]. This figure shows that the transverse flow field $\mathbf{u}_{\perp\mathbf{k}}$ or shear mode ($\lambda = \perp$) with a wave number $k < k_H^*$ is unstable, i.e. it developes vortices. On the other hand density fluctuations couple to the heat mode ($\lambda = H$), and $z_H(k)$ in Fig.2 shows that these fluctuations with $k > k_H^*$ are linearly stable, i.e. remain at thermal noise level. The stability thresholds $k_{\perp}^*$ and k_H^* are defined as the the root of $z_{\lambda}(k) = 0$ for $\lambda = \{\perp, H\}$, and are marked as black dots in the figure. Figure 2 also shows as stars the smallest wavenumbers, $2\pi/L$, allowed in a system of linear system size L, with periodic boundary conditions for $N = 50, 100, 200, \ldots, 1000$, at fixed packing fraction ϕ. In the systems

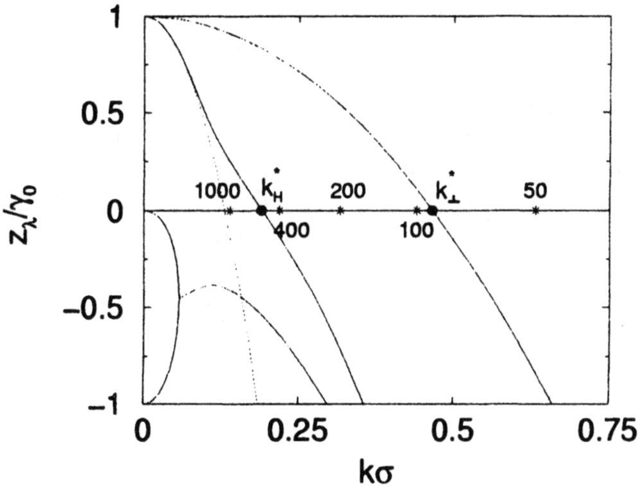

Figure 2: Dispersion relations $z_\lambda(k)$ of the Fourier modes in the IHS fluid for an area fraction of $\phi = 0.4$ and inelasticity of $\alpha = 0.85$. The stars mark the location of the minimum wave vector allowed in the system, $k_0 = 2\pi/L$ with the quoted number of particles, and the black dots the location of k_H^* and $k_\perp^*$.

with $k_H^* < k_0(N) < k_\perp^*$, vorticity grows, but density fluctuations are stable in the linear stability analysis. However, nonlinear effects induce density inhomogeneities at long times whose typical amplitude is relatively small, as it will be shown in the present paper and tested against MD simulations.

Next, we return to the Navier–Stokes equation, (7).

On the basis of the intuitive picture of the relevant instabilities in the second section, we assume that the spatial variation in the pressure are much smoother than the gradients in density and temperature. If correct, it is a reasonable approximation to neglect the term ∇p in Eq. (7), and the Navier–Stokes equation reduces to the vector Burgers equation,

$$\partial_t \mathbf{u} = -\mathbf{u} \cdot \nabla \mathbf{u} + \nu \nabla^2 \mathbf{u}. \tag{8}$$

First, we make some comments. The Burgers equation in the inviscid limit ($\nu \to 0$) has been studied [17, 18] to describe the large scale structure in the universe caused by gravitational clustering. It is also known that the Burgers equation describes the adhesion model for a perfectly elastic point particles [17, 20]. Furthermore, there is an interesting *conjecture* [20] that the large scale space time behaviour of granular gases is in the universality class of the adhesion model. If correct, then the Burgers equation (8) would be the appropriate macroscopic equation to describe the flow field in granular gases.

An interesting aside is that the Burgers equation is equivalent to the KPZ–equation, named after Kardar, Parisi and Zhang,

$$\partial_t h = \nu \nabla^2 h + \tfrac{1}{2} |\nabla h|^2 \tag{9}$$

for describing the growth of solid surfaces, where $h(\mathbf{r}, t)$ is the height function. This equivalence can be seen by applying ∇ to Eq. (9), and identifying $\nabla h = \mathbf{u}$

Next, we consider the temperature balance equation in Eq. (7), which involves two processes: heat conduction with a slow decay rate $-b_T k^2$ for small wave numbers, and a slow decrease of the ***global*** (spatially averaged) temperature $\bar{T}$ (where $\bar{a} \equiv (1/V)\int d\mathbf{r} a(\mathbf{r})$) caused by collisional damping, and partly compensated by the nonlinear viscous heating. The global $\bar{T}$ sets the typical scale of the velocity fields. Which of the two is the slowest or fastest depends on the wave numbers involved. Then a spatial average of the T-equation in (7) yields for the global temperature,

$$\partial_t \bar{T} = \bar{b}_\perp \overline{|\nabla \mathbf{u}|^2} - 2\gamma_0 \bar{\omega} \bar{T}. \tag{10}$$

We have introduced the notation $|\mathsf{A}|^2 = \sum_{\alpha\beta} |A_{\alpha\beta}|^2$, for a second rank tensor A, and we have approximated $\bar{a}(n, T) \simeq a(\bar{n}, \bar{T})$, which is allowed once the fluctuations δn and δT are small. The global temperature $\bar{T}$ sets the typical scale of the flow fields, as shown in the next section. The evolution equation for the local fluctiations $\delta T = T - \bar{T}$ will be analyzed in Ref. [7].

Spontaneous symmetry breaking. Landau–Ginzburg theory

The evolution Eqs. (3) and (9) contain the time dependent coefficients $\bar{b}_T$, $\bar{\omega}$ and $\bar{\nu}$, which are proportional to $\bar{v}_0(t) = \sqrt{2\bar{T}(t)/m}$. So, it is convenient to introduce the scaled field $\tilde{\mathbf{u}} = \mathbf{u}/\bar{v}_0$, and the scaled time τ, defined as $d\tau = \omega(\bar{T})dt$. This yields,

$$\begin{aligned}
\partial_\tau \ln \bar{T} &= \frac{4}{d}\mathcal{D}_\perp \overline{|\nabla \tilde{\mathbf{u}}|^2} - 2\gamma_0, \\
\partial_\tau \tilde{\mathbf{u}} &= \mathcal{D}_\perp \nabla^2 \tilde{\mathbf{u}} - \frac{1}{2}(\partial_\tau \ln \bar{T})\tilde{\mathbf{u}} \\
&= \gamma_0 \tilde{\mathbf{u}} + \mathcal{D}_\perp \nabla^2 \tilde{\mathbf{u}} - \frac{2}{d}\mathcal{D}_\perp \overline{|\nabla \tilde{\mathbf{u}}|^2}\tilde{\mathbf{u}}.
\end{aligned} \tag{11}$$

This equation is a closed equation for $\tilde{\mathbf{u}}$, and the physically consistent solutions need to obey the relation $\nabla \cdot \mathbf{u} = 0$. In the Navier–Stokes equation, the nonlinear convection term, $\tilde{\mathbf{u}} \cdot \nabla \tilde{\mathbf{u}}$ has been neglected. The rescaled vorticity diffusion coefficient is defined here as $\mathcal{D}_\perp = \bar{\nu}/\bar{\omega}$. The first term on the right hand side of the $\mathbf{u}$-equation represents the inestability, the second is the vorticity diffusion and the last one is the saturation effects, caused by the nonlinear viscous heating. It slows down the growth of the unstable $\mathbf{k}$-modes, and may ultimately lead to a steady state for $\tilde{\mathbf{u}}$.

The equation for $\tilde{\mathbf{u}}$ can be cast into a generic form, by introducing $\mathsf{S} = \nabla \tilde{\mathbf{u}}$, and apply ∇ to the $\mathbf{u}$-equation in (11). The result can formally be written as,

$$\begin{aligned}
\partial_\tau \mathsf{S} &= \left(\gamma_0 + \mathcal{D}_\perp \nabla^2\right)\mathsf{S} - \frac{2}{d}\mathcal{D}_\perp \overline{|\mathsf{S}|^2}\mathsf{S} \\
&= -V\delta\mathcal{H}[\mathsf{S}]/\delta\mathsf{S}^\dagger,
\end{aligned} \tag{12}$$

where $\mathcal{H}[\mathsf{S}]$ is an energy functional, defined as

$$\mathcal{H}[\mathsf{S}] = -\frac{1}{2}\gamma_0 \overline{|\mathsf{S}|^2} + \frac{1}{2}\mathcal{D}_\perp \overline{|\nabla \mathsf{S}|^2} + \frac{1}{2d}\mathcal{D}_\perp \left(\overline{|\mathsf{S}|^2}\right)^2. \tag{13}$$

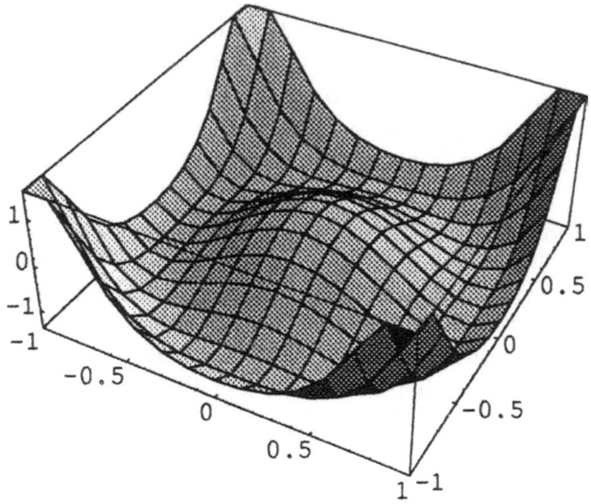

Figure 3: Landscape plot of the potential function $\mathcal{H}[\mathsf{S}]$ in the $\mathsf{S}_{\mathbf{k}_0}$ subspace.

It resembles a Landau free energy form of a *tensorial order parameter* $\mathsf{S} = \nabla\tilde{\mathbf{u}}$ with $\operatorname{tr}\mathsf{S} = \nabla\cdot\tilde{\mathbf{u}} = 0$, which is in fact the rate of deformation tensor. In our further considerations it is more convenient to use Fourier modes, with the notation $f_{\mathbf{k}} = \int_V d\mathbf{r}\, e^{-i\mathbf{k}\cdot\mathbf{r}} f(\mathbf{r})$. Moreover, $\mathbf{u}_{\mathbf{k}} = \mathbf{u}_{\perp\mathbf{k}}$, because of the assumption of incompressibility, and $\mathsf{S}_{\mathbf{k}} \equiv \mathbf{k}\tilde{\mathbf{u}}_{\perp\mathbf{k}}$. Then we obtain the equation of motion,

$$\begin{aligned}\partial_\tau \mathsf{S}_{\mathbf{k}} &= -V^2\delta\mathcal{H}[\mathsf{S}]/\delta\mathsf{S}^\dagger_{\mathbf{k}} \\ &= \left\{\gamma_0 - \mathcal{D}_\perp k^2 - \frac{2}{d}\mathcal{D}_\perp\frac{1}{V^2}\sum_{\mathbf{q}}|\mathsf{S}_{\mathbf{q}}|^2\right\}\mathsf{S}_{\mathbf{k}}\end{aligned} \tag{14}$$

with an energy functional,

$$\mathcal{H}[\mathsf{S}] = \frac{1}{2V^2}\sum_{\mathbf{k}}(-\gamma_0 + \mathcal{D}_\perp k^2)|\mathsf{S}_{\mathbf{k}}|^2 + \frac{1}{2d}\mathcal{D}_\perp\left(\frac{1}{V^2}\sum_{\mathbf{k}}|\mathsf{S}_{\mathbf{k}}|^2\right)^2, \tag{15}$$

where the wave number $\mathbf{k} = 0$ does not contribute.

These results are very interesting. If the energy functional has a minimum, then there is a fixed point solution, $\mathsf{S}_{\mathbf{k}}(\infty)$, that is approached for large times. The stationary point of (14) are found by setting the right hand side of (14) equal to zero, i.e.,

$$\left\{\gamma_0 - \mathcal{D}_\perp k^2 - \frac{2}{d}\mathcal{D}_\perp\frac{1}{V^2}\sum_{\mathbf{q}}|\mathsf{S}_{\mathbf{q}}|^2\right\}\mathsf{S}_{\mathbf{k}} = 0. \tag{16}$$

We will show that depending on the system parameter, $k_0 = 2\pi/L$, being above or below a threshold value, $k_\perp^*$, the fixed points solution, $\{\mathsf{S}_{\mathbf{k}}(\infty)\}$, which represents the order parameter has a ***vanishing*** or ***nonvanishing*** value. Here the energy functional $\mathcal{H}$[S] turns out to have schematically the shape of a Mexican hat in the subspace spanned by the S_{k_0} component, as illustrated in Fig.3, and the minimum is infinitely degenerate. In case of a fixed point solution with a nonvanishing order parameter, spontaneous fluctuations in the initial state determine which of these minima will be reached, and the symmetry of the steady state is spontaneously broken.

Consider the right hand side of (14), and observe that the expression between curly brackets is necessarily *negative* for $k > k_\perp^* = \sqrt{\gamma_0/\mathcal{D}_\perp}$, and the Fourier mode S_k decays to zero. Of the smallest possible wavenumber, $k_0 = 2\pi/L > k_\perp^*$, *all* S_k decay to zero, and there is no spontaneous symmetry breaking. The system is in a spatially homogeneous translational state with a constant velocity, $\mathbf{P}/Nm$. However, if $k < k_\perp^*$, then there exist the possibility that the expression inside brackets vanishes for a nonvanishing value of $S_k(\infty)$, i.e there is an extremum determined by the condition,

$$\frac{1}{V^2}\sum_{\mathbf{q}}|\mathsf{S}_{\mathbf{q}}|^2 = \frac{d}{2}(\gamma_0 - \mathcal{D}_\perp k^2)/\mathcal{D}_\perp. \tag{17}$$

However, if $k < k_\perp^*$, then the right hand side of Eq. (17) may become zero and even positive. There is the possibility of stationary and of exponentially growing solutions. The fixed point $\{\mathsf{S}_{\mathbf{k}}(\infty) = 0$ for any $\mathbf{k}\}$ is a saddle point. In fact, one can show [7] that all fixed point solutions with nonvanishing $\mathsf{S}_{\mathbf{k}}$ for any $|\mathbf{k}| \neq k_0$ are saddle points, and that the only nonvanishing solution $\{\mathsf{S}_{\mathbf{k}} \neq 0$ if $|\mathbf{k}| = k_0, \mathsf{S}_{\mathbf{k}} = 0$ otherwise$\}$ form a stable fixed point, which is infinitely degenerate minimum, and given schematically by the Mexican hat shape as illustrated in Fig. 3. We get a condition for $\mathbf{u}_{\perp\mathbf{k}}$ with $|\mathbf{k}| = k_0 = 2\pi/L$ from (17), given in two dimensions by

$$\frac{1}{V^2}\left(|\tilde{\mathbf{u}}_{\mathbf{k}_{0x}}|^2 + |\tilde{\mathbf{u}}_{\mathbf{k}_{0y}}|^2\right) = \frac{d}{4}(\gamma_0 - \mathcal{D}_\perp k_0^2)/\mathcal{D}_\perp k_0^2, \tag{18}$$

where $\mathbf{k}_{0x} = (k_0, 0), \mathbf{k}_{0y} = (0, k_0)$.

The stable stationary solutions in real space form an infinite set of special linear combinations of (see [7]),

$$\begin{aligned} \tilde{\mathbf{u}}_0^{st}(x) &= \hat{e}_y V_0 \cos(k_0 x + \theta_x), \\ \tilde{\mathbf{u}}_0^{st}(y) &= \hat{e}_x V_0 \cos(k_0 y + \theta_y) \end{aligned} \tag{19}$$

with an amplitude

$$V_0 = \left[(\gamma_0 - \mathcal{D}_\perp k_0^2)/\tfrac{1}{d}\mathcal{D}_\perp k_0^2\right]^{1/2}. \tag{20}$$

θ_x and θ_y are arbitrary phase constants. They are all solutions of Eq. (11). The full nonlinear equation for $\tilde{\mathbf{u}}$ also contains the nonlinear convective term $\tilde{\mathbf{u}} \cdot \nabla\tilde{\mathbf{u}}$. The two explicit solutions in (19) are the only exact stationary solutions of the full nonlinear equation with the convective terms included.

Main results for small systems

A snapshot of a typical configuration for a small system with $k_H^* < k_0 < k_\perp^*$ at long times is shown in Fig.4. It corresponds to a system at density $\phi = 0.4$, at

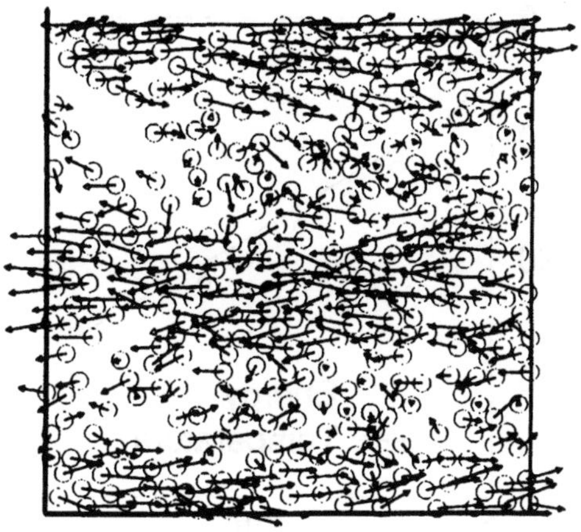

Figure 4: Configuration at $\phi = 0.4$, $\alpha = 0.85$ and $N = 400$ at $\tau = 600$. A shear velocity field is observed in agreement with the solution of the nonlinear equations given in Eq. (19).

inelasticity $\alpha = 0.85$ and with $N = 400$ particles, so that $k_0 = 0.22$ (see Fig.2). A clear shear velocity field with variation of u_x component of velocity along the y direction is observed. Individual v_x components of the velocities of the particles versus y is plotted in Fig.5. A fit to a sinusoidal curve (solid line) shows that the solution described in (19) is realized, where, in this case, only the $\tilde{\mathbf{u}}_0^{st}(y)$ is present.

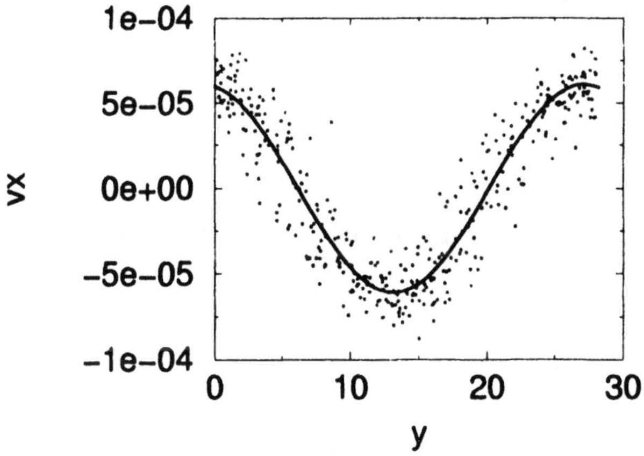

Figure 5: Shear velocity of the configuration of Fig. 4. Dots are the actual velocities of the particles, while the solid line is the fit to a sinusoidal function as described in Eq. (19).

By substituting (18) into (11), we deduce that the stationary solution of scaled velocity field $\tilde{\mathbf{u}}$ implies exponential decay of global temperature $\bar{T}$ in the long time limit:

$$\bar{T} = \bar{T}_n \exp\left[-2\mathcal{D}_\perp k_0^2 (\tau - \tau_n)\right], \tag{21}$$

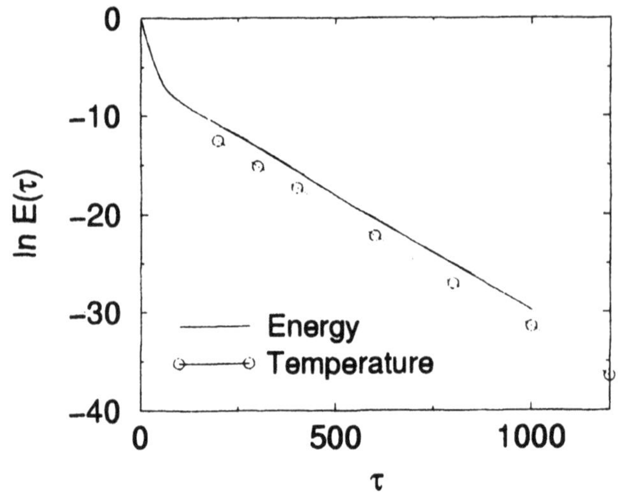

Figure 6: Energy per particle and temperature versus number of collsions per particle τ, for the same parameters than Fig. 4.

where τ_n is a scaled time which is well inside a time region where the stationary solution of $\tilde{\mathbf{u}}$ is achieved. $\bar{T}_n$ is the global temperature at τ_n.

Equation (21) gives the temperature as a function of the number of collisions per particles, τ. In order to express it in real time, t, we need the relation $t(\tau)$. From the definition of τ, $d\tau = \omega(\bar{T})\,dt$, and because ω is proportional to the square root of $\bar{T}$, we have,

$$\frac{d\tau}{dt} = \omega(\bar{T}) = \omega_0\sqrt{\frac{\bar{T}}{T_0}} = \omega_n \exp\left[-\mathcal{D}_\perp k_0^2\,(\tau - \tau_n)\right], \tag{22}$$

where $\omega_0 = \omega(T_0)$ and $\omega_n = \omega(\bar{T}_n)$ are the collision frequencies at the initial time and τ_n respectively. Moreover $\bar{T}_0$ is the initial temperature. This differential equation is easily solved:

$$t - t_n = \frac{1}{\omega_n}\frac{1}{\mathcal{D}_\perp k_0^2}\left\{\exp\left[\mathcal{D}_\perp k_0^2(\tau - \tau_n)\right] - 1\right\}, \tag{23}$$

where t_n is the time t corresponding to τ_n.

Using this relation between τ and t we can finally obtain the evolution of temperature in terms of t:

$$\bar{T} = \bar{T}_n\left(\frac{1}{\omega_n\mathcal{D}_\perp k_0^2(t - t_n) + 1}\right)^2 \simeq \frac{T_0}{(\omega_0\mathcal{D}_\perp k_0^2)^2}\,\frac{1}{t^2}. \tag{24}$$

The second expression is valid for time well inside the asymptotic time, $t \gg t_n$, and shows an ***algebraic decay*** of the temperature.

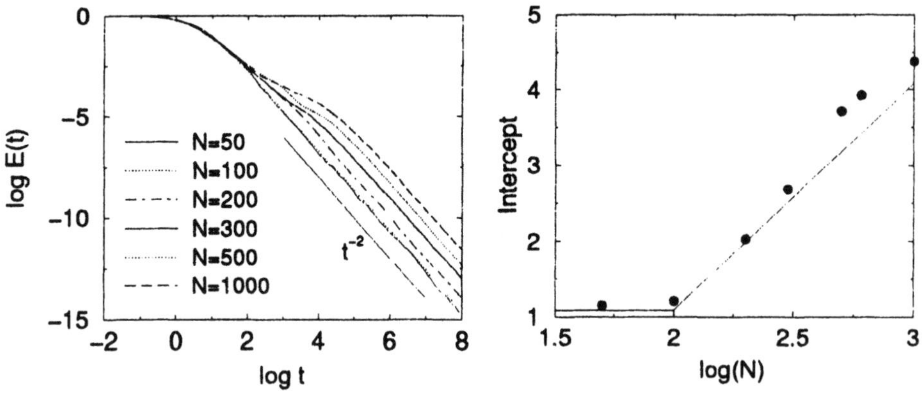

Figure 7: Left: Simulation results of the behaviour of $\log E(t)$ as a function of $\log t$. At long times (about $t > 10^4$ in the present case with $\phi = 0.4$ and $\alpha = 0.85$), a t^{-2} power law is obtained. Right: the amplitude of the t^{-2} tail depends on the number of particles in the system as N^3. Here the solid line is the result of the theory, Eq. (28), while the dots are simulation results extracted from the left panel.

Once we have $\bar{T}$, we can calculate the energy per particle, defined as,

$$E = \frac{1}{N}\left[\overline{\frac{m}{2} n\, \mathbf{u}^2 + \frac{d}{2} nT}\right]. \tag{25}$$

Under the approximation $\bar{a}(n, T) \simeq a(\bar{n}, \bar{T})$, which was used in deriving (21), we obtain

$$E = \frac{m}{2}\overline{\mathbf{u}^2} + \frac{d}{2}\bar{T} = \bar{T}\left[\overline{\tilde{\mathbf{u}}^2} + \frac{d}{2}\right]. \tag{26}$$

Substituting (19) into (26), we get

$$E \;=\; \frac{d\gamma_0}{2\mathcal{D}_\perp k_0^2}\bar{T}. \tag{27}$$

This shows that at long times, the energy and temperature are proportional. Therefore, energy decays exponentially when expressed in terms of τ (see Eqs. (21)), with the same exponent as that of the global temperature decay. This property is observed by MD symulations, as shown in Fig.6.

Moreover, from Eqs. (27) and (24), the decay of the energy per particle in terms of t in the long time limit is given by

$$E(t) \;=\; \frac{d\gamma_0 \bar{T}_n}{2\mathcal{D}_\perp k_0^2}\left(\frac{1}{\omega_n \mathcal{D}_\perp k_0^2 (t - t_n) + 1}\right)^2 \simeq \frac{d\gamma_0\, T_0}{2\omega_0^2 (\mathcal{D}_\perp k_0^2)^3}\,\frac{1}{t^2}. \tag{28}$$

Hence, if we vary the number of particles N, at fixed packing fraction ϕ, the amplitude of t^{-2}-decay of the energy is proportional to N^3 in 2-dimensional systems,

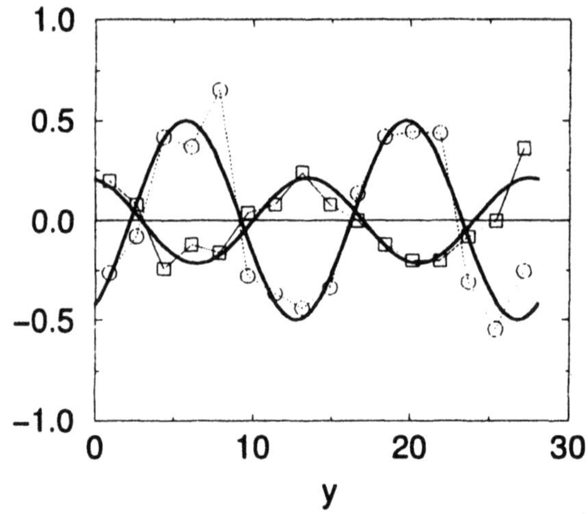

Figure 8: **Density (squares) and temperature (circles) inhomogeneities, $\delta\tilde{n}$ and $\delta\tilde{T}$ for the same parameters as in former figures and $\tau = 600$. The solid lines correspond to nonlinear fits to sinusoidal functions of double period as written in (29).**

or $N^{6/d}$ in d-dimensional systems. Comparison of Eq. (28) with simulations is presented in Fig. 7.

Finally, a systematic expansion that will be presented in [7] shows that the stationary solution of local temperature and local density are given by

$$\begin{aligned} \delta\tilde{T} \equiv \frac{T(y)}{\bar{T}} - 1 &= -\frac{\bar{b}_T V_0^2}{2d\bar{\omega}} \cos[2(k_0 y + \theta_y)], \\ \delta\tilde{n} \equiv \frac{n(y)}{\bar{n}} - 1 &= \frac{\bar{b}_T V_0^2}{2d\bar{\omega}} \frac{\frac{1}{\bar{n}}\left(\frac{\partial p}{\partial T}\right)_0}{\frac{1}{\bar{T}}\left(\frac{\partial p}{\partial n}\right)_0} \cos[2(k_0 y + \theta_y)], \end{aligned} \tag{29}$$

where $\bar{n} = N/V$, $(a(n,T))_0 = a(\bar{n},\bar{T})$, and θ_y is an arbitrary phase factor that is equal to that of Eq. (19). It means that the density and temperature inhomgeneities show a period which is half than the period of the shear velocity profile.

Figure 8 shows the observed density and temperature inhomogeneities. A fit to a sinusoidal curve supports the temperature and density profiles given by (29).

As shown, the temperature and density inhomogeneities are in opposite phase, meaning that dense regions are cold region, and vice versa. The amplitude is such that they give constant pressure, as we have assumed in the course of the paper.

Conclusions

We have presented here a analysis of the nonlinear Navier–Stokes equations, with the starting point of the incompressibility assumption. It leads to a Landau–Ginzburg

type equations. For small systems, which develop a shear instability (although the density remains stable in the linear analysis) the equations can be easily solved, leading to some important predictions:
— Exponential decay of temperature versus the number of collisions per particle τ.
— Algebraic decay of the temperature versus time t as t^{-2}.
— Proportionality relation between energy and temperature, and, therefore, algebraic decay of the energy of the system as a function of t.
— Further analysis (not presented here) leads to the result that the density and temperature inhomogeneities have a period which is half than that of the velocity profile. These results are consistent with results of a nonlinear analysis of systems close to the shear instability point [21].

The results presented here can be obtained as the lowest order in a Prandtl number series, that allows to make a more systematic expansion. It will be presented elsewhere [7].

Acknowledgements

J.W. acknowledges support of the foundation "Fundamenteel Onderzoek der Materie (FOM)", which is financially supported by the Dutch National Science Foundation (NWO). J.W. also acknowledges support of Huygens scholarship. R.B. wants to thank the hospitality of the Institute for Theoretical Physics of the University of Utrecht, where part of this work was carried out. R.B is supported by DGES-PB97-0076 (Spain).

References

[1] S.R. Nagel, Rev. Mod. Phys. **64**, 321, 1992.

[2] C.S. Campbell, Annu. Rev. Fluid Mech. **22**, 57, 1990.

[3] J.T. Jenkins and M.W. Richman, Phys. Fluids **28**, 3485, 1985. J.T. Jenkins and S.B Savage, J. Fluid Mech. **130**, 187, 1983.

[4] I. Goldhirsch and G. Zanetti, Phys. Rev. Lett. **70**, 1619, 1993. I. Goldhirsch, M-L. Tan, and G. Zanetti, J. Scient. Comp. **8**, 1, 1993.

[5] P.C. Hohenberg and B.I. Halperin, Rev. Mod. Phys. **49**, 435, 1977.

[6] R. Brito and M.H. Ernst, Europhys. Lett. **43**, 497, 1998. R. Brito and M.H. Ernst, Int. J. Mod. Phys. C **9**, 1339, 1998.

[7] J. Wakou and M.H. Ernst, to be published.

[8] J.J. Brey, J.W. Dufty, and A. Santos, J. Stat. Phys. **87**, 1051, 1997.

[9] J.A.G. Orza, R. Brito, T.P.C. van Noije, and M.H. Ernst, Int. J. Mod. Phys. C **8**, 953, 1997.

[10] P.K. Haff, J. Fluid Mech. **134**, 401 , 1983.

[11] T.P.C. van Noije, M.H. Ernst, R. Brito, and J.A.G. Orza, Phys. Rev. Lett. **79**, 411, 1997.

[12] T.P.C. van Noije, M.H. Ernst, and R. Brito, Phys. Rev. E **57**, R4891, 1998. T.P.C. van Noije and M.H. Ernst, Phys. Rev. E **61**, 1765, 2000.

[13] G.K. Batchelor, " The Theory of Homogeneous Turbulence", Cambridge University Press, 1970.

[14] U. Frisch, "Turbulence: the legacy of A.N. Kolmogorov", Cambridge University Press, 1996.

[15] P. Deltour and J.-L. Barrat, J. Phys. I France, **7**, 137, 1997.

[16] J.J. Brey, J.W. Dufty, C.S. Kim, and A. Santos, Phys. Rev. E **58**, 1051, 1998.

[17] S.F. Shandarin and Ya.B. Zeldovich, Rev. Mod. Phys. **61**, 185, 1989.

[18] M. Vergassola, B. Dubrulle, U. Frisch, and A. Noullez, Astronomy and Astrophysics **289**, 325, 1994.

[19] G.F. Carnevale, Y. Pomeau, and W.R. Young, Phys. Rev. Lett. **64**, 2913, 1990.

[20] E. Ben-Naim, S.Y. Chen, G.D. Doolen, and S. Redner, Phys. Rev. Lett. **83**, 4069, 1999.

[21] R. Soto, M. Mareschal, and M. Malek Mansour, Phys. Rev. E **62**, 3836, 2000.

The Concept of Probability in Statistical Mechanics

D. A. Lavis
Department of Mathematics
King's College, Strand, London WC2R 2LS - U.K.
Email:David.Lavis@kcl.ac.uk

1 Introduction

Thermodynamics was first formulated to describe the thermal properties of matter and, although its scope has now been enlarged, its relationship with the other main theories of physics, general and special relativity, classical and quantum mechanics and elementary particle theory, is still rather uneasy. As Sklar remarks [1, p. 4], it "is surprising that there is any place at all in this picture for a discipline such as thermodynamics" . It could be argued that it was this perception, leading to the conclusion that there is indeed no room for thermodynamics, which was the driving force behind the development of statistical mechanics. Of course, if this point has any weight, it must be seen in the historic context of the late nineteenth century, when attitudes to atomic models were rather ambivalent [1, 2]. Writing in the introduction to his *Lectures on Gas Theory* about the decline of support for atomism in continental Europe, Boltzmann [3, p. 24] remarks (presumably, sadly) that "it has been concluded that the assumption that heat is motion of the smallest particles of matter will eventually be proved false and discarded". The *energeticist* case is presented starkly by Mach:[1]

> The atom must remain a tool for representing phenomena, like functions of mathematics. Gradually, however, as the intellect ... grows in discipline, physical science will give up on its mosaic play in stones.

Maxwell in a letter to Stokes in May 1859 (quoted in [4, p. 91]), about his interest in kinetic theory, emphasizes that he had "taken to the subject for mathematical work" and that he has engaged in his calculations "as an exercise in mechanics". He also remarks that it is perhaps "absurd to ... found arguments upon measurements of strictly 'molecular' quantities before we know whether there be any molecules" (ibid, p. 93). Notwithstanding these reservations he is prepared, in a paper a few years later to mount a spirited defense of atomism [6, pp. 23–87].[2] It is significance that Cercignani [7] has chosen to subtitle his biography of Boltzmann *The Man Who Trusted Atoms*. In his paper of 1872, deriving the transport equation and H-theorem, Boltzmann [6, pp. 188–193] certainly writes as if he 'believes in' the molecules of the mechanical theory of heat. However, in his *Lectures on Gas Theory*

[1] Taken from Mach's *Popular Scientific Lectures* (1894) and quoted in [2, p. 21] .

[2] Papers by Maxwell, Clausius, Poincaré, Zermelo and Boltzmann are contained (in English translation where necessary) in the two volumes by Brush [5, 6] and will be cited accordingly.

[3][3] he takes a slightly more cautious stance. The title of the first section of the introduction is 'Mechanical analogy for the behaviour of a gas' and in that section [3, p. 21] he remarks that: "Energetics is certainly very important for science; however, up to now its concepts are still rather unclear, and its theorems not very precisely expressed, so that it cannot replace the older theory of heat".

Aside from the question of atomism versus energeticism the very existence of *statistical* mechanics could be viewed as a failure of some putative programme for the replacement of thermodynamics by a form of many-body mechanics, be it classical or quantum. An extra ingredient has been used to bridge the gap from purely mechanical concepts, like position and momentum, to the thermodynamic quantities of pressure, temperature and entropy.

2 From Kinetic Theory to Statistical Mechanics

The power of an atomistic approach was evident as early as 1738 when Bernoulli, in his *Hydrodynamics*, was able to derive Boyle's law by assuming a gas of particles all with identical velocities. He was also able to obtain the formula relating temperature to particle velocity. By the time of Clausius and Maxwell it was recognized that there would be variations in the velocities of gas particles. While Clausius [5, pp. 111–134] replaced the velocities by their average value, Maxwell [5, pp. 148–171] proposed a formula for velocity variation. Writing to Stokes in 1859 he remarks that of course his particles "have not all the same velocity, but the velocities are distributed according to the same formula as the errors are distributed in the theory of least squares".[4] The resulting formula was the, now famous, *Maxwell law* that, for a gas of N small, hard, perfectly elastic spheres acting on one another only during impact, the number of spheres whose speed[5] lies between v and $v + \mathrm{d}v$ is $f_1^{(\mathrm{M})}(v)\mathrm{d}v$, where

$$f_1^{(\mathrm{M})}(v) = \frac{4N}{\alpha^2\sqrt{\pi}}\, v^2 \exp\left(-\frac{v^2}{\alpha^2}\right).$$

It is of some interest that the word 'probability' does not occur in this part of the paper, making its appearance only at the point where Maxwell considers particle collisions. However, it would be difficult to draw a conclusion from this observation and it can be fairly said [1, p. 30] that we find in Maxwell's paper "language of sort that can be interpreted in a probabilistic or statistical vein".

When considering the origin of the insertion of probabilistic ideas into many-particle dynamics, we must examine, not only equilibrium distributions, but also the theory of transport processes, the foundations for which were laid by Maxwell [6, pp. 23–87] and developed by Boltzmann.[6] Boltzmann's analysis starts with the distribution function $f_1(\mathbf{v}, t)$ so that $f_1(\mathbf{v}, t)\mathrm{d}^3v$ is the number of particles in the volume element d^3v at the point $\mathbf{v}$ in the single-particle velocity space at

[3]Published in the same year as Zermelo's criticism [6, pp. 208–217] of his H-theorem results.

[4]Taken from *The Scientific Letters and Papers of James Clerk Maxwell* edited by P. M. Harman and quoted in [4, p. 95] .

[5]Maxwell calls v the "actual velocity" of the sphere, but in the context of his derivation it is clear that the quantity is what we would now call the speed, since he remarks that the "velocities range from 0 to ∞".

[6]Boltzmann papers on this subject began in 1868, but the most compact presentation of his work is given in his *Lectures on Gas Theory* [3].

time t.[7] To calculate how this distribution changes with time we need an expression for the number of pairs of particles with two different velocities which collide in unit time.[8] Such information will be contained in the two-particle velocity distribution function $f_2(\mathbf{v}, \mathbf{v}', t)$. The fundamental assumption that Boltzmann makes here is that there are no correlations between velocities. This means that $f_2(\mathbf{v}, \mathbf{v}', t) = f_1(\mathbf{v}, t) f_1(\mathbf{v}', t)$. This ***molecular chaos condition*** is assumed to persist for all time and enabled Boltzmann [6, pp. 188–193] to derive his transport equation, which has a solution which is independent both of time and the velocity direction and corresponds to the Maxwell distribution. However, Boltzmann aimed to prove something stronger, namely to show that the Maxwell distribution is the ***unique*** stationary solution that will be monotonically approached from any non-equilibrium distribution. He was able to do this [6, pp. 188–193] by proving the H-theorem, which established that the quantity

$$H(t) = \int \mathrm{d}^3 v f_1(\mathbf{v}, t) \ln[f_1(\mathbf{v}, t)],$$

decreases monotonically with time, with the only solution of $\mathrm{d}H/\mathrm{d}t = 0$, being the Maxwell distribution.

It is fairly clear that, up to this point, the subject under discussion is still 'kinetic theory'. Although the language and concepts of Maxwell and Boltzmann have a certain probabilistic flavour the distribution functions are still meant to be a measure of the actual number of particles with velocities in a particular range. The turning point and the perceptual change seems to have been driven by the criticisms of Boltzmann's results.

As indicated above, the scientific orthodoxy in Germany in the late nineteenth century was energetics, and this provided a constant challenge to the nascent kinetic theory. This challenge was compounded by two technical objections.

The first of these concerns the problem of reconciling the reversibility of mechanical laws and the irreversibility of natural processes as described by the second law of thermodynamics. This seems to have been first noted by Maxwell [4, p. 141], but it came to Boltzmann's attention in two papers published by Loschmidt [8].

In response to some work of Poisson, Poincaré [6, pp. 194–202] proved the *recurrence theorem* that now bears his name[9] and in a second brief paper he drew attention to what became the second problem in reconciling thermodynamics and kinetic theory [6, pp. 203–207], namely the incompatibility of the second law of thermodynamics and the mechanical theory of heat which is based on a usually recurrent dynamic system. This second paper seems to have been ignored both by Boltzmann and by Zermelo [6, pp. 208–217] who, in making a similar point elicits a reply from Boltzmann [6, pp. 218–228], followed by further dialogue ([6, pp. 229–237] and [6, pp. 238–245]).

[7] In his first presentation of the calculation in 1872 [6, pp. 188–193] the distribution was taken to be over energies, but this was modified by the time of the 1896 lectures.

[8] In his original derivation of his transport equation Boltzmann neglected the existence of particle collisions involving more than two particles.

[9] For a mechanical system there are an infinite number of ways of choosing initial conditions such that the system will return infinitely many times as close as we like to the initial position. There are also an infinite number of initial choices which do not have this property, but the latter are 'exceptional' in comparison with the former.

Boltzmann makes a number of points related both to the recurrence problem and to the question of irreversibility:

(i) If the number of molecules is infinite then the Poincaré theorem does not apply and, even for a 'small' system,[10] the recurrence time would be a number in seconds with "many trillions of digits".

(ii) In practice we would not expect a finite system to be completely isolated so again the Poincaré theorem does not apply.

(iii) The second law is, from the molecular viewpoint, a statistical law.

(iv) In order to understand irreversibility it is necessary to be able to distinguish clearly between the macroscopic and microscopic levels and to have a definition, for the dynamic system, of what is meant by a macrostate.

The importance of these replies can be seen in the fact that they have each, in different ways, led to the development of programmes for non-equilibrium statistical mechanics. With the possible exception of (iii), they are also still subjects of dispute.

The question of whether statistical mechanics applies only to systems of a large, possibly infinite, number of microsystems will be discussed below. The view that external influences are needed to achieve statistical mechanical equilibrium[11] takes two forms. The disturbance could be of the form of small random perturbations which, as envisaged by Boltzmann, will alter the trajectory of the system and prevent recurrence and reversibility. Or it could be a steady dissipation of energy. The remark by Sklar [1, p. 156] that "within the context of the dynamical theory of non-equilibrium ... the equilibrium state exists as the 'attractor' to which the dynamics of non-equilibrium drives [it]", would imply something of this sort, since isolated mechanical systems do not have attractors [9].

In his reply to Loschmidt, Boltzmann [6, pp. 188–193] argues for a statistical view of the second law. Writing in the following year a similar point is made by Maxwell. In his review of Tait's *Thermodynamics* he notes that the "truth of the second law", as a statistical theorem, was "of the nature of a strong probability ... not an absolute certainty" like dynamic laws, (quoted in [4, p. 141]). The most immediate effect of this can be seen in a change of perception of the meaning of $f_1(\mathbf{v},t)$. This function is no longer regarded as the actual distribution of the N particles of the gas. In modern terms it has become equal to $N\rho_1(\mathbf{v},t)$, where $\rho_1(\mathbf{v},t)$ is the single-particle probability density function over velocity space. From this point it is a relatively small step to change the quantity of interest to $\rho_N(\mathbf{x},\mathbf{p},t)$, the probability density function on the phase space Γ of the vector $(\mathbf{x},\mathbf{p})$, specifying the coordinates and momenta of all the degrees of freedom of the N particles. If the system is Hamiltonian and if probabilities are preserved by the Hamiltonian flow then the probability density function must satisfy Liouville's equation. This equation provides the starting point for a number of different approaches to non-equilibrium statistical mechanics, most notably the derivation of the Bogolyubov-Born-Green-Kirkwood-Yvon hierarchy of kinetic equations and the work of the Brussels-Austin School, which will be discussed in Sec. 6. However, for the moment we are interested in possible interpretations of the probability density function $\rho_N(\mathbf{x},\mathbf{p},t)$ and we must first consider the options available to us.

[10] Which he takes to be 1 c.c. of air containing 10^{18} molecules [6, pp. 218–228].

[11] Called by Sklar [1, p. 250] the *interventionist approach*.

3 Different Views of Probability

There seem to be three attitudes to the presence of statistics in statistical mechanics. The first, which tends to be that adopted by classical texts in the subject and to go with great reliance on the concept of *ensembles*, is to regard the subject algorithmically, as a procedure for arriving at answers, rather in the same way as the *replica method* is used in some calculations in critical phenomena and neural networks. Thus Tolman [10, pp. 1,2] writes:

> [The] principles of statistical mechanics are to be regarded as permitting us to make reasonable predictions as to the future of a system. ... [They consist] in abandoning the attempt to follow the precise changes in state which take place in a particular system, and in studying the behaviour of a collection or *ensemble of systems.*[12]

The second approach, is to regard probabilistic ideas rather in the way that the energeticists regarded the atom (see the quote from Mach given above), as something that a more mature science would be able to discard. The means for doing this is usually thought to be *ergodic theory*, which will be considered in detail below.

The third approach is to take a more positive attitude to the probabilistic ideas, interpreting them, by implication if not explicitly, according to one of the standard view of probability. In this context, therefore, we shall now review the different views of probability. There are many ways to subdivide these views [1, 12, 13]. The main division is between the *scientific (or objective) view* and the *subjective view.* We shall make a further subdivision of each of these categories.

3.1 The Scientific View

This may be characterized by the fact that it considers "the theory of probability as a science of the same order as geometry or theoretical mechanics", [14, p. vii]. Thus probability is an objective property. But of what? Clearly not simply of an object, like a die, without any other qualification. It seems most reasonable to define the probability as that of the outcome of a particular *experiment* with the object in question, in circumstances where some aspect of the test is *incompletely specified.* Typically this incomplete specification arises because the outcome is sensitive to initial conditions, so that it is, both in practice and in principle, impossible to fix all the aspects of the experiment necessary to determine the outcome. In the case of the throw of a die,[13] we are unable to determine which side finally lands uppermost, because we are unable to to specify exactly the initial location and orientation, and angular and linear acceleration.[14] As we have seen in relation to the comments on Tolman, given above, there is a tendency to talk of 'incomplete knowledge' rather

[12] In the same passages he also refers to 'incomplete' and 'partial' knowledge. The has led Hobson [11], an adherent of a subjectivist view of probability in statistical mechanics, to claim him as a supporter. We do not find this claim entirely convincing [12]. The passage could quite easily be read as a statement of the (presumably) uncontentious view that probability concepts are used when a system is incompletely specified.

[13] And excluding the possibility of careful dropping it with one face uppermost onto a yielding surface.

[14] Neither are we able to control the outside effects such as air resistance and wind. However, this is a separate issue leading us to the question of the role of the isolation or non-isolation of the system.

than 'incomplete specification'. However, for an objectivist such terminology would be regarded as misleading.

We now have the problem of ***defining*** probability and there are two ways in which this is done in the objective context.

The Relative Frequency Interpretation. In this tradition the probabilistic properties of a system are ***defined*** by considering the results of a large number of macroscopically identical experiments on the system. For von Mises [14, p. 29] this large number of operations or events is the ***collective***; the probability of a particular outcome is then defined as the limit of the ***relative frequency*** of this outcome in the collective, as the size of the collective increases to infinity. Probability then is, in some sense, not a property of a single experiment, but of the collective of experiments.

The Propensity Interpretation. For Popper [15, 16] (see also [13]) probability is a ***latent propensity*** of an individual system experimented on in a specified way. In keeping with his general views on the philosophy of science, the probability of the outcome of an experiment is something about which one forms a hypothesis, which is then tested by repeated experiments.

3.2 The Subjective View

According to this view the probability that someone assigns to an event is a measure of her/his degree of belief in the outcome. For some people the adoption of the viewpoint is a liberating experience. As E. T. Jaynes writes [17, p. 268]:

> As soon as we recognize that probabilities do not describe reality – only our information about reality – the gates are wide open to the optimal solution of problems of reasoning from that information.

Again we shall make two subdivisions of this viewpoint.

Degrees of Belief Interpretation. This view, which is usually associated with names of Ramsey and de Finetti, allows one to hold any coherent set of beliefs about the probabilities of the outcomes of events. Coherence is described using the ***Dutch book arguments*** which concern willingness to bet. The constraints provided by these arguments prevents one holding beliefs which would mean that one inevitably looses money and leads to probabilities which satisfy the usual rules of the probability calculus. I agree with Hobson [11, p. 33] that this type of subjectivism has no relevance for the physical sciences and it will not feature any further in our discussion.

Rational Belief Interpretation. This point of view, which is sometimes called 'objective Baysianism' has already been introduced by a quote from Jaynes. Probability is still a question of belief, but it is constrained to be rational belief, not only in the Dutch book sense, but by having taken into account in a systematic way, all the evidence available. The principle exponent of this point of view is E.

T. Jaynes.[15] His method, now usually called the ***maximum entropy method***, will be discussed in relation to statistical mechanics below. It has, however, been applied by him in other situations as well. In it simplest form, when no information on which to base our probabilities is available, the rational choice is in accord with the Keynesian ***principle of indifference*** [19]. However, the application of this principle is not straightforward. Although it leads to the choice of a uniform distribution, the variable with respect to which the distribution is uniform is not always obvious. Jaynes resolution of this problem is to argue that, if probability is to be assigned according to our state of knowledge of the system, then ***it must be assigned in the same way to equivalent problems***. The probability assignment must be invariant under all transformations between equivalent problems. He illustrates his method by proposing a solution to the Bertrand chord problem, [18, p. 131], showing that his solution is supported by a relative frequency test?

4 Ergodic Theory

Consider again a Hamiltonian system with configuration vector $\mathbf{x}$ and momentum vector $\mathbf{p}$. Then the time evolution of the system is given by vector $(\mathbf{x}(t), \mathbf{p}(t))$ in Γ, moving according to the Hamiltonian flow. Given a measure density function $\mu(\mathbf{x}, \mathbf{p}, t)$, the Hamiltonian flow is measure-preserving if $\mu(\mathbf{x}, \mathbf{p}, t)$ satisfies Liouville's equation. This will be the case for the uniform volume measure. So if γ is a subset of Γ, which is invariant under the Hamiltonian flow and of finite volume $M(\gamma)$ we can define the time-independent normalized measure function $\mu(\mathbf{x}, \mathbf{p}) = 1/M(\gamma)$, for $(\mathbf{x}, \mathbf{p}) \in \gamma$, and zero for $(\mathbf{x}, \mathbf{p}) \notin \gamma$. This mechanical system is related to a thermodynamic system through correspondences between thermodynamic quantities $\{Q_{\mathrm{T}}\}$ and mechanical phase functions $\{Q(\mathbf{x}, \mathbf{p})\}$.

The starting point for ergodic theory is to argue that Q_{T} is equal to the average $\widetilde{Q}(\mathbf{x}_0, \mathbf{p}_0, \tau)$ of Q over a period of time τ, computed along the path of the system from $(\mathbf{x}_0, \mathbf{p}_0)$. It is assumed that τ is long with respect to the microscopic correlation time, the relaxation time of macroscopic variables and the time taken to destroy purely local constants of motion. From this it is argued that the result of a measurement is effectively the infinite time average obtained in the limit $\tau \to \infty$.[16] For this to be useful it is necessary to establish that this limit exists and that it is independent of $(\mathbf{x}_0, \mathbf{p}_0)$. It was shown by Birkhoff [20] that $\lim_{\tau \to \infty} \widetilde{Q}(\mathbf{x}_0, \mathbf{p}_0, \tau) = \widehat{Q}(\mathbf{x}_0, \mathbf{p}_0)$ exists almost everywhere in γ; that is except possibly for a set of μ-measure zero. From this it follows (see e.g. [12]) that $\widehat{Q}$ is a constant of motion almost everywhere in γ. Now let $\overline{Q}$ be the average of Q over γ with respect to μ. It also follows from Birkhoff's theorem that $\overline{Q} = \overline{\widehat{Q}} = \widehat{\overline{Q}}$ and it is clear that, if $\widehat{Q}$ is a constant almost everywhere in γ, $\overline{\widehat{Q}} = \widehat{Q}$ and $\widehat{Q} = \overline{Q}$ holds almost everywhere in γ. If this is the case then $\widehat{Q}$ is a constant almost everywhere in γ. ***If $\widehat{Q} = \overline{Q}$ holds almost everywhere in γ, for all phase functions integrable over γ, then the system is said to be ergodic.*** Thus for ergodic system we can (almost) legitimately identify Q_{T} with $\overline{Q}$.

[15] For his collected papers until the date of its publication see [18]. For convenience all references to Jaynes' work will be made to this collection rather than to the original source of the paper.

[16] The obvious problem with this is that, if it were true, we should never be able to make measurements on non-equilibrium systems [1, p. 176].

It would be tempting to suppose that ergodicity has established the connection between thermodynamic quantities, defined as time averages, and phase averages, without the need to interpret the measure density function μ. We do, however, have the problem of the set of μ-measure zero. To know that this set can be neglected we must know that a measurement is never (or hardly ever) made starting at one of its points. This brings us back to assuming some sort of probabalistic interpretation for μ. We have not escaped the statistics in statistical mechanics.

Given that ergodic theory is not an escape from probability, it is still worth considering how, at least for objectivists, it can be used as a justification of the probability measure chosen. The original ***ergodic hypothesis***[17] assumed that the path of the system passed through every point of γ. It is clear both that this would be sufficient to establish ergodicity and also that it cannot be true [12]. The alternative ***quasi-ergodic hypothesis*** that the path passes arbitrarily close to every point of γ has not proved sufficient to establish ergodicity, although it is necessary.

There is, however, a condition, both necessary and sufficient, which is intuitively somewhat similar to the quasi-ergodic hypothesis. To prove the necessity of the latter we would assume that, given a particular path of the system, there exists a point in γ which has a neighbourhood not containing any points of the path. This is clearly impossible for an ergodic system since we could alter the phase average, without changing the time average, by changing the value of the phase function in the neighbourhood. The even stronger assumption that γ can be decomposed into two subsets of non-zero measure, invariant under the flow, is clearly inconsistent with ergodicity. ***Metric transitivity***, which is defined as the negation of this assumption, is thus necessary for ergodicity and it is not difficult to see that it is also sufficient.

5 Equilibrium Statistical Mechanics

This works very well; supporting, among other things, the enormous development, since the early 1970's, in the theory of phase transitions. One reason for this success is that, in spite of unresolved problems about the foundations, the superstructure is based on a few agreed propositions. Firstly on the fact that equilibrium corresponds to having a probability density function ρ which is not an explicit function of time and secondly on the form for ρ which should be used in given sets of physical circumstances.

The way that, such sets of circumstances are determined and interpreted by the subjectivists will be discussed below. We should, however, note that if the energy, given by the value of the Hamiltonian, is the only isolating constant of motion,[18] then there is general agreement that the appropriate probability density function is the one obtained by applying equal probabilities to the points of an accessible region of phase space. This leads to the microcanonical distribution and the simplest way to derive it [21] is to take the invariant set γ, defined in Sec. 4, to be the shell $E < H(\mathbf{x},\mathbf{p}) < E + \Delta E$. The distribution over the energy surface Σ_E is then induced in the limit $\Delta E \to 0$. From this the canonical distribution can be derived

[17] Usually attributed to Boltzmann, but see the translator's introduction to [3].

[18] An integral of the equations of motion does not necessarily define a surface in Γ. Only those which do can be used to reduce the dimension of a set invariant under the flow. Such an integral is called an *isolating constant of motion*.

using either the central limit theorem [21] or the method of steepest descents [22]. Both these procedures are asymptotically valid for systems with a large number of microsystems. Subjectivists do not need this limit, although "many quantities of interest are highly predictable when N is large" [11, p. 70].

So, although it is viewed in different ways according to one's view of probability, the starting problem for equilibrium statistical mechanics amounts to having a means of justifying the use of the uniform distribution over an energy shell. We see that ergodic theory will 'almost' give us such a justification if the Hamiltonian is the only isolating constant of motion, since then we might expect the energy surface to be metrically transitive. However, the problem of proving the non-existence of additional isolating constants of motion is, in general, very difficult and when they exist the form of thermodynamics differs significantly from the standard form [23, 24]. To whom is this important? Not, I think, to those like Tolman [10], who regard the object of statistical mechanics not as a single system, but an *ensemble of systems.* Choosing to model the ensemble by the microcanonical distribution is simply to include in the ensemble systems with all values of the other unknown isolating constants of motion.

As we saw above, Boltzmann aimed to justify the Maxell distribution by showing that it arose as the stationary solution to his transport equation which is attained in the limit $t \to \infty$. There are two substantial programmes which follow a similar route in aiming to show that equilibrium arises in the long-time limit from non-equilibrium situations. These are that of the Brussels-Austin School [25, 26], and that using the maximum entropy method [11, 17, 18].[19] The work of the Brussels-Austin School is most appropriately considered in the context of non-equilibrium theory in Sec. 6. However, the maximum entropy method has a form specifically for equilibrium and this we shall now discuss.

Jaynes [18, p. 416] prefers to refer to his method as "predictive statistical mechanics" and he goes on to say that:

> [It] is not a physical theory, but a form of statistical inference....instead of seeking the unattainable [it] asks a more modest question: "Given the partial information that we do in fact have, what are the best predictions we can make of observable phenomena?"

So the fact that there are *unknown* constants of motions is irrelevant, since our only task is to make predictions based on what we know.[20]

We now compare the maximum entropy formulation with a standard, objectivist, formulation for a simple problem. We consider a system with discrete energy levels $\{E_1, E_2, \ldots, E_n\}$. Then questions are posed in the following ways:

(i) *In the maximum entropy formulation:* What is the *best* probability distribution for the random variable E, the energy of the system, based on the information available to us?

(ii) *In an objectivist formulation:* Given the physical environment of the system (whether it is isolated, or in contact, in some way with its exterior), what is *the* probability distribution for E?

[19] These two approaches are discussed and compared by Dougherty [27, 28].

[20] As we see below, he argues in a similar way in relation to unknown degrees of freedom, when he discusses the non-objective nature of entropy.

For Jaynes the key to the problem is the idea of ***uncertainty***. Given an appropriate measure of uncertainty, if we choose the probability distribution which maximizes the uncertainty relative to the available information then this will be the ***best*** probability distribution because it assumes as little as possible. He shows [18, p. 16] that the unique measure of uncertainty, which satisfies some reasonable mathematical properties, is Shannon's ***information entropy***

$$S_I(p_i) = -\sum_{i=1}^{n} p_i \ln(p_i),$$

[30]. The information entropy S_I is then related to the thermodynamic entropy S_T by $S_T = k_B\{S_I\}_{Max}$. Consider the following two cases:

We know nothing about the state of the system, other than the number of energy levels.

(a) Since the system has no dynamics there is no way of 'deriving' the probability distribution. However, an ***objectivist*** will believe that there *is* a probability associated with an experiment to determine its state. In both versions of objectivism repeated experiments will be made. In the case of a relative-frequentist this will serve to define the probability; someone who holds a propensity view will have formed a hypothesis, the most reasonable being that $p_i = 1/n$, and the sequence of experiments will be used to see if the hypothesis is falsified.

(b) Using the ***maximum entropy method*** we maximize $S_I(p_i)$ subject only to the condition $p_1 + p_2 \cdots + p_n = 1$ to give the same result as that hypothesized by the objectivists.

This is the ***uniform distribution***.[21]

For the canonical distribution the objective and subjective statements of the problem differ.

(a) The objective statement here has a thermodynamic content. The system is taken to be in a heat-bath at temperature T, which is the conjugate variable to the energy E. Then the most elegant way to derive the required results in this case is, as for the Hamiltonian system described above, to use the central limit theorem [29].

(b) For the maximum entropy method the equivalent situation is to know the expectation value $\langle E\rangle$ of the energy. Then $S_I(p_i)$ is maximized subject to the normalization condition and $p_1E_1 + p_2E_2 + \cdots + p_nE_n = \langle E\rangle$ to give

$$p_i = \frac{\exp(-E_i\lambda)}{Z(\lambda)} \quad \text{where} \quad Z(\lambda) = \sum_{i=1}^{n} \exp(-E_i\lambda), \quad \langle E\rangle = -\frac{\mathrm{d}\ln(Z)}{\mathrm{d}\lambda}.$$

There are a number of problems with Jaynes' method which can be discussed in reference to this example. Two of these are:

[21] It is similar to the microcanonical distribution, which is the uniform distribution over the degenerate states corresponding to an energy level in which the system is known to be.

A conceptual problem concerns the status of entropy. It can be expressed in the following way:

- The object of interest for statistical mechanics is a system $\mathcal{O} = \{\mathcal{O}_{\rm M}, \mathcal{O}_{\rm T}\}$, where $\mathcal{O}_{\rm M}$ denotes the qualities at the micro (atomic) level and $\mathcal{O}_{\rm T}$ denotes the qualities at the macro (thermodynamic) level.
- About such a system we have a certain amount of information $\mathcal{I} = \{\mathcal{I}_{\rm M}, \mathcal{I}_{\rm T}\}$.
- For such a system we devise a model $\mathcal{M} = \{\mathcal{M}_{\rm M}, \mathcal{M}_{\rm T}\}$.

What is it that has entropy? There would probably be agreement that the entropy $S(\mathcal{O}_{\rm M})$ is not well-defined, but does the entropy $S(\mathcal{O})$ exist? Entropy is defined in terms of a probability distribution so if you believe that the distribution is an objective property of the system (including its environment) there is no problem in saying that $S(\mathcal{O})$ exists. Jaynes would deny this. For him the entropy is $S(\mathcal{I})$. He argues that you can never know what degrees of freedom a system has. You may, for example, have neglected internal degrees of freedom within your molecules which would make a contribution to the entropy. So entropy is not an 'objective' property of the system. It is 'subjective' in the sense that it is a function of the knowledge which you, the subject, have. The counter argument would go something like this. Yes, but what you are calculating is $S(\mathcal{M})$ the entropy of your model, which as long as you have carried out an exact calculation *is* the entropy of the model, however good or bad it is, for the system you are considering. The Sackur-Tetrode equation gives the correct entropy of a perfect gas in spite of the fact that perfect gases do not exist. The relation between $S(\mathcal{M})$ and $S(\mathcal{O})$ is the same as between any two other theoretical and physical quantities and doesn't lead to the rejection of the existence of $S(\mathcal{O})$.

A mathematical problem associated with Jaynes' method was first raised by Friedman and Shimony [31]. Consider the system, described above, with energy spectrum $\{E_1, \ldots, E_n\}$ and suppose and that we are first given the background datum $\mathcal{D}_0$, that contains no information apart from its structure (the number of states). Then, as we saw above, from the maximum entropy principle, the appropriate distribution is the uniform distribution $\text{Prob}[E_j|\mathcal{D}_0] = 1/n$. Suppose now the energy E is measured and let the datum be $\langle E\rangle = U$, where U is given. Referring to this new piece of datum as $\mathcal{D}_1$ and using the maximum entropy principle we now have the canonical distribution $\text{Prob}[E_j|\mathcal{D}_0 \,\text{and}\, \mathcal{D}_1] = \exp(-E_j\beta)/Z(\beta)$. Now according to the usual formula for conditional probabilities (Bayes' Theorem)

$$\text{Prob}[E_j|\mathcal{D}_0] = \sum_{\mathcal{D}_1} \text{Prob}[E_j|\mathcal{D}_0 \,\text{and}\, \mathcal{D}_1]\text{Prob}[\mathcal{D}_1|\mathcal{D}_0]$$

and since $\mathcal{D}_1$ varies over all values of β it can be supposed to have a probability density function $p(\beta)$, giving

$$\frac{1}{n} = \int \mathrm{d}\beta p(\beta)\frac{\exp(-E_j\beta)}{Z(\beta)}, \qquad 1 \le j \le n.$$

But for $j = 1$ and all $p(\beta)$, except $p(\beta) = \delta^{\rm D}(\beta)$

$$\int \mathrm{d}\beta p(\beta)\frac{\exp(-E_1\beta)}{Z(\beta)} > \int \mathrm{d}\beta p(\beta)\frac{\exp(-E_1\beta)}{n\exp(-E_1\beta)} = \frac{1}{n}.$$

This problem has generated a lot of discussion (see [32]). The response by Jaynes [18, p. 250] was that "if $\mathcal{D}_1$ is a statement about a probability distribution on the sample space $\Sigma = \{E_1, \ldots, E_n\}$, then it can be used as a constraint when maximizing entropy but not as conditioning statement in Bayes' theorem, since it is not a statement about an event in Σ. On the other hand, if $\mathcal{D}_1$ is a statement defining an event in the sample space Σ^m of m trials, then the converse is the case".

6 Non-Equilibrium Statistical Mechanics

In the final chapter of his seminal work on the philosophical foundations of statistical mechanics Sklar [1] begins a summing up of the current state of the area by stating that, in his opinion, "most important questions still remain unanswered in very fundamental and important ways". Although, as we have seen, this is to some extent true for equilibrium it is more evidently the case for non-equilibrium.

It seems to be the case that the attempt to remove the statistics from statistical mechanics (via ergodic theory and associated dynamic analysis) is now at a dead-end. So when we are considering the way a system behaves over time the "evolution we describe ... will be that of a probability distribution over microstates of systems compatible with the macroconstraints defining the systems of interest" [1, p. 261]. There, therefore, remains the question of the interpretation of probability. Or, at least, whether you want to embrace a subjective view of probability. Because, as we shall see, if you do that you will be able to develop a type of solution to the problem of entropy increase and the evolution to equilibrium which would not make sense to an objectivist. On the other hand most of the approaches to this problem proposed by objectivists could be regarded as 'interpretation-free'.[22]

One group who would probably disagree with Sklar as to the unresolved nature of the problem of irreversiblity are those like Lebowitz [33] and Bricmont [34], who believe that the problem was solved in a satisfactory way by Boltzmann, and that current problems are caused by the fact that he has been misunderstood. The 'successful' explanation is based on the implementation of the procedure[23] for defining macrostates by dividing the phase space into small cells. This *course graining* approach works very well, in the sense that it gives a clear (possible) physical insight into the mechanism at work in irreversiblity. The usual objection is to the rather arbitrary nature of the course graining procedure. This is acknowledged by Lebowitz who says that while "this specification of the macroscopic state clearly contains some arbitrariness, this need not concern us too much, since all the statements we are going to make about the evolution of [the macrostate] are independent of the precise definition as long as there is a large separation between the macro and microscales" [33, pp. 33–34].

Lebowitz's article elicited a number of letters in Physics Today, two of which are of particular interest since they represent the main competing schools in non-equilibrium theory. The first, from Barnum *et al.* [36] criticizes Boltzmann's ideas from the perspective of "Shannon's statistical information and Edwin Jaynes' principle of maximum entropy". The criticism here, as I understand it, is not so much

[22] This is a view I expressed [12] with regard to the work of Progogine. It was subsequently endorsed by Dougherty [27].

[23] More fully developed by Paul and Tatiana Ehrenfest [35].

of course-graining *per se* as of the underlying philosophy. In fact Jaynes is on the whole favourable to Boltzmann's approach, giving an account of it, together with the comment that, in "Boltzmann's method of most probable distribution, we have already the essential mathematical content of the principle of maximum entropy" [18, p. 227]. However, this kind of approach is not the way Jaynes seems to prefer to describe increase of entropy. The following is the account given in [18, p. 27].

6.1 A Subjectivist Approach to Non-Equilibrium

Entropy is a measure of uncertainty or lack of information. As time passes our information about the system becomes out of date. There is a loss of information, which is an increase in uncertainty (entropy). This perception is realized in the following way. Suppose we have a set $\{\Omega_1(t),\ldots,\Omega_m(t)\}$ of time-dependent observables related respectively to the phase functions $\{\omega_1(\mathbf{x},\mathbf{p};t),\ldots,\omega_m(\mathbf{x},\mathbf{p};t)\}$ by

$$\Omega_j(t) = \langle\omega_j(\mathbf{x},\mathbf{p};t)\rangle = \int_\Gamma \rho(\mathbf{x},\mathbf{p};t)\omega_j(\mathbf{x},\mathbf{p};t)\mathrm{d}\Gamma.$$

Measurements are made of these observables at the time t_0 with the results $\{\overline{\Omega}_1(t_0),\ldots,\overline{\Omega}_m(t_0)\}$. The probability density function $\rho(\mathbf{x},\mathbf{p};t_0) = \rho_0(\mathbf{x},\mathbf{p};t_0)$ is the one which maximizes

$$S(\rho(t_0)) = -k_\mathrm{B}\int_\Gamma \rho(\mathbf{x},\mathbf{p};t_0)\ln\{\rho(\mathbf{x},\mathbf{p};t_0)\}\mathrm{d}\Gamma,$$

subject to the constraints

$$\overline{\Omega}_j(t_0) = \int_\Gamma \rho(\mathbf{x},\mathbf{p};t_0)\omega_j(\mathbf{x},\mathbf{p};t_0)\mathrm{d}\Gamma.$$

The probability density function evolves according to Liouville's equation and at a later time t is given by $\rho_0(\mathbf{x},\mathbf{p};t)$. According to our state of knowledge our best predictions for the observables at time t are now given by

$$\Omega_j(t) = \int_\Gamma \rho_0(\mathbf{x},\mathbf{p};t)\omega_j(\mathbf{x},\mathbf{p};t)\mathrm{d}\Gamma.$$

Using these predicted values as new constraints we derive a new probability density function $\rho(\mathbf{x},\mathbf{p};t)$ which maximizes $S(\rho(t))$. It is clear that $S(\rho(t_0)) = S(\rho_0(t_0)) = S(\rho_0(t)) \leq S(\rho(t))$. This approach, even more clearly that does the equilibrium treatment, highlights the fact that entropy is to be regarded, not as an objective property of the system but as dependent upon our knowledge of the system. It is also somewhat more limited that the usual statement of the second law. This can be seen if we consider a number of instances of time later that t_0. Suppose $t_0 < t < t'$. Then using the analysis given above $S(\rho(t_0)) \leq S(\rho(t))$ and $S(\rho(t_0)) \leq S(\rho(t'))$, but we know nothing about the relative sizes of $S(\rho(t))$ and $S(\rho(t'))$. Entropy has not been shown to be monotonically increasing. This aspect of Jaynes' programme was discussed in detail by Lavis and Milligan [32].

6.2 An Objectivist Approach to Non-Equilibrium

The second letter responding to Lebowitz's article in Physics Today is from Driebe [37], a member of Progogine's Brussels-Austin group. His criticism of the Boltzmann/Lebowitz approach is more radical than that of Barnum *et al.* [36]. He makes two points of particular interest for the present discussion: (i) "Irreversibility is not to be found on the level of trajectories or wave-functions, but is instead manifest on the level of probability distributions". (ii) "Many degrees of freedom is not a necessary condition for irreversible behaviour. It is the chaotic dynamics, associated with positive Lyapunov exponents or Poincaré resonances, that causes the system to behave irreversibly". At first sight (i) appears to be simply a restatement of the quote from Sklar, given above, about the need to use, as our element of interest, the probability density function rather than the trajectory. However, I think something more fundamental is implied. Before discussing this question we give a very brief summary of the methods of Prigogine and co-workers.[24]

The subject of interest is the evolution of a set of observable macroscopic quantities, which are taken to be the expectation values $\langle Q_i(t)\rangle$ of phase functions $Q_i(\mathbf{x},\mathbf{p},t)$, $i = 1,2,\ldots$. Now phase functions corresponding to observables are functions of only a small number of variables,[25] so the probability density function contains a great deal of unwanted detail. The method is to show that, relative to any particular Q_i, the probability density function ρ can be split into two parts $\rho = \rho_1 + \rho_2$, with $\langle Q_i(t)\rangle = \langle Q_i(t)\rangle_1 + \langle Q_i(t)\rangle_2$, so that, the unwanted detail is in ρ_2 with $\langle Q_i(t)\rangle_2$ vanishing identically and $\langle Q_i(t)\rangle_1$ reproducing the unique equilibrium value, corresponding to the thermodynamic quantity, in the limit $t \to \infty$. This procedure can be seen as a "series of successive contractions of the description of a many-body system" [26, p. 689]. For it to work it is necessary that the system has a large number of degrees of freedom and that it satisfies some level of mixing. The latter would certainly be the case if it were a K-system, that is to say chaotic (possesses a positive Lyapunov exponent for almost all initial conditions), [9, p. 262]. This is the point made by Driebe [37].[26] Returning to his comment concerning trajectories and probabilities density functions; it is illuminating to see them in the context of remarks by Prigogine to the effect that we must "eliminate the notion of trajectory from our microscopic description. This actually corresponds to a realistic description: no measurement, no computation leads strictly to a point, to the consideration of a *unique* trajectory. We shall always face a *set* of trajectories", [38, p. 60].[27] I think what is being referred to here is the 'sensitivity to initial conditions' which is present in chaotic systems. This means that, even in principle, we cannot specify the initial conditions with sufficient accuracy to know that the evolution corresponds to the flow along a particular *single* trajectory. However, it seems to me, that there is a conceptual difference between that and the "elimination of the notion of a trajectory".

[24] For detailed accounts see Prigogine [25] or Balescu [26].

[25] This is the intuition underlying the use of the Boltzmann transport equation and the early terms in the Bogolyubov-Born-Green-Kirkwood-Yvon hierarchy.

[26] His reference to the irrelevance of "many degrees of freedom" is not, I think, a contradiction of Balescu [26, p. 689]. It is certainly the case that some systems with a few degrees of freedom behave irreversibly, but this doesn't mean that they are models for thermodynamic behaviour.

[27] The English translation of this passage is taken from [34].

7 Conclusions

We have seen that programmes for the foundation of statistical mechanics can be based both of subjective and objective views of probability.

The subjective programme, based on the maximum entropy principle of Jaynes, is coherent and mathematically complete. It does, however, have some drawbacks, to which we have referred. (i) Because of it reliance on data collecting it does not seem to be able to take into account non-quantitative information. (ii) Because the probability density function is a property both of the system and our knowledge of the system, thermodynamic quantities, most particularly the entropy, are also properties both of the system and our knowledge of the system. This second point is closely related to the lack of a clear distinction between systems and models of systems.

Objective programmes are much more technically difficult. Equilibrium calculations either rely on some kind of justification for the microcanonical distribution or are the consequence of showing that equilibrium is achieved as the long-time limit from non-equilibrium. The most developed programme for the latter is the work of the Brussels-Austin School which needs the system to be chaotic.

References

[1] L. Sklar, "Physics and Chance", Cambridge U.P., 1993.

[2] Y.M. Guttmann, "The Concept of Probability in Statistical Physics, Cambridge U. P., 1999.

[3] L. Boltzmann, "Lectures on Gas Theory", English translation by S. G. Brush 1964,California U. P., 1896.

[4] P.M. Harman, "The Natural Philosophy of James Clerk Maxwell", Cambridge U. P., 1998.

[5] S.G. Brush, "Kinetic Theory", Vol. 1, Pergamon, 1965.

[6] S.G. Brush, "Kinetic Theory", Vol. 2, Pergamon, 1966.

[7] C. Cercignani, "Ludwig Boltzmann: The Man Who Trusted Atoms", Oxford U. P., 1998.

[8] J. Loschmidt, Wiener Beri. **73**, 139, 1876; Wiener Beri. **75**, 67, 1877.

[9] E. Ott, "Chaos in Dynamical Systems", (C.U.P.).

[10] R.C. Tolman, "The Principles of Statistical Mechanics", Oxford U. P., 1938.

[11] A. Hobson, "Concepts in Statistical Mechanics", Gordon and Breach, 1971.

[12] D.A. Lavis, Brit. J. Phil. Sci. **28**, 255-279, 1977.

[13] D.A. Gillies, "An Objective Theory of Probability", Methuen, 1973.

[14] R. Von Mises, "Probability, Statistics and Truth", George, Allen and Unwin, 1957.

[15] K.R. Popper, Brit., J. Phil. Sci. **10**, 1959, 25–42.

[16] K.R. Popper, "The Logic of Scientific Discovery", Hutchinson, 1959.

[17] W.T. Grandy and P.W. Milonni (Editors), "Physics and Probability: Essays in Honour of E. T. Jaynes", Cambridge U. P., 1993.

[18] E.T. Jaynes, "Papers on Probability, Statistics and Statistical Physics", Edited by R. D. Rosenkratz, Reidel, 1983.

[19] J.M. Keynes, "A Treatise on Probability", Macmillan.

[20] G.D. Birkhoff, Proc. Nat. Ac. Sci. **17**, 1931, 656–660.

[21] A.I. Khinchin, "The Mathematical Foundations of Statistical Mechanics", Dover, 1949.

[22] R. Kubo, "Statistical Mechanics", North-Holland, 1965.

[23] R.M. Lewis, Arch. Rat. Mech. Anal.**5**, 1960, 355.

[24] H. Grad, Comm. Pure and App. Maths. **5**, 1952, 455–494.

[25] I. Prigogine, "Non-Equilibrium Statistical Mechanics", Interscience-Wiley.

[26] R. Balescu, "Equilibrium and Non-Equilibrium Statistical Mechanics", Wiley, 1975.

[27] J.P. Dougherty, Stud. Hist. Phil. Sci. **24**, 1993, 843–866.

[28] J.P. Dougherty, Phil. Trans. Roy. Soc. A, **346**, 1994, 259–305.

[29] A.I. Khinchin, "Analytical Foundations of Physical Statistics", Dover, 1961.

[30] C.E. Shannon, and W. Weaver, "The Mathematical Theory of Communication", Illinois U. P., 1964.

[31] J. Friedman and A.J. Shimony Stat. Phys. **3**, 1971, 381.

[32] D.A. Lavis and P.J. Milligan, Brit. J. Phil. Sci. **36**, 1985, 193–210.

[33] J.L. Lebowitz, Physics Today, September, 1993. 32–38.

[34] J. Bricmont, Physicalia Mag. **17**, 1995, 159–208.

[35] P.Ehrenfest, and T., "The Conceptual Foundations of the Statistical Approach in Mechanics", 1912; English translation Cornell U. P., 1959.

[36] H. Barnum, C.M. Caves, C. Fuchs and R. Schack, Physics Today, November 1994, 11–13.

[37] D.J. Driebe, Physics Today, November 1994, 13–15.

[38] I. Progogine, "Les Lois du Chaos, Flammarion, 1994.

The "Mass Boom": The Effect of the Expansion of the Universe on the Fundamental "Constants"

A. Alfonso-Faus
E.U.I.T. Aeronáutica
Plaza Cardenal Cisneros s/n, Madrid 28040 España
Email:aalfonso@euita.upm.es

1 Introduction

Perhaps the most analyzed case for the possibility of time-varying "constants" corresponds to Newton's gravitational constant G. Much work has been done in the past. A few names that have been involved in this research are Eddington [1], Dirac [2], Brans and Dicke [3], Hoyle and Narlikar [4], Narlikar [5], Canuto [6], Adams [7], Alfonso-Faus [8] etc. At present the general feeling is that there is no evidence for a changing G. But the vast majority of this work has been done leaving the speed of light constant. One exception is the work of Petit [9], Alfonso-Faus [10], Belinchón [11] and Belinchón and Alfonso-Faus [12] using the conservation equations to keep G/c^2 constant. Hence under this view, a time varying G implies a time varying speed of light c. We proceed here to further analyze this approach.

We put together four basic principles, Lorentz invariance and general covariance, Mach, Equivalence and causality. We then incorporate the experience we have in quantum processes to match the principles with this evidence. The result is a contracting quantum world, as $1/a(t)$, where $a(t)$ is the cosmic scale factor that we identify with the size of the Universe. Both effects, the expansion of the Universe and the contraction of the quantum world go together. The speed of light varies also as $1/a(t)$ and the gravitational constant as t^{-1}, the same time dependence that Dirac [2] proposed. Our theory is not of the type of a Large Number Hypothesis, as proposed by Dirac [2]. In fact we prove that there are dimensionless parameters that are very large or very small and that they are not time varying.

We revise Newton's laws in the light of these findings. We predict that there is a gravitational torque on all orbiting bodies, as is the case for the Moon. Next we solve the cosmological equations. We find the solution $a(t)^2 = t$ for the cosmic scale factor valid at all times, for the radiation as well as the matter dominated Universe. Here one factor $a(t)$ comes from the expansion of the Universe, and the other $a(t)$ factor comes from the contraction of the quantum world so that cosmological and quantum sizes relate in a linear way with time.

We determine a set of units consistent with the above findings. In this theory Planck's length and mass are constant. Since the speed of light is not constant Planck's time is not constant either: it varies as the cosmic scale factor. Then one

Frontiers of Fundamental Physics 4, Edited by Sidharth and Altaisky,
Kluwer Academic/Plenum Publishers, New York, 2001

can go much further back in time than the usual 10^{-43} seconds, up to about 10^{-104} seconds.

We find two different Planck´s constants, one for the LAB and another one for cosmology (gravity included) that have about the same value today (at least not different by many orders of magnitude). So we conjecture a Universe with a bimetric property, with two tics for each world: a constant quantum tic, of the order of 10^{-23} seconds, and a constant cosmological tic, of the order of 10^{-104} seconds, that have a ratio of about the constant number of particles in the Universe.

We find a chain of masses, all proportional to the expansion of the Universe, from the gravity quanta (Alfonso-Faus [18]) up to the mass of the Universe. The missing mass, the dark matter in the Universe, is identified with one empty link in this chain.

With this mass boom cosmological model we solve the most important problems of standard cosmology: fine-tuning, Planck's, lambda, flatness, horizon, magnetic monopoles, causality, entropy, age of stars versus age of the Universe. Today there is only one known solution to some of these problems: inflation [13, 14]. Here we present a model that is an alternative to the inflationary theories.

Finally we state some definite predictions and conclude that the cosmological model based on a *mass boom* concept is a valid complement to the standard big bang model.

2 Four Principles Together

In one of the most important principles in theoretical physics, relativity, there is covariance and, of course, Lorentz invariance. Allowing the gravitational constant G to vary with time requires the speed of light to vary with time too. This condition comes from the field equations and the application of self-similarity to them (Belinchón [11], Belinchón and Alfonso-Faus [12]) and it is as follows

$$\frac{G}{c^2} = const. \tag{1}$$

Condition (1) has been pointed out elsewhere [9, 10].The present state of the art on a possible time variation for G, i.e. the experimental results concluding that G is constant, has to be modified to consider that both, G and c, may be varying with time.

The relativistic Einstein-Lorentz transformations should be invariant with respect to time, i.e., with respect to the cosmic scale factor of the Universe $a(t)$. This requires that the ratio v/c appearing in the transformations must be constant with time:

$$\frac{v}{c} = const. \tag{2}$$

By keeping v/c constant we do not have the braking of local Lorentz invariance that is the case presented by Moffat [15], Albrecht and Magueijo [16] and Barrow [17].

It is well known from general relativity that the speed v varies with the cosmic scale factor $a(t)$ as

$$v = \frac{const}{a(t)} \tag{3}$$

hence, in a certain system of units, one must have

$$c = \frac{1}{a(t)} \tag{4}$$

At the first tic of time t_1 we make $a(t_1) = 1$, the initial cosmic scale factor that will be Planck's length. At this time the speed of light is $c(t_1) = 1$.To preserve Lorentz invariance against the expansion of the Universe one must have the speed of light varying as in (4), and from (1) the gravitational constant, in a certain system of units, varying as

$$G = c^2 = \frac{1}{a(t)^2} \tag{5}$$

This choice of $G = c^2$ and the identification of the speed of light to Planck's constant will also be the result of taking Planck's mass and length as the units of mass and length.

The consideration that the relativistic energy of any particle mc^2 is proportional to the gravitational potential energy of this particle, relative to the rest of the Universe $GMm/a(t)$, gives a form of Mach's principle as

$$\kappa \frac{GMm}{a(t)} = mc^2 \tag{6}$$

where the dimensionless constant κ is of order one. This principle should be an important ingredient in general relativity. The principle of Equivalence is of fundamental importance too: the basis of general relativity. We will now prove that both principles, Mach and Equivalence together, imply causality at any time.

We impose that physical laws must not change with time, i.e., with the expansion of the Universe. The time dilation of a clock near a gravitating mass M of size any fraction of the mass of the observable Universe must not be affected by expansion, i.e., cosmological time. Since this dilation depends on v/c, v being in this case the gravitational equivalent $GMt/a(t)^2$, i.e., gravitational acceleration times time, then one has the following form for the principle of Equivalence:

$$\frac{GM}{a(t)^2} \frac{t}{c} = const. \tag{7}$$

Hence, Mach's principle (6) and the Equivalence principle (7) together imply $a(t) \propto ct$. Initial conditions impose that the proportionality constant be one. Then,

$$a(t) = ct \tag{8}$$

This is the causality principle, the expression that the rate of the expansion of the universe is of the order of the speed of light, ensuring causal connection always:

$$\frac{d[2a(t)]}{dt} = c \tag{9}$$

We conclude that the fulfillment of the principle of causality (8) identifies both principles, Mach's and Equivalence.

From the requirement of Lorentz invariance (4) and causality (8) we obtain in a certain system of units:

$$a(t) = t^{1/2} \tag{10}$$

It is interesting to note that the invariance of the space-time interval implies that the homogeneity of space (conservation of linear momentum) and the homogeneity of time (conservation of energy) go together. In this theory the total energy of the Universe is conserved: it is zero, as can be interpreted from Mach's principle (6). We now find the relation between length and time, units l_1 and t_1, from (4) and (10):

$$c = \frac{1}{a(t)} = \frac{1}{t^{1/2}} \tag{11}$$

Using t_0 as the age of the universe and c_o the present speed of light we get

$$l_1^2 = c_0^2 t_0 t_1 \tag{12}$$

This is a cosmological relation between the units of length and time consistent with the Lorentz invariance (4), Mach's principle (6) and the Equivalence principle (7) that together imply causality as in (8).

One can check if Planck's units of length and time can be considered to have a cosmological meaning. It is seen that this is not the case: one of them does not comply with the cosmological condition (12). It will later be seen that in this theory Planck's length and mass are constant but Planck's time is not constant: it varies as the cosmic expansion factor $a(t)$. Hence one can go back in time much further than the present Planck's time of 10^{-43} seconds. The usual Planck's mass and length, and a much smaller Planck's time, characterize the Planck's era in this theory: 10^{-104} seconds.

Now, from Mach's principle (6) and Lorentz invariance (4) we get in a certain system of units

$$\kappa G M = c \text{ i.e. } \kappa M c = 1 \tag{13}$$

This is the conservation of the maximum possible linear momentum in the Universe. Expressing the mass of the Universe M as the product of a number of particles N times their typical mass m, and using (4) we get

$$\kappa M c = \frac{\kappa N m}{a(t)} = 1 \tag{14}$$

There is strong evidence in favor of a constant number of particles N at the present time. Let us consider only baryons N_p and photons N_γ . The number of photons, as long as they are blackbody distributed as observed, is constant and of value:

$$N_\gamma \approx 8 \left(\frac{a(t)}{\lambda} \right)^3 = const. \approx 2 \cdot 10^{88} \tag{15}$$

This relation holds for blackbody radiation in a box of size $a(t)$ with photons of typical wavelength λ. For the Universe $\lambda \propto a(t)$, the stretching due to expansion. Since we know the age of the Universe that is close to $1.5 \cdot 10^{10}$ years we have the present value for the cosmic scale factor from (8) $a(t_0) = 1.4 \cdot 10^{28} cm$. The cosmic microwave background radiation has at present a typical wavelength of $0.1 cm$. Then we get for the typical wavelength:

$$\lambda \approx \frac{a(t)}{1.4 \cdot 10^{29}} \tag{16}$$

Hence, N_γ is constant. On the other hand the known deuterium abundance in the Universe, interpreted as primordial, implies that the ratio

$$\eta = \frac{N_p}{N_\gamma} \tag{17}$$

is a constant of the order of 10^{-9} or 10^{-10}. Hence, N_p is also constant as long as the radiation is blackbody, and has a value between 10^{78} and 10^{79}. Then one has from (14)

$$m_p = \frac{1}{\kappa} \frac{a(t)}{N_p} \tag{18}$$

The mass of the particles increases with time. One needs a mechanism for that. The best one is gravitation itself: the emission of negative gravity quanta (Alfonso-Faus [18]) explains this increase in the mass of the Universe, a mass BOOM.

Time variations are such that one has,

$$m_p c = \frac{1}{\kappa} \frac{1}{N_p} = const. \tag{19}$$

No more principles are invoked. We see that momentum is conserved, and that at a cosmological scale the total energy is conserved, it is zero always. Here we have found $mc = const.$, a characteristic of the condition for the conservation of momentum. Since m is proportional to $a(t)$ and the speed is $v = const/a(t)$ one has $mv = const$. The kinetic energy or the rest energy is not constant. They are affected by the expansion of the Universe. We can see that in he relativistic equation

$$m^2 c^4 = p^2 c^2 + m_0^2 c^4 \tag{20}$$

where the constancy of v/c, as required by Lorentz invariance, and the constancy of $m_0 c$ implies

$$p = const. \qquad mc = const. \tag{21}$$

Hence, the expansion of the Universe conserves momentum.

We can now check the effect of the expansion of the Universe on the gravitational forces between any two elementary masses m. The force is proportional to the product Gmm and from (5) and (18) one has $Gmm \propto c^2 m^2 = const$. Hence, there is no change in the processes known to occur in stars etc. as far as gravitation is concerned. Next we will impose that there is no change with time in the nuclear processes, despite the time varying condition for m and c.

3 Quantum Mechanics

A theory with time varying constants must keep an eye on both: cosmology and quantum physics. In the previous section we have taken care of cosmology. Now we analyze the quantum world and fit the merging between them.

Radioactive decay is used for time analysis. To keep its validity and uniformity in time one has to have the rates constant. All quantum rates are proportional to the time that light takes to cross the nuclear sizes. Then one must have

$$\tau = \frac{\hbar}{mc^2} = const. \tag{22}$$

where $\hbar$ is Planck's constant. This is the same as to say that the atomic clocks have a constant rate. But to keep the rates constant is not enough. One has to ensure that the nuclear processes are the same as time goes. Then the nuclear forces must remain constant with time, not affected by the expansion of the Universe. This is consistent with radioactive evidence and stellar evolution. Then,

$$\frac{mc^2}{r_p} = const. \tag{23}$$

where r_p is the size of a nucleon. Since we know the constancy of momentum mc we get from (22) and (23)

$$\hbar \propto c \propto r_p \tag{24}$$

These proportionalities imply that the quantum world goes as the speed of light. If the expansion of he Universe forces this speed to decrease as $c = 1/a(t)$ so does Planck's "constant" and the nuclear sizes. Expansion of the Universe and contraction of the quantum world go together. This has an important effect on the calculations of the cosmological red shifts. Spectra are measured in the lab, subject to contraction, so that the effective red shift has to take into account that the stretching of wavelengths is measured relative to the lab sizes. Then the effective expansion parameter must be

$$a(t)_{eff} \propto \frac{a(t)}{r_p} \propto a(t)^2 = t \tag{25}$$

Hence, the measured Hubble constant H_0 has to be related to the effective expansion:

$$H_0 = \frac{a(t)'_{eff}}{a(t)_{eff}} = \frac{1}{t} \tag{26}$$

The conclusion is that the Hubble time is exactly the age of the Universe in this theory. If we take a value for H_0 of 70 $Km/sec/Mpc$ we get an age for the Universe of $1.43 \cdot 10^{10}$ years, consistent with the age of he older stars.

Another way to see it is to consider the percentage of red shift when the source is moving at speed v, that is v/c. Then, since v is proportional to distance that varies as $a(t)$ the ratio v/c varies as $a(t)^2 = t$, which is the variation of the effective $a(t)$. Besides, the deceleration parameter q is now

$$q = -\frac{a(t)''_{eff}a(t)_{eff}}{\left(a(t)'_{eff}\right)^2} = 0 \tag{27}$$

4 Newton's Laws

We will reinterpret Newton's laws in the light of this theory. Using $mc = const$, m the relativistic mass now, one has for Newton's 2nd law:

$$\vec{f} = \frac{dm\vec{v}}{dt} = mc\frac{d\left(\frac{\vec{v}}{c}\right)}{t} \tag{28}$$

Newton's 1st law, with $f = 0$, gives $v/c = const$. The constancy of the velocity is seen here in terms of the speed of light. This ensures Lorentz invariance against the expansion of the Universe. The angular momentum equation is

$$\vec{r} \wedge \vec{f} = \vec{r} \wedge \frac{dm\vec{v}}{dt} = \frac{d(\vec{r} \wedge m\vec{v})}{dt} \tag{29}$$

Assuming gravity to produce a central acceleration then one has

$$\vec{r} \wedge \frac{d\vec{v}}{dt} \qquad \text{i.e.} \qquad \vec{r} \wedge \vec{v} = const. \tag{30}$$

that is Kepler's 2nd law, the law of areas. The angular mometum equation (29) is then

$$\vec{r} \wedge \vec{f} = \vec{r} \wedge \vec{v}\frac{dm}{dt} \tag{31}$$

which means that there is a gravitational force in the direction of the orbital velocities due to the mass BOOM:

$$\vec{f} = \vec{v}\frac{dm}{dt} \tag{32}$$

Hence there is a gravitational torque on the orbiting bodies. An example is the Moon, accepting that its recession is not all due to the tidal effect. This effect may explain by itself the expansion of the Universe, and is then related to the *mass boom* concept. For near circular orbits one has

$$v^2 = \frac{GM}{r} \qquad\qquad vr = const. \tag{33}$$

These relations give for the orbital speeds an sizes:

$$v \propto \frac{1}{a(t)} \qquad\qquad r \propto a(t) \tag{34}$$

These are the same time variations as the expansion rate of the Universe and its size.

From (34) we see that the orbital period goes as

$$T \propto \frac{r}{v} \propto a(t)^2 = t \tag{35}$$

Hence, in this theory the length of the year increases with cosmological time.

5 Conjecture: There Are Two Planck´s Constants In Nature

The contraction of the quantum world with $m_p c$ constant implies a Planck´s "constant"

$$\hbar = c = \frac{1}{a(t)} \tag{36}$$

This is a "local" or LAB result that gives a Hubble "constant" $H = 1/t$ with an acceptable age for the Universe and allows for nucleosynthesis, two important cosmological conditions. There is another different result from self-similarity relations (Belinchón and Alfonso-Faus [12]),

$$\hbar_g c = const. \tag{37}$$

The point is that all the relations between cosmological and quantum parameters, as presented by Sidharth [19] are consistent with the use of a gravitational wavelength λ_g, as introduced by Alfonso-Faus [18], instead of the Compton wavelength:

$$\sigma_g \approx \lambda_g^2 \approx \frac{Gm}{c^2} \cdot a(t) \tag{38}$$

The gravitational wavelength of any mass m is about the geometrical mean of its gravitational radius and the size of the Universe. For a fundamental particle $m = a(t)/N_p$ and using $G = c^2$ we get Eddington´s relation:

$$a(t) \approx \sqrt{N_p}\lambda_g \tag{39}$$

By equating the Compton wavelength to the gravitational size we get for the Planck´s "constant" now

$$\hbar_g = \frac{a(t)}{N_p^{3/2}} \tag{40}$$

Both Planck´s "constants" (36) and (40) have about the same value today, but they vary very differently with the expansion of the Universe. There is then a conjecture of a bimetric in the Universe: one local and one cosmological. The two tics are 10^{-23} seconds for the quantum world and 10^{-104} seconds for cosmology.

6 Einstein´s Cosmological Equations

The General Relativity field equations can be written as

$$G^{ij} = 8\pi \frac{G(t)}{c^4(t)} T^{ij} + \Lambda(t) g^{ij} \tag{41}$$

Taking a homogeneous and isotropic Universe as a fluid of mass density ρ and pressure p, one has for the energy-momentum tensor (e.g.Weinberg [20])

$$T^{ij} = (\rho c^2 + p) u^i u^j + p g^{ij} \tag{42}$$

and using the Robertson-Walker metric (e.g. Weinberg [20])

$$ds^2 = c(t)^2 - a(t)^2\left(\frac{dr^2}{1-kr^2} + r^2(d\theta^2 + \sin^2\theta d\phi^2)\right) \tag{43}$$

The two cosmological equations with a pressure term due to gravity quanta are:

$$\begin{aligned}\left(\frac{a(t)'}{a(t)}\right)^2 + 2\frac{a(t)''}{a(t)} + 8\pi\frac{G}{c^2}p + \frac{kc^2}{a(t)^2} &= \Lambda c^2\\ \left(\frac{a(t)'}{a(t)}\right)^2 - \frac{8\pi}{3}\frac{G}{c^2}\rho + \frac{kc^2}{a(t)^2} &= \frac{1}{3}\Lambda c^2\end{aligned} \tag{44}$$

In a certain system of units the cosmological equations have the solution $a(t) = t^{1/2}$ that is what we have obtained from first principles in (10). Substituting this solution into (44) we get the two relations

$$\begin{aligned}-\frac{1}{4t^2} + 8\pi\frac{G}{c^2}p + \frac{kc^2}{a(t)^2} &= \Lambda c^2\\ \frac{1}{4t^2} - \frac{8\pi}{3}\frac{G}{c^2}\rho + \frac{kc^2}{a(t)^2} &= \frac{1}{3}\Lambda c^2\end{aligned} \tag{45}$$

Now defining the dimensionless parameters as used in the literature Ω_p, Ω_m, Ω_k, Ω_Λ,

$$\begin{aligned}\Omega_p &= 8\pi\frac{G}{c^2}\frac{p}{H^2} & \Omega_m &= \frac{8\pi}{3}\frac{G}{c^2}\frac{\rho}{H^2}\\ \Omega_k &= \frac{kc^2}{a(t)^2H^2} & \Omega_\Lambda &= \frac{1}{3}\frac{\Lambda c^2}{H^2}\end{aligned} \tag{46}$$

we get from (45)

$$\begin{aligned}-1 + 3\Omega_p + \Omega_k &= 3\Omega_\Lambda\\ 1 - \Omega_m + \Omega_k &= \Omega_\Lambda\end{aligned} \tag{47}$$

Here we have the most important features of this cosmology. The gravity quanta equation of state is not known yet. If we assume a relation $p = w\rho c^2$ then we get a pressure parameter $\Omega_p = w\Omega_m$ and from (47)

$$\begin{aligned}\Omega_m &= \frac{4}{3(w+1)} + \frac{2}{3(w+1)}\Omega_k\\ \Omega_\Lambda &= \frac{3w-1}{3(w+1)} + \frac{3w+1}{3(w+1)}\Omega_k\end{aligned} \tag{48}$$

Using the most recent experimental values for the cosmological parameters ([21]) from MAXIMA, BOOMERANG, COBE AND SNeIA, $\Omega_m = 0.35$ $\Omega_\Lambda = 0.75$, we get from (48) $\Omega_k = 0.1$ and $w = 3$. The equation of state for the gravity quanta (Alfonso-Faus [18]) is then

$$p = 3\rho c^2 \tag{49}$$

If one uses $w = 1/3$ with $\Omega_k \approx 0$ the result is $\Omega_m \approx 1$ and $\Omega_\Lambda \approx 0$.

7 Dark Matter And The Mass Boom

The mass boom concept is an effect of the expansion of the Universe, or the other way around. The proposed mechanism is the emission of negative gravity quanta by all masses (Alfonso-Faus [18]). It implies that all masses very proportionally to the scale factor $a(t)$:

Gravity quanta

$$\frac{a(t)}{N_p^{3/2}} \approx 10^{-64} gr. \tag{50}$$

Equivalent mass of a CMBR photon

$$\frac{10^{29} a(t)}{N_p^{3/2}} \approx 10^{-35} gr. \tag{51}$$

Planck´s mass

$$\frac{a(t)}{N_p^{3/4}} \approx 10^{-5} gr. \tag{52}$$

Missing link

$$\frac{a(t)}{10^{29}} \approx 10^{27} gr. \tag{53}$$

Universe

$$a(t) \approx 10^{56} gr. \tag{54}$$

These five masses differ by a scaling factor of about 10^{30}. We conjecture that the dark matter in the Universe, the missing mass, is composed of millimeter black holes (about 10^{29} in total) that emit thermal radiation in equilibrium with the CMBR.

8 System Of Units

In quantum cosmology usually one works with Planck's units, defined as

$$\begin{aligned} l_* &= \left(\frac{G\hbar}{c^3}\right)^{1/2} \\ m_* &= \left(\frac{\hbar c}{G}\right)^{1/2} \\ t_* &= \left(\frac{G\hbar}{c^5}\right)^{1/2} \end{aligned} \tag{55}$$

In this theory we have $G = c^2$ and $\hbar \propto c$. Hence Planck's length and mass are constants and can be taken as units. By doing so we get $\hbar = c$ exactly, and Planck's density is a constant too

$$\rho_* \approx \frac{m_*}{l_*^3} = const. \approx 5 \cdot 10^{93} \frac{gr}{cm^3} \tag{56}$$

while Planck's time is not constant:

$$t_* = \left(\frac{G\hbar}{c^3}\right)^{1/2} \frac{1}{c} = \frac{1}{c} = a(t) \tag{57}$$

Since Planck's length and mass are constant, we can take them as units. Then, we have the relation (12) that contains two units and we have determined one of them. The unit of time is then

$$t_1 = \frac{G\hbar}{c^5 t} \approx 6 \cdot 10^{-105} \,\text{sec} \tag{58}$$

or equivalently using Mach's principle (6) and causality (8)

$$t_1 = \frac{\hbar}{m_p c^2} \frac{1}{\kappa N_p} \tag{59}$$

We see that the cosmological unit of time is just the basic nuclear time interval $\hbar/m_p c^2$ divided by the number of particles in the Universe. In this unit the present values for the age of the Universe and the cosmic expansion factor are

$$t_0 \approx 10^{122} \qquad a(t) \approx 10^{61} \tag{60}$$

Now, the first fluctuation in the Universe should have Planck's density (56). Since the density of the Universe in this theory varies as $M/a(t)^3 \approx 1/a(t)^2 = 1/t$, if we divide Planck's density by 10^{122} we get a density of about $10^{-29}\ gr/cm^3$ which is about the present density of the Universe. We have solved Planck's problem and have explained the present density of the Universe, originally created as a Planck's fluctuation. In standard cosmology this is not the case: an initial Planck's density implies a present density many orders of magnitude lower than observed, a very embarrassing problem for cosmology.

9 Problems Solved

The worse problem present in standard cosmology is fine-tuning. If we extrapolate back in time some properties that we observe today in the Universe. they become more and more precise in value so that close to the standard Planck's time, $10^{-43} sec$, the value obtained is extremely well defined, within one part in about 10^{60}. It comes to say that the Universe was highly improbable. The fine-tuning problem is solved in our model. Inflation solves some of the problems, and is the only theory that does so at present. We have here an alternative.

Planck's problem is related to the initial density out of a fluctuation. Planck's density (56) is about $5 \cdot 10^{93} gr/cm3$ and if we expand the Universe from then to the present time we have to divide this density by $a(t_0)^3 \approx 10^{183}$ that gives $5 \cdot 10^{-90} gr/cm^3$, to be compared with the presently observed value of about $10^{-29} gr/cm^3$. Due to the mass increase with time, in our model we have to divide only by $a(t_0)^2 \approx 10^{122}$ which gives a present density of $5 \cdot 10^{-29}$ very close to the observed one. Hence, we get rid of this problem.

The lambda problem is related to the value of the lambda "constant" needed initially in the Universe by particle theorists, a value of about 10^{122} to be compared with the value close to one observed today. The discrepancy has forced to speculate if the constant is exactly zero always. But we have here an alternative. According to our findings the lambda term varies inversely proportional to time. Hence the problem is solved because $a(t_0)^2 = t \approx 10^{122}$. The lambda "constant" is a function of time here and meets the requirement of particle physicists as well as the present observations.

The flatness problem in standard cosmology comes from the cosmological equations (44) when going back in time. The curvature term $kc^2/a(t)^2$ becomes negligible compared to the others. An almost flat time today implies an extremely flat Universe in the past. This is not so in our theory. Since $c = 1/a(t)$, the curvature term is always of the same order of magnitude as the rest of the terms in the equations. Flatness is then the same now and at the beginning of the Universe.

We have already addressed the horizon problem. With the relation $a(t) = ct$ one does not have this problem either: both the horizon and causality problems are solved and all parts in the Universe are causally connected always.

Another problem is related to the magnetic monopoles. Gauge field theories predict the existence of magnetic monopoles, giving a present monopole density far in excess of the observed density of the Universe. At the GUT epoch the horizon size is $2ct$ (with one monopole in it) , but since in our theory is $c = 1/a(t)$, the horizon size is then 10^{61} times larger. The initial density is much lower and no monopoles should be observed today.

Finally, the entropy problem. The present value is very high, usually taken as the number of photons in the Universe. With the units defined here the entropy as given by the photons is about 1 . As given by the gravity quanta the entropy is about 10^{29}, the number of millimeter blackholes we propose as constituents of the dark matter in the Universe.

10 Prediction

Our prediction refers to the Zeeman effect. The splitting of spectral lines of an atom under a magnetic field is proportional, from the point of view of energy, to c^2 in our theory. Since energies go as $mc^2 \propto c$ there is a factor of c in the splitting. If c varies with time the Zeeman splitting has a time variation of one part in $3 \cdot 10^{10}$ per year. So, one is able to falsify this theory by the possible time variations in the length of the year and the Zeeman effect.

11 Conclusions

We have constructed a cosmological model free from the main problems of the big-bang: fine-tuning, Planck's, lambda, flatness, horizon, magnetic monopoles, causality, entropy. It is an alternative to inflation.

The model has definite predictions: time variations in the length of the year and the Zeeman effect.

Two conjectures are advanced here: the existence of two Planck´s constants, one for the quantum world and one for cosmology. A bimetric with two different tics is foreseen. The second conjecture is the proposal to explain the dark matter in the Universe as composed of millimeter black holes in thermal equilibrium with the CMBR.

We have followed the method of preserving the basic principles in physics: Lorentz invariance, general covariance, Mach, Equivalence and causality. The model is unique and well defined. It is a complement to the standard big-bang model, allowing for time variations in some of the physical constants, and keeping gravitational and nuclear processes unchanged with time.

There is no acceleration in the expansion of the Universe: the conclusions from the SNeIa (Perlmutter [22]) may change as seen from the point of view of this model.

References

[1] A. Eddington, "The Expanding Universe", Cambridge Univ. Press, 1933.

[2] P.A.M. Dirac, Nature, 164, 637, 1937.

[3] C. Brans, & R.H. Dicke, Phys. Rev., 124, 925, 1961.

[4] F. Hoyle, & J.V. Narlikar, Proc. Roy. Soc. A270, 334, 1962.

[5] J.V. Narlikar, Nature, 242, 135, 1973.

[6] V. Canuto, et al., Phys. Rev. (Ser. 3), D16, 1643, 1977.

[7] P.J. Adams, Interntl. J. of Theor. Physics, 22, 421, 1983.

[8] A. Alfonso-Faus, Interntl. J. of Theor. Physics, 25. No. 3, 293, 1986.

[9] J.P. Petit, Mod. Phys. Let. 3, 16, 1733, 1988.

[10] A. Alfonso-Faus, Proc. E.R.E. Inst.de Astrof. de Canarias, 1987.

[11] J.A. Belinchón, Interntl. J. of Theor. Physics, 39 No 6, 1669, 2000.

[12] J.A. Belinchón. and A. Alfonso-Faus, Interntl. J. of Modern Phys. D

[13] A. Guth, Phys. Rev. D23 347, 1981.

[14] A. Linde, Phys. Lett. B 108, 1220, 1982.

[15] J. Moffat, Inter. J. of Phys. D, Vol. 2, No. 3, 1993, 351-365.

[16] A. Albrecht and J. Magueijo, Phys. Rev. D 59 043516, 1999.

[17] J. Barrow, Phys. Rev. D 59 043515, 1999.

[18] A. Alfonso-Faus, A., Physics Essays,Vol.12,No.4,December 1999.

[19] B.G. Sidharth, astro-ph/9904088, 2000 arXiv:Physics/0004002-3-4, 1999.

[20] S. Weinberg, "Gravitation and Cosmology", NY, John Wiley & Sons, 1992.

[21] A. Balbi, "Cosmology and Particle Physics",Proc.of the CAPP 2000 Conference, eds. J. Garcia-Bellido et al., Verbier, Switzerland, 2000 and Astro-ph/0011202

[22] S. Perlmutter, et al., Astro. Jour. 517,1999, 565-586.

Do Virtual Field Quanta Follow Geodesics?

Munawar Karim
Department of Physics
St. John Fisher College, Rochester, NY 14618, U.S.A.
Email:karim@sjfc.edu

1 Introduction

We address this question within the framework of quantum fields in curved space. We compute the effects of gravitational curvature on phenomena which appear because of quantum fluctuations of the vacuum field, specifically the Casimir force. The Casimir force is a purely quantum electrodynamic phenomenon, it is the result of zero point fluctutations of the vacuum field [1]. Experiments have provided tentative verification of the Casimir force [2]. Because of its nature the Casimir force provides a laboratory where concepts of quantum fields in curved space may be tried out both computationally and experimentally, as a precursor to attempts at quantizing gravity. In this note I will describe our derivation of an expression for the Casimir force in the Schwarzschild metric.

2 Casimir force in Schwarzschild metric

The Casimir force/area as derived by Casimir is:

$$C = -\frac{\pi^{\in}ƫ\text{J}}{\in\Delta/\lceil^{\Delta}} \quad (1)$$

where 'd' is the gap between a pair of plane, parallel conductors. This expression may be re-written as the vacuum expectation value of the stress tensor [4]:

$$\langle T\rangle^{\mu\nu} = (\frac{1}{4}\eta^{\mu\nu} - \hat{x}^{\mu}\hat{x}^{\nu})\frac{\pi^{2}\hbar c}{180d^{4}} \quad (2)$$

Identifying the Casimir force as an element of a stress or energy-momentum tensor opens up computational advantages; since in this form the expression is Lorentz invariant and divergence-free. It is renormalized by ignoring the infinite vacuum energy. Our computational approach is to start with the renormalized expression and to use the equivalence principle to re-write the stress tensor in terms of the elements of the Schwarzschild, instead of the Minkowski metric. Properties such as zero (now covariant) divergence are maintained as they must, since they imply local conservation of energy and momentum. We will show that, when written in this form:

$$\langle T\rangle^{\mu\nu}_{G} = (\frac{1}{4}g^{\mu\nu} - \hat{\chi}^{\mu}\hat{\chi}^{\nu})\frac{\pi^{2}\hbar c}{180d^{4}} \quad (3)$$

Frontiers of Fundamental Physics 4, Edited by Sidharth and Altaisky,
Kluwer Academic/Plenum Publishers, New York, 2001

it may be possible to actually measure the Casimir force in the Schwarzschild metric. We proceed by reducing the stress tensor elements to scalar components of forces. First we calculate a local tetrad by identifying the time-like four-velocity of the observer with respect to the Casimir plates, with the time-like Killing vector of the Schwarzschild metric. Having done this we construct a complete set of orthonormal tetrads. These are:

$$\begin{aligned}
\hat{\chi}_t^\mu &= [-\frac{1}{\sqrt{g_{0,0}}}, 0, 0, 0] \\
\hat{\chi}_r^\mu &= [0, \frac{1}{\sqrt{g_{1,1}}}, 0, 0] \\
\hat{\chi}_\theta^\mu &= [0, 0, \frac{1}{\sqrt{g_{2,2}}}, 0] \\
\hat{\chi}_\phi^\mu &= [0, 0, 0, \frac{1}{\sqrt{g_{3,3}}}]
\end{aligned} \tag{4}$$

The four-momenta are integrals of the energy-momentum tensor over a three-surface:

$$P^\mu = \int_\Sigma \langle T \rangle_G^{\mu\nu} d^3\Sigma_\nu \tag{5}$$

The three-surfaces $d^3\Sigma_\nu$ are:

$$\begin{aligned}
d^3\Sigma_1 &= \epsilon_{1,j,k,0} A^j B^k C^0 = \sqrt{g_{2,2}} d\theta \sqrt{g_{3,3}} d\phi \sqrt{-g_{0,0}} dt \hat{r} \\
d^3\Sigma_3 &= \epsilon_{3,j,k,0} A^j B^k C^0 = \sqrt{g_{1,1}} dr \sqrt{g_{2,2}} d\theta \sqrt{-g_{0,0}} dt \hat{\phi}
\end{aligned} \tag{6}$$

The forces are calculated from the four-momenta:

$$F^1 = \frac{dP^1}{dt} \text{ and } F^3 = \frac{dP^3}{dt} \tag{7}$$

The time variable needs to be modified; along the r- and $\phi-$ directions; it will be:

$$dt = \sqrt{\frac{g_{1,1}}{-g_{0,0}}} dr \text{ and } dt = \sqrt{\frac{g_{3,3}}{-g_{0,0}}} d\phi \tag{8}$$

The corresponding forces are:

$$F^1 = \frac{dP^1}{dt} = C\pi\mathcal{R}^{\in}(\infty - \frac{\alpha}{\in\nabla_\infty})\heartsuit \text{ and } \mathcal{F}^{\ni} = \frac{\lceil\mathcal{P}^{\ni}}{\lceil\sqcup} = C\pi\mathcal{R}^{\in}\frac{\infty}{\nabla_\infty^{\in}}(\infty + \frac{\alpha}{\in\nabla_\infty})\phi \tag{9}$$

where we have used r_1 to denote the location of the apparatus. Although we have used Schwarzschild polar coordinates, using other coordinates such as isotropic and harmonic, yield the same result, affirming its robustness. It can be seen that the general relativistic contributions appear as first order effects in $\frac{\alpha}{r}$. Although this appears unexpected, it is not really so. It may be understood by drawing an analogy with the time-delay of a light pulse, a first order effect in $\frac{\alpha}{r}$, between points in a Schwarzschild metric. In this example the points in question are located across the gap between the two plates. The time-delay between the plates is increased along

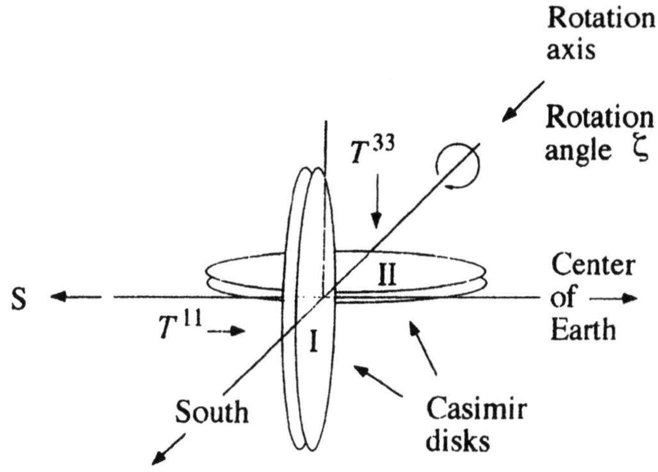

Figure 1

the radial direction (the plates appear further apart), less so in the tangent plane; as a consequence there is a decrease, relative to the flat-space value, in the Casimir force in the two directions. The general relativistic contribution can be extracted by measuring the difference between the radial and tangential elements of the stress tensor.

3 Explicit expressions

We observe that the general relativistic contribution can be extracted using the radial and one of the tangential components; we choose this to be the ϕ component. We go over to a local Cartesian frame, The x- and y- components of the force are:

$$\begin{aligned} F^x &= C\pi\mathcal{R}^{\in}[(\infty - \frac{\alpha}{\in\nabla_\infty})\rfloor\wr\int^{\in}\zeta + (\infty + \frac{\alpha}{\in\nabla_\infty})\int\rangle\backslash^{\in}\zeta) \\ F^y &= C\pi\mathcal{R}^{\in}[(\infty - \frac{\alpha}{\in\nabla_\infty})\int\rangle\backslash^{\in}\zeta + (\infty + \frac{\alpha}{\in\nabla_\infty})\rfloor\wr\int^{\in}\zeta) \end{aligned} \quad (10)$$

where ζ is the polar angle (see figure). There are two mutually orthogonal pairs of Casimir disks, each sensitive to only the r- or $\phi-$ component of the stress tensor. The apparatus is aligned north-south on Earth's surface and located on the equator $(\theta = 0)$. The two force components are then:

$$F^x = C\pi\mathcal{R}^{\in}[(\infty - \frac{\alpha}{\in\nabla_\infty}) \quad \dashv\backslash\lceil \quad \mathcal{F}^{\dagger} = C\pi\mathcal{R}^{\in}[(\infty + \frac{\alpha}{\in\nabla_\infty}) \quad (11)$$

Their difference gives the general relativistic contribution:

$$\frac{F^x - F^y}{\pi R^2} = -C\frac{\alpha}{\nabla_\infty}\rfloor\wr\int\in\zeta \equiv C_g\rfloor\wr\int\in\zeta \quad (12)$$

The Galactic center is the strongest Schwarzschild source affecting space-time in the vicinity of Earth. Using it and a point of observation on Earth's surface $\frac{\alpha}{r_1} \approx 6 \times 10^{-}7$. Thus

$$\frac{F^x - F^y}{\pi R^2} = \mathcal{C}_{\mathcal{G}} \rfloor \wr \int \in \zeta \approx \infty \prime^{-} \mathcal{G} \rfloor \wr \int \in \zeta \tag{13}$$

Or

$$\frac{1}{\mathcal{C}} \frac{F^x - F^y}{\pi R^2} \approx 10^{-} 6 cos 2\zeta \tag{14}$$

The Casimir force in Schwarzschild metric is then:

$$\mathcal{C}_{\mathcal{G}} = -\frac{\alpha}{\nabla_{\infty}} \mathcal{C} \tag{15}$$

This is our central result [3]. The Casimir force is smaller in the Schwarzschild metric.

4 Experiment

The apparatus consists of two pairs of Casimir disks. Springs hold apart each pair to counter the Casimir force, with a gap of 8μm. The disks are electrically insulated. A change in the Casimir force causes a change in the gap, which shows up as a change in the capacitance. The two pairs form two capacitors in an A.C bridge. The bridge detects an imbalance, or force difference, between the two disk pairs.

The apparatus is located at the equator, aligned north-south. As Earth rotates on its axis with respect to the Galactic center, the orientation of the disk pairs changes with a 12-hour period. The sensitivity of the apparatus can be estimated by calculating the noise from amplifiers and Brownian sources. The apparatus will be cooled to 77K. Using transistors with noise figures of $\approx 10^{-}9V/\sqrt{Hz}$, low-dissipation aluminum alloy 5056, an integration time of 10^5 secs is sufficient to detect a differential change in the Casimir force of 10^{-13}; more than sufficient to detect the expected signal of 10^{-6} using homodyne methods with a reference of frequency 1/12 hours.

5 Interpretation

The experiment measures the effect of gravitational curvature on zero point fluctutations of the vacuum field. A positive result will extend the interpretation of the energy-momentum tensor in the Einstein equation to include zero-point energy and in the process confirm that virtual field quanta follow geodesics. The decrease in the Casimir force may be interpreted as the momentum loss because of inelastic scattering of virtual photons off the classical gravitational field. Virtual photons transfer quanta to the gravitational continuum.

References

[1] H.G.B. Casimir, Proc. K. Ned. Akad. Wet.,**51**, 793, 1948.

[2] S. K. Lamoreaux, Phys. Rev. Lett. **78** (1), 5, 1997, U. Mohideen and A. Roy, Phys. Rev. Lett.,**81** (21), 4549, 1998.

[3] M. Karim and A.H. Bokhari, "Frontier Tests of QED and Physics of the Vacuum", Eds. E. Zavattini, D. Bakalov and C. Rizzo, Heron Press, Sofia, 1998, p. 405; M. Karim, A.H. Bokhari and B.J. Ahmedov, Class. Quantum Grav. **17**, 2459, 2000.

[4] L.S. Brown and G.J. Maclay, Phys. Rev. **184** (5), 1272, 1969.

Masses and fields in Microdynamics: a possible foundation for dynamic gravity

W.A. Hofer
Dept. of Physics and Astronomy, University College London
Gower Street, London WC1E 6BT, UK Email:w.hofer@ucl.ac.uk

1 In search of new foundations

A few years ago physicists thought that all the fundamental puzzles of their field were essentially solved. The experiments in high energy physics, designed to test the standard model, gave agreement between theoretical predictions and experimental findings to the n-th decimal, all the fundamental interactions seemed to be known, the full variety of atomic behaviour explained; and the whole history of our Universe appeared to be at the brink of final disclosure. The first cracks in this monumental edifice of quantum field theory and general relativity appeared on the level of atoms and, independently, on the level of galaxies. On the level of atoms it was the remarkable success of experimental methods in condensed matter physics. These methods, most prominently the scanning tunnelling microscope, allowed to study the behaviour of atoms in real time and real space. Henceforth, chemical processes could not only be mathematically described, they could actually be observed. Implicitly, this made a reinterpretation of quantum mechanics necessary. Because if I can study the behaviour of a *single* atom (which is by now routinely done), how can this experiment be the result of statistics? And if this atom is real, in what sense precisely is the atom not, which we treat in quantum mechanics? Exactly the same question can be asked in experiments with single photons or electrons. Also these experiments are by now routine in many laboratories around the world. The experimental situation allows only one answer: quantum mechanics is an algorithm and essentially incomplete. But then we may immediately ask: Is there a complete theory, and is there a reality behind the mathematical objects in quantum mechanics? Since quantum mechanics is the basis for quantum field theory, these questions are important for the whole of microphysics. Also, in principle, for high energy physics. So what, we may ask, is the reality behind a neutrino, a quark, a Higgs boson?

On the level of galaxies the standard model is based on the experimental fact of an isotropic background radiation in the range of 160 GHz. This radiation is equal to the radiation of a black body with $T = 2.726$ K [1]. It is thought to originate from the birth of our Universe in the "big bang", some 10 to 20 billion years ago. On closer scrutiny this model encountered several obstacles. The cosmic background radiation, for example, is constant in all directions. The fact is problematic, because the early universe could not be completely homogeneous. To reconcile facts with the theory, one either has to assume a varying velocity of light (in contrast with special relativity), or a peculiar way, how the universe initially expanded. Today, the most serious objection against its validity is motion of galactic mass. According to model calculations the observations can only be accounted for, if more than 90 percent of a galaxy's mass is unobservable [2, 3]. But this, in turn, contradicts the calculations of the upper limit of baryon mass in the big bang model. One either has to resort to ad hoc hypotheses about the qualities of this "dark matter", or concede, that the big bang model is flawed. But if it is fundamentally flawed, then the cosmic radiation background must have other origins. And this, then, could be the point of departure of a completely different theory about our Universe.

2 Alternative formulations

The problems sketched have only surfaced in the last ten to twenty years of the Twentieth century. Limiting our discussion to theories amending or extending quantum mechanics, these fall essentially into two categories: (i) Theories aimed at reconciling the microscopic with the macroscopic domain. (ii) Theories aimed at recovering a physical reality behind the mathematical formulations.

It is a constitutional feature of the theories type (i) that they accept the existing formulations. The aim is therefore not so much an extension of our knowledge, than limiting the application of existing mathematical expressions. In this sense, they provide a subset of the unrestricted original frameworks. Examples of this type are: The consistent histories approach due to Griffiths, Hartle, and Omnes [4, 5, 6]; the Ghirardi-Rimini-Weber model [7]; or the many-world concept [8]. None of these models develops a new physical reality beyond quantum mechanics, which could in principle be measured. In this sense, none of the models is suitable to answer the principal question: What is the physical reality of an atom, photon, electron?

The same deficiency probably afflicts all theories based on classical mechanics. Because the main addition, in quantum mechanics, is the uncertainty principle. Mathematically this is expressed by non-commuting variables. Since this condition inhibits any observation of systems with a higher degree of precision, it is theoretically necessary to postulate some hidden variables, which describe fundamental reality. Mechanics is not developed beyond point-like objects and density distributions. In particular the notion of a phase is not a mechanical concept. But as phases play a major role in field physics - and thus, generally, in microphysics - their omission is a severe, probably decisive, limitation. This is true before and apart from any analysis of the actual models used in quantum mechanics. In this sense Bohm's theory [9], if it is interpreted as a theory of real particles with real trajectories, pays the price that also the non-local connections, inherent in quantum mechanics, must be real. Even though it is therefore a theory of type (ii), it cannot be consistently formulated as such.

Another theory of type (ii), stochastic electrodynamics [10], focuses on electrodynamics and photons. Classical electrodynamics in this concept is extended by a zero-point field which carries the statistical uncertainty inherent in quantum mechanics. Even though there has recently been new interest in the theory, it is for all practical purposes restricted to photons. It thus has nothing to say, so far, about atoms, or electrons.

The most recent theory type (ii), microdynamics [11, 12, 13, 14], focuses on wave properties of matter. These waves are thought to be real. It could be shown that electrodynamics and quantum mechanics are limiting cases of the same general model of matter in microphysics. In this theory the reality behind quantum mechanics is the reality of fields and densities. Contrary to other theories, microdynamics also allows to develop a dynamic model of atoms [15].

In the initial concept, the model of hydrogen required an ad-hoc hypothesis: the existence of fields inside the atom, which are due to density oscillations of the proton. In this sense the density oscillations were postulated in addition to the electrostatic charge of the proton. It was only realized recently, that these oscillations may play a far more fundamental role and in fact be related to the very existence of electric charge. A line of research, we shall pursue in the rest of this paper. As shall be seen shortly, the concept can even be used to bridge the gap between electrodynamics and gravitation in microphysics. It thus appears to be at the bottom of what could well become a completely new field: gravity dynamics.

3 The nature of charge

Since the discovery of the electron by J. J. Thomson [16] the concept of electric charge has remained nearly unchanged. Apart from Lorentz' extended electron [17], or Abraham's electromagnetic electron [18], the charge of an electron remained a point like entity, in one way or another related to electron mass [19, 20]. In atomic nuclei we think of charge as a smeared out region of space, which is structured by the elementary constituents of nuclear particles, the quarks [21].

But experiments on the quantum Hall effect[22, 23], performed around 1980, suggested the existence of "fractional charge" of electrons. Although this effect has later been explained on the basis of standard theory [24], its implications are worth a more thorough analysis. Because it cannot be excluded that the same feature, fractional or even continuous charge, will show up in other experiments, especially since experimental practice more and more focuses on the properties of single particles. And in this case the conventional picture, which is based on discrete and unchangeable charge of particles, may soon prove too narrow a frame of reference. It seems therefore justified, at this point, to analyse the very nature of charge itself. A nature, which would reveal itself as an answer to the question: What is charge?

With this problem in mind, we reanalyse the fundamental equations of intrinsic particle properties [12]. The consequences of this analysis are developed in two directions. First, we determine the interface between mechanic and electromagnetic properties of matter, where we find that only one fundamental constant describes it: Planck's constant $\hbar$. And second, we compute the fields of interaction within a hydrogen atom, where we detect oscillations of the proton density of mass as their source. Finally, the implications of our results in view of unifying gravity and quantum theory are discussed and a new model of gravity waves derived, which is open to experimental tests.

4 The origin of dynamic charge

The intrinsic vector field $\mathbf{E}(\mathbf{r},t)$, the momentum density $\mathbf{p}(\mathbf{r},t)$, and the scalar field $\phi(\mathbf{r},t)$ of a particle are described by (see [12], Eq. (18)):

$$\mathbf{E}(\mathbf{r},t) = -\nabla \frac{1}{\bar{\sigma}}\,\phi(\mathbf{r},t) + \frac{1}{\bar{\sigma}}\frac{\partial}{\partial t}\,\mathbf{p}(\mathbf{r},t) \tag{1}$$

Here $\bar{\sigma}$ is a dimensional constant introduced for reasons of consistency. Rewriting the equation with the help of the definitions:

$$\beta := \frac{1}{\bar{\sigma}} \qquad \beta\phi(\mathbf{r},t) := \phi(\mathbf{r},t) \tag{2}$$

we obtain the classical equation for the electric field, where in place of a vector potential $\mathbf{A}(\mathbf{r},t)$ we have the momentum density $\mathbf{p}(\mathbf{r},t)$. This similarity, as already noticed, bears on the Lorentz gauge as an expression of the energy principle ([12] Eqs. (26) - (28)).

$$\mathbf{E}(\mathbf{r},t) = -\nabla\,\phi(\mathbf{r},t) + \beta\,\frac{\partial}{\partial t}\,\mathbf{p}(\mathbf{r},t) \tag{3}$$

Note that β describes the interface between dynamic and electromagnetic properties of the particle. Taking the gradient of (3) and using the continuity equation for $\mathbf{p}(\mathbf{r},t)$:

$$\nabla\,\mathbf{p}(\mathbf{r},t) + \frac{\partial}{\partial t}\rho(\mathbf{r},t) = 0 \tag{4}$$

where $\rho(\mathbf{r},t)$ is the density of mass, we get the Poisson equation with an additional term. And if we include the source equation for the electric field $\mathbf{E}(\mathbf{r},t)$:

$$\nabla\,\mathbf{E}(\mathbf{r},t) = \sigma(\mathbf{r},t), \tag{5}$$

with $\sigma(\mathbf{r},t)$ being the density of charge, ϵ set to 1 for convenience, we end up with the modified Poisson equation:

$$\Delta\phi(\mathbf{r},t) = -\underbrace{\sigma(\mathbf{r},t)}_{static\ charge} - \underbrace{\beta\,\frac{\partial^2}{\partial t^2}\rho(\mathbf{r},t)}_{dynamic\ charge} \tag{6}$$

The first term in (6) is the classical term in electrostatics. The second term does not have a classical analogue, it is an essentially novel source of the scalar field ϕ, its novelty is due to the fact, that no dynamic interpretation of the vector potential $\mathbf{A}(\mathbf{r},t)$ exists, whereas, in the current framework, $\mathbf{p}(\mathbf{r},t)$ has a dynamic meaning: that of momentum density.

To appreciate the importance of the new term, think of an aggregation of mass in a state of oscillation. In this case the second derivative of ρ is a periodic function, which is, by virtue of Eq. (6), equal to periodic charge. Then this dynamic charge gives rise to a periodic scalar field ϕ. This field appears as a field of charge in periodic oscillations: hence its name, dynamic charge. We demonstrate the implications of Eq. (6) on an easy example: the radial oscillations of a proton. The treatment is confined to monopole oscillations, although the results can easily be generalised to any multipole. Let a proton's radius be a function of time, so that $r_p = r_p(t)$ will be:

$$r_p(t) = R_p + d \cdot \sin \omega_H t \tag{7}$$

Here R_p is the original radius, d the oscillation amplitude, and ω_H its frequency. Then the volume of the proton V_p and, consequently, its density of mass ρ_p depend on time. In first order approximation we get:

$$\rho_p(t) = \frac{3M_p}{4\pi}\left(R_p + d \sin \omega_H t\right)^{-3} \approx \rho_0 \left(1 - x \sin \omega_H t\right) \qquad x := \frac{3d}{R_p} \tag{8}$$

The Poisson equation for the dynamic contribution to proton charge then reads:

$$\Delta\phi(\mathbf{r}, t) = -\beta x \rho_0 \omega_H^2 \sin \omega_H t \tag{9}$$

Integrating over the volume of the proton we find for the dynamic charge of the oscillating proton the expression:

$$q_D(t) = \int_{V_p} d^3 r \beta x \rho_0 \omega_H^2 \sin \omega_H t = \beta x M_p \omega_H^2 \sin \omega_H t \tag{10}$$

This charge gives rise to a periodic field within the hydrogen atom, as already analysed in some detail and in a slightly different context [15]. We shall turn to the calculation of a hydrogen's fields of interaction in the following sections. But in order to fully appreciate the meaning of the dynamic aspect it is necessary to digress at this point and to turn to the discussion of electromagnetic units.

5 Natural electromagnetic units

By virtue of the Poisson equation (6) dynamic charge must be dimensionally equal to static charge, which for a proton is + e. But since it is, in the current framework, based on dynamic variables, the choice of β also defines the interface between dynamic and electromagnetic units. From (10) we get, dimensionally:

$$[e] = [\beta][M_p \omega_H^2] \quad \Rightarrow \quad [\beta] = \left[\frac{e}{M_p \omega_H^2}\right] \tag{11}$$

The unit of β is therefore, in SI units:

$$[\beta] = C \cdot \frac{s^2}{kg} = C \cdot \frac{m^2}{J} \qquad [SI] \tag{12}$$

We define now the ***natural system of electromagnetic units*** by setting β equal to 1. Thus:

$$[\beta] := 1 \qquad \Rightarrow \qquad [C] = \frac{J}{m^2} \tag{13}$$

The unit of charge C is then energy per unit area of a surface. Why, it could be asked, should this definition make sense? Because, would be the answer, it is the only suitable definition, if electrostatic interactions are accomplished by photons. Suppose a $\delta^3(\mathbf{r} - \mathbf{r}')$ like region around $\mathbf{r}'$ is the origin of photons interacting with another $\delta^3(\mathbf{r} - \mathbf{r}'')$ like region around $\mathbf{r}''$. Then $\mathbf{r}'$ is the location of charge. Due to the geometry of the problem the interaction energy will decrease with the square of $|\mathbf{r}' - \mathbf{r}''|$. What remains constant, and thus characterises the charge at $\mathbf{r}'$, is only the interaction energy per surface unit. Thus the definition, which applies to all r^{-2} like interactions, also, in principle, to gravity.

Returning to the question of natural units, we find that all the other electromagnetic units follow straightforward from the fundamental equations [12]. However, if we analyse the units in Lorentz' force equation, we observe, at first glance, an inconsistency.

$$\mathbf{F}_L = q\,(\mathbf{E} + \mathbf{u} \times \mathbf{B}) \tag{14}$$

The unit on the left, Newton, is not equal to the unit on the right. As a first step to solve the problem we include the dielectric constant ϵ^{-1} in the equation, since this is the conventional definition of the electric field $\mathbf{E}$. Then we have:

$$[\mathbf{F}_L] = \frac{Nm}{m^2}\left(\frac{m^4}{N}\frac{N}{m^3} + \frac{m}{s}\frac{Ns}{m^4}\right) = N + N \cdot \frac{N}{m^4} \tag{15}$$

Interestingly, now the second term, which describes the magnetic forces, is wrong in the same manner, the first term was before we included the dielectric units. It seems thus, that the dimensional problem can be solved by a constant η, which is dimensionally equal to ϵ, and by rewriting the force equation (14) in the following manner:

$$\mathbf{F}_L = \frac{q}{\eta}\,(\mathbf{E} + \mathbf{u} \times \mathbf{B}) \qquad [\eta] = Nm^{-4} = Cm^{-3} = [\sigma] \tag{16}$$

The modification of (14) has an implicit meaning, which is worth being emphasise. It is common knowledge in special relativity, that electric and magnetic fields are only different aspects of a situation. They are part of a common field tensor $F_{\mu\nu}$ and transform into each other by Lorentz transformations. From this point of view the treatment of electric and magnetic fields in the SI, where we end up with two different constants (ϵ, μ), seems to go against the requirement of simplicity. On the other hand, the approach in quantum field theory, where one employs in general only a dimensionless constant at the interface to electrodynamics, the finestructure constant α, is over the mark. Because the information, whether we deal with the electromagnetic or the mechanic aspect of a situation, is lost. The natural system, although not completely free of difficulties, as seen further down, seems a suitable compromise. Different aspects of the intrinsic properties, and which are generally electromagnetic, are not distinguished, no scaling is necessary between $\mathbf{p}, \mathbf{E}$ and $\mathbf{B}$. The only constant necessary is at the interface to mechanic properties, which is η. This also holds for the fields of radiation, which we can describe by:

$$\phi_{Rad}(\mathbf{r}, t) = \frac{1}{8\pi\eta}\left(\mathbf{E}^2 + c^2\mathbf{B}^2\right) \tag{17}$$

Note that in the natural system the usage, or the omission, of η ultimately determines, whether a variable is to be interpreted as an electromagnetic or a mechanic property. Forces and energies are mechanic, whereas momentum density is not. The numerical value of η has to be determined by explicit calculations. This will be done in the next sections. Comparing with existing systems we note three distinct advantages: (i) The system reflects the dynamic origin of fields, and it is based on only three fundamental units: m, kg, s. A separate definition of the current is therefore obsolete. (ii) There is a clear cut interface between mechanics (forces, energies), and electrodynamics (fields of motion). (iii) The system provides a common framework for macroscopic and microscopic processes.

6 Interactions in hydrogen

Returning to proton oscillations let us first restate the main differences between a free electron and an electron in a hydrogen atom [15]: (i) The frequency of the hydrogen system is constant ω_H, as is the frequency of the electron wave. It is thought to arise from the oscillation properties of a proton. (ii) Due to this feature the wave equation of momentum density $\mathbf{p}(\mathbf{r}, t)$ is not homogeneous, but inhomogeneous:

$$\Delta\mathbf{p}(\mathbf{r}, t) - \frac{1}{u^2}\frac{\partial^2}{\partial t^2}\mathbf{p}(\mathbf{r}, t) = \mathbf{f}(t)\delta^3(\mathbf{r}) \tag{18}$$

for a proton at $\mathbf{r} = 0$ of the coordinate system. The source term is related to nuclear oscillations. We do not solve (18) directly, but use the energy principle to simplify the problem. From a free electron it is known that the total intrinsic energy density, the sum of a kinetic component ϕ_K and a field component ϕ_{EM} is a constant of motion [12]:

$$\phi_K(\mathbf{r}) + \phi_{EM}(\mathbf{r}) = \rho_0 u^2 \tag{19}$$

where u is the velocity of the electron and ρ_0 its density amplitude. We adopt this notion of energy conservation also for the hydrogen electron, we only modify it to account for the spherical setup:

$$\phi_K(\mathbf{r}) + \phi_{EM}(\mathbf{r}) = \frac{\rho_0}{r^2}u^2 \tag{20}$$

The radial velocity of the electron has discrete levels. Due to the boundary values problem at the atomic radius, it depends on the principal quantum number n. From the treatment of hydrogen we recall for u_n and ρ_0 the results [15]:

$$u_n = \frac{\omega_H R_H}{2\pi n} \qquad \rho_0 = \frac{M_e}{2\pi R_H} \tag{21}$$

where R_H is the radius of the hydrogen atom and M_e the mass of an electron. Since ρ_0 includes the kinetic as well as the field components of electron "mass", e.g. in Eq. (20), we can define a momentum density $\mathbf{p}_0(\mathbf{r}, t)$, which equally includes both components. As the velocity $u_n = u_n(t)$ of the electron wave in hydrogen is periodic, the momentum density $\mathbf{p}_0(\mathbf{r}, t)$ is given by:

$$\mathbf{u}_n(t) = u_n \cos\omega_H t\, \mathbf{e}^r \qquad \Longrightarrow \qquad \mathbf{p}_0(\mathbf{r}, t) = \frac{\rho_0 u_n}{r^2} \cos\omega_H t\, \mathbf{e}^r \tag{22}$$

The combination of kinetic and field components in the variables has a physical background: it bears on the result that photons change both components of an electron wave [12]. With these definitions we can use the relation between the electric field and the change of momentum, although now this equation refers to both components:

$$\mathbf{E}_0(\mathbf{r}, t) = \frac{\partial}{\partial t}\mathbf{p}_0(\mathbf{r}, t) = -\frac{\rho_0 u_n}{r^2}\omega_H \sin\omega_H t\, \mathbf{e}^r \tag{23}$$

Note that charge, by definition, is included in the electric field itself. Integrating the dynamic charge of a proton from Eq. (10) and accounting for flow conservation in our spherical setup, the field of a proton will be:

$$\mathbf{E}_0(\mathbf{r}, t) = \frac{q_D}{r^2} = \frac{M_p \omega_H^2}{r^2} x \,\sin\omega_H t\, \mathbf{e}^r \tag{24}$$

Apart from a phase factor the two expressions must be equal. Recalling the values of u_n and ρ_0 from (21), the amplitude x of proton oscillation can be computed. We obtain:

$$x = \frac{3d}{R_p} = \frac{M_e}{(2\pi)^2 M_p} \cdot \frac{1}{n} \tag{25}$$

In the highest state of excitation, which for the dynamic model is $n = 1$, the amplitude is less than 10^{-5} times the proton radius: Oscillations are therefore comparatively small. This result indicates that the scale of energies within the proton is much higher than within the electron, say. The result is therefore well in keeping with existing nuclear models. For higher n, and thus lower excitation energy, the amplitude becomes smaller and vanishes for $n \to \infty$.

It is helpful to consider the different energy components within the hydrogen atom at a single state, say $n = 1$, to understand, how the electron is actually bound to the proton. The energy of the electron consists of two components.

$$\phi_K(\mathbf{r}, t) = \frac{\rho_0 u_1^2}{r^2} \sin^2 k_1 r \cos^2 \omega_H t \tag{26}$$

is the kinetic component of electron energy (k_1 is now the wavevector of the wave). As in the free case, the kinetic component is accompanied by an intrinsic field, which accounts for the energy principle (i.e. the requirement, that total energy density at a given point is a constant of motion). Thus:

$$\phi_{EM}(\mathbf{r}, t) = \frac{\rho_0 u_1^2}{r^2} \cos^2 k_1 r \cos^2 \omega_H t \tag{27}$$

is the field component. The two components together make up for the energy of the electron. Integrating over the volume of the atom and a single period τ of the oscillation, we obtain:

$$W_{el} = \frac{1}{\tau} \int_0^\tau dt \int_{V_H} d^3r \, (\phi_K(\mathbf{r}, t) + \phi_{EM}(\mathbf{r}, t)) = \frac{1}{2} M_e u_1^2 \tag{28}$$

This is the energy of the electron in the hydrogen atom. W_{el} is equal to 13.6 eV. The binding energy of the electron is the *energy difference* between a free electron of velocity u_1 and an electron in a hydrogen atom at the same velocity. Since the energy of the free electron W_{free} is:

$$W_{free} = \hbar \omega_H = M_e u_1^2 \tag{29}$$

the energy difference ΔW or the binding energy comes to:

$$\Delta W = W_{free} - W_{el} = \frac{1}{2} M_e u_1^2 \tag{30}$$

This value is also equal to 13.6 eV. It is, furthermore, the energy contained in the photon field $\phi_{Rad}(\mathbf{r}, t)$ of the proton's radiation

$$W_{Rad} = \Delta W = \frac{1}{\tau} \int_0^\tau dt \int_{V_H} d^3r \, \phi_{Rad}(\mathbf{r}, t) = \frac{1}{2} M_e u_1^2 \tag{31}$$

This energy has to be gained by the electron in order to be freed from its bond, it is the ionization energy of hydrogen. However, in the dynamic picture the electron is not thought to move as a point particle in the static field of a central proton charge, the electron is, in this model, a dynamic and oscillating structure, which emits and absorbs energy constantly via the photon field of the central proton. In a very limited sense, the picture is still a statistical one, since the computation of energies involves the average over a full period.

7 The meaning of η

The last problem, we have to solve, is the determination of η, the coupling constant between electromagnetic and mechanic variables. To this end we compute the energy of the radiation field W_{Rad}, using Eqs. (17), (24), and (25). From (17) and (24) we obtain:

$$\phi_{Rad}(r,t) = \frac{1}{8\pi\eta}\mathbf{E}^2 = \frac{1}{8\pi\eta} \cdot \frac{M_p^2 \omega_H^4}{r^4} x^2 \sin^2 \omega_H t \tag{32}$$

Integrating over one period and the volume of the atom this gives:

$$W_{Rad} = \frac{1}{\tau}\int_0^{\tau} dt \int_{R_p}^{R_H} 4\pi r^2 dr\, \phi_{Rad}(\mathbf{r},t) \approx -\frac{1}{4\eta} \cdot \frac{M_p^2 \omega_H^4 x^2}{R_p} \tag{33}$$

provided R_p, the radius of the proton is much smaller than the radius of the atom. With the help of (25), and remembering that W_{Rad} for $n = 1$ equals half the electron's free energy $\hbar\omega_H$, this finally leads to:

$$W_{Rad} = \frac{1}{4\eta} \cdot \frac{M_p^2 \omega_H^4 x^2}{R_p} = \frac{1}{2}\hbar\omega_H \tag{34}$$

$$\eta = \frac{M_e^2 \nu_H^3}{2hR_p} = \frac{1.78 \times 10^{20}}{R_p} \tag{35}$$

since the frequency ν_H of the hydrogen atom equals 6.57×10^{15} Hz. Then η can be calculated in terms of the proton radius R_p. This radius has to be inferred from experimental data, the currently most likely parametrisation being [26]:

$$\frac{\rho_p(r)}{\rho_{p,0}} = \frac{1}{1 + e^{(r-1.07)/0.55}} \tag{36}$$

radii in fm. If the radius of a proton is defined as the radius, where the density $\rho_{p,0}$ has decreased to $\rho_{p,0}/e$, with e the Euler number, then the value is between 1.3 and 1.4 fm. Computing 4π the inverse of η, we get, numerically:

$$\begin{aligned} \frac{4\pi}{\eta} &= 0.92 \times 10^{-34} \qquad (R_p = 1.3fm) \\ &= 0.99 \times 10^{-34} \qquad (R_p = 1.4fm) \\ &= 1.06 \times 10^{-34} \qquad (R_p = 1.5fm) \end{aligned} \tag{37}$$

Numerically, this value is equal to the numerical value of Planck's constant $\hbar$ [27]:

$$\hbar_{UIP} = 1.0546 \times 10^{-34} \tag{38}$$

Given the conceptual difference in computing the radius the agreement seems remarkable. Note that this is a genuine derivation of $\hbar$, because nuclear forces and radii fall completely outside the scope of the theory in its present form. If measurements of R_p were any different, then we would be faced, at this point, with a meaningless numerical value. Reversing the argument it can be said, that the correct value - or rather the meaningful value - is a strong argument for the correctness of our theoretical assumptions. For the following, we redefine the symbol $\hbar$:

$$\hbar := 1.0546 \times 10^{-34}[N^{-1}m^{4}] \tag{39}$$

Then we can rewrite the equations for $\mathbf{F}$, the Lorentz force, for $\mathbf{L}$, angular momentum related to this force, and ϕ_{Rad}, the radiation energy density of a photon in a very suggestive form:

$$\mathbf{F} = \hbar q \left(\frac{\mathbf{E}}{4\pi} + \mathbf{u} \times \frac{\mathbf{B}}{4\pi}\right) \qquad \mathbf{L} = \hbar q \, \mathbf{r} \times \left(\frac{\mathbf{E}}{4\pi} + \mathbf{u} \times \frac{\mathbf{B}}{4\pi}\right) \tag{40}$$

$$\phi_{Rad} = \frac{\hbar}{2}\left[\left(\frac{\mathbf{E}}{4\pi}\right)^2 + c^2\left(\frac{\mathbf{B}}{4\pi}\right)^2\right] \tag{41}$$

Every calculation of mechanic properties involves a multiplication by $\hbar$. Since $\hbar$ is a scaling constant, the term "quantization", commonly used in this context, is misleading. Furthermore, it is completely irrelevant, whether we compute an integral property (the force in (40)), or a density (ϕ_{Rad} in (41), a force density can also be obtained by replacing charge q by a density value). What is, in a sense, discontinuous, is only the mass contained in the shell of the atom. But this mass depends, as does the amplitude of $\phi_{Rad}(r, t)$, on the mass of the atomic nucleus. Thus the only discontinuity left on the fundamental level, is the mass of atomic nuclei. That the energy spectrum of atoms is discrete, is a trivial observation in view of boundary conditions and finite radii.

All our calculations so far focus on single atoms. To get the values of mechanic variables in SI units used in macrophysics, we have to include the scaling between the atomic domain and the domain of everyday measurements. Without proof, we assume this value to be N_A, Avogadro's number. The scale can be made plausible from solid state physics, where statistics on the properties of single electrons generally involve a number of N_A particles in a volume of unit dimensions [28]. And a dimensionless constant does not show up in any dimensional analysis.

8 Dynamic gravity fields

The dynamic fields of mass in motion, and the interpretation of charge as a dynamic feature of mass has far-reaching consequences. In general relativity the connection between electrodynamics and gravity is obtained only via the energy stress tensor T_{ab}, according to Einstein's equation [29]:

$$G_{ab} = 8\pi T_{ab} \tag{42}$$

Here T_{ab} is related in a complicated manner to the electromagnetic fields at a given position. G_{ab} describes the curvature of spacetime at the same position, which in turn is related to gravity. There is no direct way, in Einstein's theory, from the electromagnetic to the gravitational properties of matter. Indirect routes have been explored in the past [30, 31, 32, 33], focusing on the notion of black holes and a length scale considerably below the atomic domain. To date, the only connection between microphysics and gravity is thought to exist in extreme environments, like the first few seconds after the big bang or the vicinity of a black hole. The interpretation of electric charge as a dynamic feature of mass, elaborated in the previous sections, allows a connection in standard situations. For the first time we may ask, whether the gravitational aspect of matter (= its mass) and the electromagnetic aspect (= its charge) are only different scales of the same physical feature: its oscillations.

Oscillations couple strongly to the environment of matter, therefore transport, dissipation and radiation effects should in principle alter the picture in a dynamic way. The fundamental question, in any such theory, is exactly the opposite one posed in static concepts. Static theories have difficulties with the question: How do things change? Dynamic theories, on the other hand, must answer the following one: How can things remain stable?

An answer to this puzzle could lie in the frequency, thus the time scale involved. It is well known that biological organisms utilise mainly electromagnetic or chemical interactions, the time scale for the life of these organisms is in the range of days to years. The frequencies involved in the interactions are around 10^{14} to 10^{15} Hz. The frequency is also the main parameter for the amplitude of the field of dynamic charge. From Eq. (24) we get:

$$|\mathbf{E}| = \frac{q_D}{r^2} \approx \frac{M\omega^2}{r^2} \tag{43}$$

The coupling between mass and gravity is much weaker than between charge and its electrostatic field. We have two ways, in principle, to estimate the difference: either we set mass and charge equal to 1 and compute the fields. Then the ratio between gravity and electrostatics is $\epsilon_0 G \approx 10^{-22}$. For the frequency of the dynamic gravity field we get consequently:

$$\omega_G \approx 10^{-11} \omega_E \tag{44}$$

Or we compute the ratio of the fields from the static equations, accounting for physical units by a constant k_u, which shall be purely dimensional. Then we have:

$$|\mathbf{E}| = \frac{e}{4\pi\epsilon_0 r^2} := M\frac{\omega_E^2}{r^2} \qquad |\mathbf{G}| = G\frac{M}{r^2} := M\frac{\omega_G^2}{r^2} \qquad \omega_G = \sqrt{k_u \frac{4\pi\epsilon_0 GM}{e}}\omega_E \tag{45}$$

At present, we have no way to remove this ambiguity about the exact numerical value. Consequently, the scale η between the two frequencies will be:

$$\omega_G = \eta \cdot \omega_E \qquad 10^{-14} \leq \eta \leq 10^{-11} \tag{46}$$

Here ω_E is the characteristic electromagnetic frequency, ω_G its gravitational counterpart. The hypothesis can in principle be tested. If ν_G, the frequency of gravity radiation, is about η times the frequency of proton oscillation, we get:

$$\nu_G \approx 10\text{Hz} - 10\text{kHz} \tag{47}$$

The frequency is in the same range as the theoretical results based on general relativity [34]. Also the implications are the same: if fields of this frequency range exist in space, we would attribute these fields to stellar gravity. However, since there is no difference between electromagnetic waves and gravitational waves, apart from their frequency, we would also attribute electromagnetic waves in this range to gravity. To estimate the intensity of this, hypothetical, field, we use Eq. (23):

$$\mathbf{G}_S(\mathbf{r}, t) = \frac{\partial}{\partial t}\mathbf{p}_E(\mathbf{r}, t) \tag{48}$$

Here $\mathbf{G}_S$ is the solar gravity field. The momentum density and its derivative can be inferred from centrifugal acceleration.

$$\frac{\partial}{\partial t}\mathbf{p}_E(\mathbf{r}, t) = \rho_E \, a_C \, \mathbf{e}^r \qquad \rho_E = \frac{3M_E}{4\pi R_E^3} \qquad a_C = \omega_E^2 \, R_O \tag{49}$$

where R_O is the earth's orbital radius and where we have assumed isotropic distribution of terrestrial mass. Then Eq. (32) leads to:

$$\phi_G(r = R_O) = \frac{\hbar}{2}\left(\frac{G_S}{4\pi}\right)^2 = \frac{\hbar}{2}\left(\frac{3M_E R_O}{4R_E^3 \tau_E^2}\right)^2 \tag{50}$$

Note the occurrence of Planck's constant also in this equation, although all masses and distances are astronomical. The intensity of the field, if calculated from (50), is very small. To give it in common measures, we compute the flow of gravitational energy through a surface element at the earth's position. In SI units we get:

$$J_G(R_O) = \phi_G(R_O) \cdot N_A \cdot c \approx 70mW/m^2 \tag{51}$$

Compared to radiation in the near visible range - the solar radiation amounts to over 300 Watt/m^2 [35] - the value seems rather small. But considering, that also radiation in the visible range could have an impact on terrestrial motion, the intensity of the gravity waves could be, in fact, much higher. Concerning the detectors of gravity waves, by now in operation at several locations around the world [34], the hypothesis involves a conjecture: not, that gravity waves have not been detected so far, because their intensity is so small, but because they are so ubiquitous. The background noise, all experimental groups report as a major obstacle, could be the main experimental feature of gravity waves. If this is the case, then an unambiguous detection is not possible on earth. To that end, electromagnetic radiation in space is the only possible source for analysis.

Coming back to the question, how the solar system in a dynamic theory of gravity could be stable, we find that the frequency scale between gravity and electricity is very different. If the typical lifetime of a primitive organism based on chemical interactions is in the range of days, then the typical lifetime of an organism based on gravity would be in the range of billion years. This value is in the same order of magnitude as current estimates in cosmology about the typical lifetime of a small yellow star like our sun.

Acknowledgements

Thanks are due to Dr. Sidharth and the B M Birla Foundation for the invitation. Financial support by the University College London is gratefully acknowledged.

References

[1] J. Mather et al., Ap. J. Lett., **354**, L37, 1990.

[2] V.C. Rubin, "Bright Galaxies, Dark Matters", Springer/AIP Press, New York, 1997.

[3] V.C. Rubin, Scientific American Presents, **9**, 106, 1998.

[4] R.B. Griffiths, J. Statist. Phys., **36**, 219, 1984.

[5] J.B. Hartle, Phys. Rev. D, **44**, 3173, 1991.

[6] R. Omnes, Rev. Mod. Phys., **64**, 339, 1992.

[7] G.C. Ghirardi, A. Rimini, and T. Weber, Phys. Rev. D, **34**, 470, 1986.

[8] H. Everett, Rev. Mod. Phys., **29**, 454, 1957.

[9] D. Bohm, Phys. Rev., **85**, 166 and 180, 1952.

[10] L. de la Pena and A.M. Cetto, "The Quantum Dice - An Introduction to Stochastic Electrodynamics", Kluwer, Dordrecht, 1996.

[11] W.A. Hofer, "Beyond Uncertainty: internal structures of electrons and photons", in B.G. Sidharth and A. Burinskij (eds.), Frontiers of Fundamental Physics, Universities Press, Hyderabad, 1999, p. 44.

[12] W.A. Hofer, Physica A, **256**, 178, 1998.

[13] W.A. Hofer, "A realist view of the electron: recent advances and unsolved problems", in V.A. Dvoeglazov (ed.), Photon: Old Problems in Light of New Ideas, Nova Science, Huntington, NY, 2000.

[14] W.A. Hofer, "Measurements in quantum physics: towards a physical picture of relevant processes", proceedings of the VIth Wigner Symposium, August 16-22, 1999.

[15] W.A. Hofer, "A dynamic model of atoms: structure, internal interactions and photon emissions of hydrogen", quant-ph/9801044.

[16] J. J. Thomson, Phil. Mag., **44**, 293, 1897.

[17] H. A. Lorentz, Proc. R. Acad. Sci. Amsterdam, **6**, 809, 1904.

[18] M. Abraham, Ann. Physik, **10**, 105, 1903.

[19] D. Bender et al., Phys. Rev D, **30**, 515, 1984.

[20] J. Keller and Z. Oziewicz, "The Theory of the Electron", UNAM Mexico, 1997.

[21] L. Montanet et al., Phys. Rev. D, **50**, 1173, 1994.

[22] D. C. Tsui, H. L. Störmer, and A. C. Grossard, Phys. Rev. Lett., **48**, 1559, 1982.

[23] R. E. Prange and S. H. Girvin, "The Quantum Hall effect", Springer-Verlag New York, 1990.

[24] R. B. Laughlin, Phys. Rev. Lett., **50**, 1395, 1983.

[25] J. D. Jackson, "Classical Electrodynamics", 3rd edition, Wiley & Sons, New York, 1999, p.779.

[26] R. Eisberg and R. Resnick, "Quantum Physics", 2nd edition, Wiley & Sons, New York, 1985, p.517.

[27] UIPAP docoment UIP 20. Physica A, **93**, 1, 1978.

[28] N. W. Ashcroft and N. D. Mermin, "Solid State Physics", Holt-Saunders, Philadeslphia ,1974, p.4.

[29] R.M. Wald, "General Relativity", University of Chicago Press, Chicago, 1984, p.72.

[30] R.M. Wald, Phys. Rev. D, **6**, 1476, 1972.

[31] S.W. Hawking, Commun. Math. Phys., **43**, 199, 1975.

[32] R.M. Wald, Commun. Math. Phys., **45**, 9, 1975.

[33] B.G. Sidharth, Indian J. Pure Appl. Phys., **35**, 456, 1997.

[34] For an overview of the current experimental and theoretical state on gravity waves see: Sydney Meshkov (ed.), Gravitational Waves, AIP Conference Proceedings 523, AIP New York, 2000.

[35] D. Morrison and T. Owen, "The Planetary System", 2nd edition, Addison Wesley, New York, 1996, 243.

Consistent Equation of Classical Gravitation to Quantum Limit and Beyond

Shantilal G. Goradia
Senior Engineer, Cook Nuclear Station
Bridgman, MI 49106 (USA)
Email: Shantlalg@juno.com

1 Introduction

General Relativity makes a distinction between mass and space. Mass tells space how to curve and space tells mass how to move. Newtonian gravity equation makes a distinction between them by having its numerator as mass effect and its denominator as inverse square law space effect at macroscopic approximation. At microscopic distances it makes sense to substitute surface-to-surface distance between two nucleons for center-to-center distance between them to account for the mass space distinction, keeping in mind the smallest distance between coupled nucleons is Planck length. Any distance less than Planck makes no sense in the classical world. When we calculate the force between two nucleons of one femtometer diameter each, separated by a surface-to-surface distance of Planck length, we get the force that matches well known nuclear force i.e. 10^{40} times the value of the force of gravitation "g" calculated by assuming the Newtonian center-to-center distance of 1 femtometer. What we get is what is described as the nuclear force in scientific literature. This leads to the question: Is the nuclear force (well recognized secondary effect of color force) high intensity gravitation?

2 Analysis

Consider the following equations **(1)** and **(2)**. The notation d_n in Equation **(2)** is the diameter of the nucleons in question.

$$\textbf{Newtonian} \quad F_N = Gm_1 x m_2/D^2 \tag{1}$$

$$\textbf{Proposed} \quad F_P = Gm_1 x m_2/(D - d_n)^2 \tag{2}$$

The notation D in the denominator in Newton's equation **(1)** of the gravitational force denotes the separating distance between the centers of mass of the particles in question. The validity of equation **(1)** has been verified for distances as low as a few centimeters. Its validity is not verified when the subatomic separating distance between nuclei of atoms is a few femtometers **(fm)**. Newton's equation is an approximation that explains macroscopic observations. If Newton meant his equation to hold true at microscopic distances, he would have explained the binding energy. Newtonian physics implies point masses and action between points. A point has no mass. I am asserting that, instead, the classically deterministic Newtonian gravity originates at the surfaces of nucleons, not at a central point within the nucleons. I am not addressing electrons and coulomb forces in this paper. This paper is dedicated to the investigation of the nuclear force alone at this stage. The deviation from inverse square logic resulting from our proposal is insignificant. The consequences of mass and space microscopic distinction are enormous. The proposed correction is the injection of d_n. Equation (2) is good for all distances greater than Planck length (10^{-35} meters = 10^{-20} fm). I require the Planck length as a lower bound so as to include the dominant first order quantum effect in this classical model. The relative strength of the proposed equation is the ratio obtained by dividing equation (2) by equation (1), which is

$$\textbf{The ratio} \quad F_P/F_N = D^2/(D - d_n)^2. \qquad (3)$$

3 Strength of Gravity at Short Range per Eqn. (2)

When D is very large compared to d_n, D^2 is almost equal to $(D - d_n)^2$. The diameter of atoms is hundreds of millions of times greater than the diameter of nucleons located at the center of atoms. The force of gravitation calculated by these two equations between the nucleons of two adjoining atoms is practically the same, because the ratio $D^2/(D - d_n)^2$ is almost equal to one. When D is small compared to d_n, the force of attraction calculated by equation (2) will be significantly greater than that calculated by equation (1). The following results bring home the concept. If we call the force of gravitation calculated by equation(1) "g", the force of gravitation calculated by equation (2) would be higher by the ratio $D^2/(D - d_n)^2$. When D exceeds d_n, by Planck length (10^{-20} fm), one obtains the ratio:

$$\begin{aligned} D^2/(D - d_n)^2 &= (d_n + \text{Planck length})^2/(\text{Planck length})^2 \\ &= (1 + 10^{-20})^2/(10^{-20})^2 \quad \text{(All lenghts in femtometers)} \\ &= 10^{40}. \quad (d_n = 1 \text{ femtometer}) \end{aligned}$$

At surface to surface separations of 1, 2, 3, 4 and 10 femtometers, the calculated nuclear forces rapidly diminish to 4.0, 2.1, 1.77, 1.56 and 1.23 times the gravitational forces respectively and match Newtonian gravitation at 1000 femtometers as tabulated below. My calculations meet the observed boundary values. Nucleon deformation is neglected.

Separating Distance		Nuclear Force / Gravity
One	Planck length, 10^{-20} fm	10^{40}
1	Femtometer	4.0
2	Femtometers	2.1
3	Femtometers	1.77
4	Femtometers	1.56
5	Femtometers	1.44
6	Femtometers	1.36
7	Femtometers	1.31
8	Femtometers	1.26
9	Femtometers	1.23
10	Femtometers	1.21
15	Femtometers	1.15
20	Femtometers	1.11
25	Femtometers	1.09
50	Femtometers	1.04
100	Femtometers	1.02
1000	Femtometers	1.00

4 Analogy

Considering a hollow metal sphere containing smaller metal balls rumbling inside the sphere, a probe inside the sphere or close to the outside surface would detect non-central high intensity, indeterministic noise with intensity increasing with distance from the center. For a distant listener, the sound would be of deterministic nature originating from the center of the sphere with its intensity decreasing with distance. Indeterministic high intensity noise at a short range is deterministic low intensity sound at large distances. What this analogy brings home is that the color force is potentially the high intensity, non-central, classically indeterministic interaction. The gravity is potentially the low intensity, macroscopically central, Newtonian deterministic manifestation of the same fundamental interaction. I am taking the liberty to use the prevailing view that the nuclear force is the secondary effect of the color force to reach the following conclusion. This view does not need to be reestablished.

5 Conjectures

(A) We do not have quantum gravity: gravity is potentially not a separate fundamental interaction of Nature.

(B) Rutherford's scattering experiments showed nuclear forces as far as 10 femtometers [2]: not that they do not exist beyond that range. At higher distances they are too weak to detect.

(C) There is no proof of a central force detected inside the nucleons.

(D) Despite its theoretical justification, Yukawa potential does not predict the observations.

(E) There is no feature of nuclear force that distinguishes it from gravitation.

(F) Einstein attempted to explain nuclear force in terms of gravity [1].

(G) The Standard Model does not incorporate gravity.

6 Conclusions

Newtonian gravitation is potentially a deterministic manifestation of the classically indeterminate color forces addressed in QCD. The prevailing view is that the nuclear force is the secondary effect of the color force. The proposed theory connects the nuclear force with gravitation in one common equation. The combined contributions of these two clearly imply that gravitation is the Newtonian deterministic manifestation of the classically indeterminate color forces.

7 Acknowledgements.

I am grateful to Professor Fridolin Weber (University of Notre Dame) for his comments following the presentation of the concepts.

References

[1] Tian Yu Cao, "The conceptual foundation of quantum field theory", Boston University, (QC174.45.A1C646 1999) Page 85, "Does quantum field theory need a foundation?"
See discussion referring to Nobel Prize Winner Sheldon Glashow.

[2] L.R.B. Elton, "Introductory Nuclear Theory", D. Sc., F. Inst. P, Professor of Physics, Battersea College of Technology, Second Edition, 1966. Section 1.7, Nuclear Forces.

[3] S.G. Goradia, http://ar.Xiv.org/abs/math-ph/0009025, 9/15/00.

Index

MIX
Papier aus verantwortungsvollen Quellen
Paper from responsible sources
FSC® C105338

If you have any concerns about our products,
you can contact us on
ProductSafety@springernature.com

In case Publisher is established outside the EU,
the EU authorized representative is:
Springer Nature Customer Service Center GmbH
Europaplatz 3, 69115 Heidelberg, Germany

Printed by Libri Plureos GmbH
in Hamburg, Germany